Introduction to Reliability Engineering

Second Edition

D0162187

Introduction to Reliability Engineering

Second Edition

E. E. Lewis

Department of Mechanical Engineering
Northwestern University
Evanston, Illinois
July, 1994

John Wiley & Sons, Inc.

ACQUISITIONS EDITOR   Cliff Robichaud

ASSISTANT EDITOR   Catherine Beckham

MARKETING MANAGER   Laura Eckley

PRODUCTION EDITOR   Cathey Ronda

ASSISTANT PRODUCTION EDITOR   Michelle Comparetti

DESIGNER   Harry Nolan

MANUFACTURING MANAGER   Dorothy Sinclair

ILLUSTRATION COORDINATOR   Gene Aiello

Recognizing the importance of preserving what has been written, it is a policy of John Wiley & Sons, Inc. to have books of enduring value published in the United States printed on acid-free paper, and we exert our best efforts to that end.

Library of Congress Cataloging in Publication Data:
Lewis, E. E. (Elmer Eugene), 1938-
    Introduction to reliability engineering / E. E. Lewis. — 2nd ed.
        p.   cm.
    Includes bibliographical references.
    ISBN 0-471-01833-3 (cloth : alk. paper)
    1. Reliability (Engineering)   I. Title.
TA169.L47   1996
620'.00452—dc20
                                                                95-13533
                                                                    CIP

10 9 8 7 6 5 4

To Ann, Elizabeth, and Paul

About the Author

Elmer E. Lewis is Professor and Chairman of the Department of Mechanical Engineering, Northwestern University. He received his B.S. in Engineering Physics and an M.S. and Ph.D. in Nuclear Engineering at the University of Illinois, Urbana. He served as a Captain in the U.S. Army and as a Ford Foundation Fellow and Assistant Professor at MIT before joining Northwestern's faculty. He has also held appointments as Visiting Professor at the University of Stuttgart and Guest Scientist at the Nuclear Research Center at Karlsruhe, Germany, and has served as a consultant to Argonne and Los Alamos National Laboratories and to a number of industrial firms. A Fellow of the American Nuclear Society, and winner of its Arthur Holly Compton Award, Professor Lewis is actively involved in research in the areas of reliability modeling, radiation transport, and the physics and safety of nuclear systems. He is the author or co-author of over 125 publications including the Wiley books, *Nuclear Power Reactor Safety* and *Computational Methods of Neutron Transport.*

Preface

The object of this text is to provide an elementary and reasonably self-contained overview of reliability engineering that is suitable for an upper level undergraduate or first year graduate course for students of any engineering discipline. The materials reflect the inherently interdisciplinary character of reliability considerations and the central role played by probability and statistical analysis in presenting reliability principals and practices.

The examples and exercises are drawn from a variety of engineering fields. They can be understood, however, with only the knowledge from the physics, chemistry, and basic engineering courses contained in the first years of nearly all engineering curricula. Likewise, the reader is presumed to have completed only the standard mathematics sequence, through ordinary differential equations, required of most engineering students. No prior knowledge of probability or statistics is assumed; the development of the required concepts is contained within the text.

Since the mid 1980s when the bulk of the manuscript for the first edition was written, at least two major changes have taken place that are incorporated into this new edition. The first is the increased industrial emphasis on quality in the product development cycle and its relationship to reliability. The second is the rapid advances that have taken place in personal computer software.

In implementing quality goals, a number of methodologies are finding increased use which also have a strong impact on reliability considerations. Two of the most important of these, the Taguchi and the Six Sigma methodologies, are examined in some detail in Chapter 4. Quality loss functions and capability indices, robust design, design of experiments, and statistical process control are taken up in the context of reliability and safety engineering.

The power and user-friendliness of spread sheet and graphics software available on personal computers has obviated most of the statistical tables and nonlinear graph papers contained in the first edition for the analysis of reliability data. Tedious table look-ups and the use of subjective visual fitting of curves to data may be replaced by the use of the statistical formulas and least-square fitting incorporated into present-day spread sheets. While the materials in this text may be understood without specific reference to spread sheet terminologies, the analysis of data is greatly facilitated by their use.

For each appropriate example in this edition, the necessary steps for obtaining a spread sheet solution are indicated explicitly using Excel-4*, augmented in some cases with the Cricket Graph†. These two programs were chosen because they are widely available, and instructions for their use are virtually identical on Macintosh and on IBM compatible computers. Thus no computer-specific information is required. Spread sheet fundamentals are not included in the text since instruction can be found in many readily available books and manuals; an extra class meeting or two should suffice to give students with no previous experience adequate background to master the examples and exercises in this text. Some instructors may wish to utilize more focused statistical software packages. This should cause no difficulty.

A number of additional improvements have been incorporated into the new edition. In Chapter 7, a more comprehensive treatment has been included to tie the load-capacity approach to reliability to failure rate methodology. In Chapter 9, more emphasis is placed on the limitations and trade-offs encountered in the use of active and standby redundant systems. A section has also been added to Chapter 12 dealing with safety hazards of products and equipment to provide a more balanced treatment of potential hazards arising from the use of consumer, as well as, industrial products. Finally, the text now contains over 100 solved examples, and well over 200 exercises, many of which are new. The answers to the odd-numbered exercises are given at the end of the book.

The text contains somewhat more material than can be treated in detail in a normal one-semester undergraduate course, providing some latitude in the topics that may be emphasized. If the students have had some previous exposure to elementary probability, Chapters 2 and 3 can be telescoped because those probability concepts that are more specific to reliability analysis are set fourth in Chapter 6. The statistical treatment of data contained mainly in Chapters 5 and 8 is essential to a well-rounded course in reliability engineering; nevertheless, the materials in the remaining chapters may be covered independently. Likewise, the quality methodologies contained in Chapter 4 are important, but need not be covered to comprehend the materials in the latter chapters. Similarly, the quantitative analysis of the effects of load and capacities contained in Chapter 7 is critical to the understanding of failure mechanisms, but the reliability systems considerations concentrated in Chap-

---

* Excell is a registered trademark of the Microsoft Corporation.
† Cricket Graph is copyright © by Cricket Software.

ters 9 through 11 may be read independently of it. Finally, the system safety analysis contained in Chapter 12 may be understood without first covering the Markov analysis methods developed in Chapter 11.

In addition to the continued thanks owed to the students and colleagues who provided their advice and assistance with the first addition, I would like to acknowledge the help of the Northwestern students and colleagues of other institutions, who have ferreted out errors in the first edition and made constructive criticisms and suggestions for improvements. George Coons of the Motorola Corporation has been particularly helpful in providing materials and suggestions related to the treatment of quality issues, and Jim Lookabaugh of Northwestern designed the data acquisition system and obtained the light bulb reliability results that serve as the basis for several examples in Chapters 5 and 8. Finally, I would like to express my appreciation for the continued understanding of my wife and children while I monopolized the family computer.

Evanston, Illinois   Elmer E. Lewis

Contents

CHAPTER 1

Introduction

"When an engineer, following the safety regulations of the Coast Guard or the Federal Aviation Agency, translates the laws of physics into the specifications of a steamboat boiler or the design of a jet airliner, he is mixing science with a great many other considerations all relating to the purposes to be served. And it is always purposes in the plural—a series of compromises of various considerations, such as speed, safety, economy and so on."

D. K. Price, The Scientific Estate, 1968

1.1 RELIABILITY DEFINED

The emerging world economy is escalating the demand to improve the performance of products and systems while at the same time reducing their cost. The concomitant requirement to minimize the probability of failures, whether those failures simply increase costs and irritation or gravely threaten the public safety, is also placing increased emphasis on reliability. The formal body of knowledge that has been developed for analyzing such failures and minimizing their occurrence cuts across virtually all engineering disciplines, providing the rich variety of contexts in which reliability considerations appear. Indeed, deeper insight into failures and their prevention is to be gained by comparing and contrasting the reliability characteristics of systems of differing characteristics: computers, electromechanical machinery, energy conversion systems, chemical and materials processing plants, and structures, to name a few.

In the broadest sense, reliability is associated with dependability, with successful operation, and with the absence of breakdowns or failures. It is necessary for engineering analysis, however, to define reliability quantitatively as a probability. Thus reliability is defined as the probability that a system will perform its intended function for a specified period of time under a given

1

set of conditions. System is used here in a generic sense so that the definition of reliability is also applicable to all varieties of products, subsystems, equipment, components and parts.

A product or system is said to fail when it ceases to perform its intended function. When there is a total cessation of function—an engine stops running, a structure collapses, a piece of communication equipment goes dead—the system has clearly failed. Often, however, it is necessary to define failure quantitatively in order to take into account the more subtle forms of failure; through deterioration or instability of function. Thus a motor that is no longer capable of delivering a specified torque, a structure that exceeds a specified deflection, or an amplifier that falls below a stipulated gain has failed. Intermittent operation or excessive drift in electronic equipment and the machine tool production of out-of-tolerance parts may also be defined as failures.

The way in which time is specified in the definition of reliability may also vary considerably, depending on the nature of the system under consideration. For example, in an intermittently operated system one must specify whether calendar time or the number of hours of operation is to be used. If the operation is cyclic, such as that of a switch, time is likely to be cast in terms of the number of operations. If reliability is to be specified in terms of calendar time, it may also be necessary to specify the frequency of starts and stops and the ratio of operating to total time.

In addition to reliability itself, other quantities are used to characterize the reliability of a system. The mean time to failure and failure rate are examples, and in the case of repairable systems, so also are the availability and mean time to repair. The definition of these and other terms will be introduced as needed.

1.2   PERFORMANCE, COST, AND RELIABILITY

Much of engineering endeavor is concerned with designing and building products for improved performance. We strive for lighter and therefore faster aircraft, for thermodynamically more efficient energy conversion devices, for faster computers and for larger, longer-lasting structures. The pursuit of such objectives, however, often requires designs incorporating features that more often than not may tend to be less reliable than older, lower-performance systems. The trade-offs between performance, reliability, and cost are often subtle, involving loading, system complexity, and the employment of new materials and concepts.

Load is most often used in the mechanical sense of the stress on a structure. But here we interpret it more generally so that it also may be the thermal load caused by high temperature, the electrical load on a generator, or even the information load on a telecommunications system. Whatever the nature of the load on a system or its components may be, performance is frequently improved through increased loading. Thus by decreasing the weight of an aircraft, we increase the stress levels in its structure; by going to higher—thermodynamically more efficient—temperatures we are forced to operate materials under conditions in which there are heat-induced losses of strength and more rapid corrosion. By allowing for ever-increasing flows of information in communications systems, we approach the frequency limits at which switching or other digital circuits may operate.

Approaches to the physical limits of systems or their components to improve performance increases the number of failures unless appropriate countermeasures are taken. Thus specifications for a purer material, tighter dimensional tolerance, and a host of other measures are required to reduce uncertainty in the performance limits, and thereby permit one to operate close to those limits without incurring an unacceptable probability of exceeding them. But in the process of doing so, the cost of the system is likely to increase. Even then, adverse environmental conditions, product deterioration, and manufacturing flaws all lead to higher failure probabilities in systems operating near their limit loads.

System performance may often be increased at the expense of increased complexity; the complexity usually being measured by the number of required components or parts. Once again, reliability will be decreased unless compensating measures are taken, for it may be shown that if nothing else is changed, reliability decreases with each added component. In these situations reliability can only be maintained if component reliability is increased or if component redundancy is built into the system. But each of these remedies, in turn, must be measured against the incurred costs.

Probably the greatest improvements in performance have come through the introduction of entirely new technologies. For, in contrast to the trade-offs faced with increased loading or complexity, more fundamental advances may have the potential for both improved performance and greater reliability. Certainly the history of technology is a study of such advances; the replacement of wood by metals in machinery and structures, the replacement of piston with jet aircraft engines, and the replacement of vacuum tubes with solid-state electronics all led to fundamental advances in both performance and reliability while costs were reduced. Any product in which these trade-offs are overcome with increased performance and reliability, without a commensurate cost increase, constitutes a significant technological advance.

With any major advance, however, reliability may be diminished, particularly in the early stages of the introduction of new technology. The engineering community must proceed through a learning experience to reduce the uncertainties in the limits in loading on the new product, to understand its susceptibilities to adverse environments, to predict deterioration with age, and to perfect the procedures for fabrication, manufacture, and construction. Thus in the transition from wood to iron, the problem of dry rot was eliminated, but failure modes associated with brittle fracture had to be understood. In replacing vacuum tubes with solid-state electronics the ramifications of reliability loss with increasing ambient temperature had to be appreciated.

Whether in the implementation of new concepts or in the application of existing technologies, the way trade-offs are made between reliability, performance and cost, and the criteria on which they are based is deeply imbedded in the essence of engineering practice. For the considerations and criteria are as varied as the uses to which technology is put. The following examples illustrate this point.

Consider a race car. If one looks at the history of automobile racing at the Indianapolis 500 from year to year, one finds that the performance is continually improving, if measured as the average speed of the qualifying cars. At the same time, the reliability of these cars, measured as the probability that they will finish the race, remains uniformly low at less than 50%.* This should not be surprising, for in this situation performance is everything, and a high probability of breakdown must be tolerated if there is to be any chance of winning the race.

At the opposite extreme is the design of a commercial airliner, where mechanical breakdown could well result in a catastrophic accident. In this case reliability is the overriding design consideration; degraded speed, payload, and fuel economy are accepted in order to maintain a very small probability of catastrophic failure. An intermediate example might be in the design of a military aircraft, for here the trade-off to be achieved between reliability and performance is more equally balanced. Reducing reliability may again be expected to increase the incidence of fatal accidents. Nevertheless, if the performance of the aircraft is not sufficiently high, the number of losses in combat may negate the aircraft's mission, with a concomitant loss of life.

In contrast to these life or death implications, reliability of many products may be viewed primarily in economic terms. The design of a piece of machinery, for example, may involve trade-offs between the increased capital costs entailed if high reliability is to be achieved, and the increased costs of repair and of lost production that will be incurred from lower reliability. Even here more subtle issues come into play. For consumer products, the higher initial price that may be required for a more reliable item must be carefully weighed against the purchaser's annoyance with the possible failure of a less reliable item as well as the cost of replacement or repair. For these wide classes of products it is illuminating to place reliability within the wider context of product quality.

1.3   QUALITY, RELIABILITY, AND SAFETY

In competitive markets there is little tolerance for poorly designed and/or shoddily constructed products. Thus over the last decade increasing emphasis has been placed on product quality improvement as manufacturers have striven to satisfy customer demands. In very general terms quality may be defined as the totality of features and characteristics of a product or service that bear on its ability to satisfy given needs. Thus, while product quality and reliability invariably are considered to be closely linked, the definition of quality implies performance optimization and cost minimization as well. Therefore it is important to delineate carefully the relationships between

* R. D. Haviland, *Engineering Reliability and Long Life Design,* Van Nostrand, New York, 1964, p. 114.

quality, reliability, and safety. We approach this task by viewing the three concepts within the framework of the design and manufacturing processes, which are at the heart of the engineering enterprise.

In the product development cycle, careful market analysis is first needed to determine the desired performance characteristics and quantify them as design criteria. In some cases the criteria are upper limits, such as on fuel consumption and emissions, and in others they are lower limits, such as on acceleration and power. Still others must fall within a narrow range of a specified target value, such as the brightness of a video monitor or the release pressure of a door latch. In conceptual or system design, creativity is brought to the fore to formulate the best system concept and configuration for achieving the desired performance characteristics at an acceptable cost. Detailed design is then carried out to implement the concept. The result is normally a set of working drawings and specifications from which prototypes are built. In designing and building prototypes, many studies are carried out to optimize the performance characteristics.

If a suitable concept has been developed and the optimization of the detailed design is successful, the resulting prototype should have performance characteristics that are highly desirable to the customer. In this process the costs that eventually will be incurred in production must also be minimized. The design may then be said to be of high quality, or more precisely of high characteristic quality. Building a prototype that functions with highly desirable performance characteristics, however, is not in and of itself sufficient to assure that the product is of high quality; the product must also exhibit low variability in the performance characteristics.

The customer who purchases an engine with highly optimized performance characteristics, for example, will expect those characteristics to remain close to their target values as the engine is operated under a wide variety of environmental conditions of temperature, humidity, dust, and so on. Likewise, satisfaction will not be long lived if the performance characteristics deteriorate prematurely with age and/or use. Finally, the customer is not going to buy the prototype, but a mass produced engine. Thus each engine must be very nearly identical to the optimized prototype if a reputation of high quality is to be maintained; variability or imperfections in the production process that lead to significant variability in the performance characteristics should not be tolerated. Even a few "lemons" will damage a product's reputation for high quality.

To summarize, two criteria must be satisfied to achieve high quality. First, the product design must result in a set of performance characteristics that are highly optimized to customer desires. Second, these performance characteristics must be robust. That is, the characteristics must not be susceptible to any of the three major causes of performance variability: (1) variability or defects in the manufacturing process, (2) variability in the operating environment, and (3) deterioration resulting from wear or aging.

In what we shall refer to as product dependability, our primary concern is in maintaining the performance characteristics in the face of manufacturing variability, adverse environments, and product deterioration. In this context we may distinguish between quality, reliability, and safety. Any variability of performance characteristics concerning the target values entails a loss of quality. Reliability engineering is primarily concerned with variability that is so severe as to cause product failure, and safety engineering is focused on those failures that create hazards.

To illustrate these relationships consider an automatic transmission for an automobile. Among the performance characteristics that have been optimized for customer satisfaction are the speeds at which gears automatically shift. The quality goal is then to produce every transmission so that the shift takes place at as near as possible to the optimum speed, under all environmental conditions, regardless of the age of the transmission and independently of where in the production run it was produced. In reality, these effects will result in some variability in the shift speeds and other performance characteristics. With increased variability, however, quality is lost. The driver will become increasingly displeased if the variability in shift speed is large enough to cause the engine to race before shifting, or low enough that it grinds from operating in the higher gear at too low a speed. With even wider variability the transmission may fail altogether, by one of a number of modes, for example by sticking in either the higher or lower gear, or by some more catastrophic mode, such as seizure.

Just as failures studied in reliability engineering may be viewed as extreme cases of the performance variability closely associated with quality loss, safety analysis deals with the subset of failure modes that may be hazardous. Consider again our engine example. If it is a lawn mower engine, most failure modes will simply cause the engine to stop and have no safety consequences. A safety problem will exist only if the failure mode can cause the fuel to catch fire, the blades to fly off or some other hazardous consequence. Conversely, if the engine is for a single-engine aircraft, reliability and safety considerations clearly are one and the same.

In reliability engineering the primary focus is on failures and their prevention. The foregoing example, however, makes clear the intimate relationship among quality loss, performance variability, and failure. Moreover, as will become clearer in succeeding chapters, there is a close correlation between the three causes of performance variability and the three failure modes categories that permeate reliability and safety engineering. Variability due to manufacturing processes tends to lead to failures concentrated early in product life. In the reliability community these are referred to as early or infant mortality failures. The variability caused by the operating environment leads to failures designated as random, since they tend to occur at a rate which is independent of the product's age. Finally, product deterioration leads to failures concentrated at longer times, and is referred to in the reliability community as aging or wear failures.

The common pocket calculator provides a simple example of the classes of variability and of failure. Loose manufacturing tolerances and imprecise quality control may cause faulty electrical connections, misaligned keys or other imperfections that are most likely to cause failures early in the design life of the calculator. Inadvertently stepping on the calculator, dropping it in water, or leaving it next to a strong magnet may expose it to environmental stress beyond which it can be expected to tolerate. The ensuing failure will have little correlation to how long the calculator has been used, for these are random events that might occur at any time during the design life. Finally, with use and the passage of time, the calculator key contacts are likely to become inoperable, the casing may become brittle and crack, or other components may eventually cause the calculator to fail from age. To be sure, these three failure mode classes often subtly interact. Nevertheless they provide a useful framework within which we can view the quality, reliability, and safety considerations taken up in succeeding chapters.

The focus of the activities of quality, reliability, and safety engineers respectively, differ significantly as a result of the nature and amount of data that is available. This may be understood by relating the performance characteristics to the types of data that engineers working in each of these areas must deal with frequently. Quality engineers must relate the product performance characteristics back to the design specifications and parameters that are directly measurable; the dimensions, material compositions, electrical properties and so on. Their task includes both setting those parameters and tolerances so as to produce the desired performance characteristics with a minimum of variability, and insuring that the production processes conform to the goals. Thus corresponding to each performance characteristic there are likely to be many parameters that must be held to close conformance. With modern instrumentation, data on the multitude of parameters and their variability may be generated during the production process. The problem is to digest the vast amounts of raw data and put it to useful purposes rather than being overwhelmed by it. The processes of robust design and statistical quality control deal with utilizing data to decrease performance characteristic variability.

Reliability data is more difficult to obtain, for it is acquired through observing the failure of products or their components. Most commonly, this requires life testing, in which a number of items are tested until a significant number of failures occur. Unfortunately, such tests are often very expensive, since they are destructive, and to obtain meaningful statistics substantial numbers of the test specimens must fail. They are also time consuming, since unless unbiased acceleration methods are available to greatly compress the time to failure, the test time may be comparable or longer to the normal product life. Reliability data, of course, is also collected from field failures once a product is put into use. But this is a lagging indicator and is not nearly as useful as results obtained earlier in the development process. It is imperative that the reliability engineer be able to relate failure data back to performance characteristic variability and to the design parameters and tolerances. For then quality measures can be focused on those product characteristics that most enhance reliability.

The paucity of data is even more severe for the safety engineer, for with most products, safety hazards are caused by only a small fraction of the failures.

Conversely, systems whose failures by their very nature cause the threat of injury or death are designed with safety margins and maintenance and retirement policies such that failures are rare. In either case, if an acceptable measure of safety is to be achieved, the prevention of hazardous failures must rely heavily on more qualitative methods. Hazardous design characteristics must be eliminated before statistically significant data bases of injuries or death are allowed to develop. Thus the study of past accidents and of potential unanticipated uses or environments, along with failure modes and effects analysis and various other "what if" techniques find extensive use in identifying potential hazards and eliminating them. Careful attention must also be paid to field reports for signs of hazards incurred through product use—or misuse—for often it is only through careful detective work that hazards can be identified and eliminated.

1.4 PREVIEW

In the following two chapters we first introduce a number of concepts related to probability and sampling. The rudiments of the discrete and continuous random variables are then covered, and the distribution functions used in later discussion are presented. With this mathematical apparatus in place, we turn, in Chapter 4, to a quantitative examination of quality and its relationships to reliability. We deal first with the Taguchi methodology for the measure and improvement of quality, and then discuss statistical process control within the framework of the Six Sigma criteria. Chapter 5 is concerned with elementary methods for the statistical analysis of data. Emphasis is placed on graphical methods, particularly probability plotting methods, which are easily used in conjunction with widely available personal computer spread sheets. Classical point estimate and confidence intervals are also introduced, as are the elements of control charting.

In Chapter 6 we investigate reliability and its relationship to failure rates and other phenomena where time is the primary variable. The bathtub curve is introduced, and the relationships of reliability to failure modes, component failures, and replacements is discussed. In contrast, Chapter 7 concerns the relationships between reliability, the loading on a system, and its capacity to withstand those loads. This entails, among other things, an exposition of the probabilistic treatment of safety factors and design margins. The treatment of repetitive loading allows the time dependence of failure rates on loading, capacity and deterioration to be treated explicitly.

In Chapter 8 we return to the statistical analysis of data, but this time with emphasis on working within the limitations frequently encountered by the reliability engineer. After reliability growth and environmental stress testing are reviewed, the probability plotting methods introduced earlier are used to treat product life testing methods. Both single and multiple censoring and the various forms of accelerated testing are discussed.

Chapters 9 through 11 deal with the reliability of more complex systems. In Chapter 9 redundancy in the form of active and standby parallel systems is introduced, limitations—such as common mode failures—are examined, and the incorporation of redundancy into more complex systems is presented. Chapter 10 concentrates on maintained systems, examining the effects of both preventive and corrective maintenance and then focusing on maintainability and availability concepts for repairable system. In Chapter 11 the treatment of complex systems and their failures is brought together through an introduction to continuous-time Markov analysis.

Chapter 12 concludes the text with an introduction to system safety analysis. After discussions of the nature of hazards caused by equipment failures and by human error, quantitative methods for safety analysis are reviewed. The construction and analysis of fault tree analysis methods are then treated in some detail.

Bibliography

Brockley, D. (ed.), *Engineering Safety*, McGraw-Hill, London, 1992.

Green, A. E., and A. J. Bourne, *Reliability Technology*, Wiley, NY, 1972.

Haviland, R. D., *Engineering Reliability and Long Life Design*, Van Nostrand, New York, 1964.

Kapur, K. C., and L. R. Lamberson, *Reliability in Engineering Design*, Wiley, NY, 1977.

McCormick, N. J., *Reliability and Risk Analysis*, Academic Press, NY, 1981.

Mitra, A., *Fundamentals of Quality Control and Improvement*, Macmillan, NY, 1993.

Smith, D. J., *Reliability, Maintainability and Risk*, 4th ed., Butterworth-Heinemann, Oxford, 1993.

CHAPTER 2

Probability and Sampling

"Probability is the very guide to life."

Thomas Hobbes, 1588—1679

2.1 INTRODUCTION

Fundamental to all reliability considerations is an understanding of probability, for reliability is defined as just the *probability* that a system will not fail under some specified set of circumstances. In this chapter we define probability and discuss the logic by which probabilities can be combined and manipulated. We then examine sampling techniques by which the results of tests or experiments can be used to estimate probabilities. Although quite elementary, the notions presented will be shown to have immediate applicability to a variety of reliability considerations ranging from the relationship of the reliability of a system to its components to the common acceptance criteria used in quality control.

2.2 PROBABILITY CONCEPTS

We shall denote the probability of an event, say a failure, X, as $P\{X\}$. This probability has the following interpretation. Suppose that we perform an experiment in which we test a large number of items, for example, light bulbs. The probability that a light bulb fails the test is just the relative frequency with which failure occurs when a very large number of bulbs are tested. Thus, if N is the number of bulbs tested and n is the number of failures, we may define the probability formally as

$$P\{X\} = \lim_{N \to \infty} \frac{n}{N}.$$ (2.1)

Equation 2.1 is an empirical definition of probability. In some situations symmetry or other theoretical arguments also may be used to define probability. For example, one often assumes that the probability of a coin flip resulting in "heads" is 1/2. Closer to reliability considerations, if one has two pieces of equipment, A and B, which are chosen from a lot of equipment of the same design and manufacture, one may assume that the probability that A fails before B is 1/2. If the hypothesis is doubted in either case, one must verify that the coin is true or that the pieces of equipment are identical by performing a large number of tests to which Eq. 2.1 may be applied.

Probability Axioms

Clearly, the probability must satisfy

$$0 \leqslant P\{X\} \leqslant 1. \tag{2.2}$$

Now suppose that we denote the event not X by $\tilde{X}$. In our light-bulb example, where X indicates failure, $\tilde{X}$ then indicates that the light bulb passes the test. Obviously, the probability of passing the test, $P\{\tilde{X}\}$, must satisfy

$$P\{\tilde{X}\} = 1 - P\{X\}. \tag{2.3}$$

Equations 2.2 and 2.3 constitute two of the three axioms of probability theory. Before stating the third axiom we must discuss combinations of events.

We denote by $X \cap Y$ the event that both X and Y take place. Then, clearly $X \cap Y = Y \cap X$. The probability that both X and Y take place is denoted by $P\{X \cap Y\}$. The combined event $X \cap Y$ may be understood by the use of a Venn diagram shown in Fig. 2.1a. The area of the square is equal to one. The circular areas indicated as X and Y are, respectively, the probabilities $P\{X\}$ and $P\{Y\}$. The probability of both X and Y occurring, $P\{X \cap Y\}$, is indicated by the cross-hatched area. For this reason $X \cap Y$ is referred to as the intersection of X and Y, or simply as X *and* Y.

Suppose that one event, say X, is dependent on the second event, Y. We define the conditional probability of event X, given event Y as $P\{X|Y\}$. The third axiom of probability theory is

$$P\{X \cap Y\} = P\{X|Y\}P\{Y\}. \tag{2.4}$$

That is, the probability that both X and Y will occur is just the probability that Y occurs times the conditional probability that X occurs, given the occur-

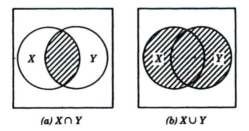

(a) $X \cap Y$ (b) $X \cup Y$

FIGURE 2.1 Venn diagrams for the intersection and union of two events.

rence of Y. Provided that the probability that Y occurs is greater than zero, Eq. 2.4 may be written as a definition of the conditional probability:

$$P\{X|Y\} = \frac{P\{X \cap Y\}}{P\{Y\}}. \tag{2.5}$$

Note that we can reverse the ordering of events X and Y, by considering the probability $P\{X \cap Y\}$ in terms of the conditional probability of Y, given the occurrence of X. Then, instead of Eq. 2.4, we have

$$P\{X \cap Y\} = P\{Y|X\}P\{X\}. \tag{2.6}$$

An important property that we will sometimes assume is that two or more events, say X and Y, are mutually independent. For events to be independent, the probability of one occurring cannot depend on the fact that the other is either occurring or not occurring. Thus

$$P\{X|Y\} = P\{X\} \tag{2.7}$$

if X and Y are independent, and Eq. 2.4 becomes

$$P\{X \cap Y\} = P\{X\}P\{Y\}. \tag{2.8}$$

This is the definition of independence, that the probability of two events both occurring is just the product of the probabilities of each of the events occurring. Situations also arise in which events are mutually exclusive. That is, if X occurs, then Y cannot, and conversely. Thus $P\{X|Y\} = 0$ and $P\{Y|X\} = 0$; or more simply, for mutually exclusive events

$$P\{X \cap Y\} = 0. \tag{2.9}$$

With the three probability axioms and the definitions of independence in hand, we may now consider the situation where either X or Y or both may occur. This is referred to as the union of X and Y or simply $X \cup Y$. The probability $P\{X \cup Y\}$ is most easily conceptualized from the Venn diagram shown in Fig. 2.1b, where the union of X and Y is just the area of the overlapping circles indicated by cross hatching. From the cross-hatched area it is clear that

$$P\{X \cup Y\} = P\{X\} + P\{Y\} - P\{X \cap Y\}. \tag{2.10}$$

If we may assume that the events X and Y are independent of one another, we may insert Eq. 2.8 to obtain

$$P\{X \cup Y\} = P\{X\} + P\{Y\} - P\{X\}P\{Y\}. \tag{2.11}$$

Conversely, for mutually exclusive events, Eqs. 2.9 and 2.10 yield

$$P\{X \cup Y\} = P\{X\} + P\{Y\}. \tag{2.12}$$

EXAMPLE 2.1

Two circuit breakers of the same design each have a failure-to-open-on-demand probability of 0.02. The breakers are placed in series so that both must fail to open in order for the circuit breaker system to fail. What is the probability of system failure (*a*) if the failures are independent, and (*b*) if the probability of a second failure is 0.1, given the failure of the first? (*c*) In part *a* what is the probability of one or more breaker failures on demand? (*d*) In part *b* what is the probability of one or more failures on demand?

Solution $X \equiv$ failure of first circuit breaker
$Y \equiv$ failure of second circuit breaker
$P\{X\} = P\{Y\} = 0.02$

(*a*) $P\{X \cap Y\} = P\{X\}P\{Y\} = 0.0004.$

(*b*) $P\{Y|X\} = 0.1$
$P\{X \cap Y\} = P\{Y|X\}P\{X\} = 0.1 \times 0.02 = 0.002.$

(*c*) $P\{X \cup Y\} = P\{X\} + P\{Y\} - P\{X\}P\{Y\}$
$= 0.02 + 0.02 - (0.02)^2 = 0.0396.$

(*d*) $P\{X \cup Y\} = P\{X\} + P\{Y\} - P\{Y|X\}P\{X\}$
$= 0.02 + 0.02 - 0.1 \times 0.02 = 0.038.$

Combinations of Events

The foregoing equations state the axioms of probability and provide us with the means of combining two events. The procedures for combining events may be extended to three or more events, and the relationships may again be presented graphically as Venn diagrams. For example, in Fig. 2.2*a* and *b* are shown, respectively, the intersection of *X*, *Y*, and *Z*, $X \cap Y \cap Z$; and the union of *X*, *Y*, and *Z*, $X \cup Y \cup Z$. The probabilities $P\{X \cap Y \cap Z\}$ and $P\{X \cup Y \cup Z\}$ may again be interpreted as the cross-hatched areas.

The following observations are often useful in dealing with combinations of two or more events. Whenever we have a probability of a union of events, it may be reduced to an expression involving only the probabilities of the individual events and their intersection. Equation 2.10 is an example of this. Similarly, probabilities of more complicated combinations involving unions and intersections may be reduced to expressions involving only probabilities of intersections. The intersections of events, however, may be eliminated only by expressing them in terms of conditional probabilities, as in Eq. 2.6, or if

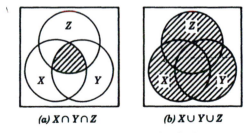

(a) $X \cap Y \cap Z$ (b) $X \cup Y \cup Z$

FIGURE 2.2 Venn diagrams for the intersection and union of three events.

TABLE 2.1 Rules of Boolean Algebra[a]

Mathematical symbolism	Designation
(1a) $X \cap Y = Y \cap X$	Commutative law
(1b) $X \cup Y = Y \cup X$	
(2a) $X \cap (Y \cap Z) = (X \cap Y) \cap Z$	Associative law
(2b) $X \cup (Y \cup Z) = (X \cup Y) \cup Z$	
(3a) $X \cap (Y \cup Z) = (X \cap Y) \cup (X \cap Z)$	Distributive law
(3b) $X \cup (Y \cap Z) = (X \cup Y) \cap (X \cup Z)$	
(4a) $X \cap X = X$	Idempotent law
(4b) $X \cup X = X$	
(5a) $X \cap (X \cup Y) = X$	Law of absorption
(5b) $X \cup (X \cap Y) = X$	
(6a) $X \cap \bar{X} = \phi^b$	Complementation
(6b) $X \cup \bar{X} = I^b$	
(6c) $(\bar{\bar{X}}) = X$	
(7a) $\overline{(X \cap Y)} = \bar{X} \cup \bar{Y}$	de Morgan's theorem
(7b) $\overline{(X \cup Y)} = \bar{X} \cap \bar{Y}$	
(8a) $\phi \cap X = \phi$	Operations with I
(8b) $\phi \cup X = X$	
(8c) $I \cap X = X$	
(8d) $I \cup X = I$	
(9a) $X \cup (\bar{X} \cap Y) = X \cup Y$	These relationships are unnamed.
(9b) $\bar{X} \cap (X \cup \bar{Y}) = \bar{X} \cap \bar{Y} = \overline{(X \cup Y)}$	

[a] Adapted from H. R. Roberts, W. E. Vesley, D. F. Haastand, and F. F. Goldberg, *Fault tree Handbook*, NUREG-0492, U.S. Nuclear Regulatory Commission, 1981.

[b] ϕ = null set; I = universal set.

the independence may be assumed, they may be expressed in terms of the probabilities of individual events as in Eq. 2.8.

The treatment of combinations of events is streamlined by using the rules of Boolean algebra listed in Table 2.1. If two combinations of events are equal according to these rules, their probabilities are equal. Thus since according to Rule 1a, $X \cap Y = Y \cap X$, we also have $P\{X \cap Y\} = P\{Y \cap X\}$. The communicative and associative rules are obvious. The remaining rules may be verified from a Venn diagram. For example, in Fig. 2.3a and b, respectively, we show the distributive laws for $X \cap (Y \cup Z)$ and $X \cup (Y \cap Z)$. Note that

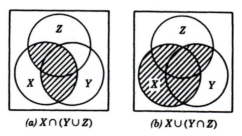

(a) $X \cap (Y \cup Z)$ *(b)* $X \cup (Y \cap Z)$

FIGURE 2.3 Venn diagrams for combinations of three events.

in Table 2.1, ϕ is used to represent the null event for which $P\{\phi\} = 0$, and I is used to represent the universal event for which $P\{I\} = 1$.

Probabilities of combinations involving more than two events may be reduced sums of the probabilities of intersections of events. If the events are also independent, the intersection probabilities may further be reduced to products of probabilities. These properties are best illustrated with the following two examples.

EXAMPLE 2.2

Express $P\{X \cap (Y \cup Z)\}$ in terms of the probabilities of intersections of X, Y, and Z. Then assume that X, Y, and Z are independent events and express the result in terms of $P\{X\}$, $P\{Y\}$, and $P\{Z\}$.

Solution   Rule 3a: $P\{X \cap (Y \cup Z)\} = P\{(X \cap Y) \cup (X \cap Z)\}$
This is the union of two composites $X \cap Y$ and $Y \cap Z$. Therefore from Eq. 2.10:
$P\{X \cap (Y \cup Z)\} = P\{X \cap Y\} + P\{X \cap Z\} - P\{(X \cap Y) \cap (X \cap Z)\}$.
Associative rules 2a and 2b allow us to eliminate the parenthesis from the last term by first writing $(X \cap Y) \cap (X \cap Z) = (Y \cap X) \cap (X \cap Z)$ and then using law 4a to obtain
$(Y \cap X) \cap (X \cap Z) = Y \cap (X \cap X) \cap Z = Y \cap X \cap Z = X \cap Y \cap Z$.
Utilizing these intermediate results, we have
$P\{X \cap (Y \cup Z)\} = P\{X \cap Y\} + P\{X \cap Z\} - P\{X \cap Y \cap Z\}$.
If the events are independent, we may employ Eq. 2.8 to write
$P\{X \cap (Y \cup Z)\} = P\{X\}P\{Y\} + P\{X\}P\{Z\} - P\{X\}P\{Y\}P\{Z\}$.

EXAMPLE 2.3

Repeat Example 2.2 for $P\{X \cup Y \cup Z\}$.

Solution   From the associative law, $P\{X \cup Y \cup Z\} = P\{X \cup (Y \cup Z)\}$
Since this is the union of event X and $(Y \cup Z)$, we use Eq. 2.10 to obtain
$P\{X \cup Y \cup Z\} = P\{X\} + P\{Y \cup Z\} - P\{X \cap (Y \cup Z)\}$
and again to expand the second term on the right as
$P\{Y \cup Z\} = P\{Y\} + P\{Z\} - P\{Y \cap Z\}$.
Finally, we may apply the result from Example 2.2 to the last term, yielding
$P\{X \cup Y \cup Z\} = P\{X\} + P\{Y\} + P\{Z\} - P\{X \cap Y\}$
$\qquad\qquad - P\{X \cap Z\} - P\{Y \cap Z\} + P\{X \cap Y \cap Z\}$.
Applying the product rule for the intersections of independent events, we have
$P\{X \cup Y \cup Z\} = P\{X\} + P\{Y\} + P\{Z\} - P\{X\}P\{Y\}$
$\qquad\qquad - P\{X\}P\{Z\} - P\{Y\}P\{Z\} + P\{X\}P\{Y\}P\{Z\}$

In the following chapters we will have occasion to deal with intersections and unions of large numbers of n independent events: $X_1, X_2, X_3 \ldots X_n$. For intersections, the treatment is straightforward through the repeated application of the product rule:

$$P\{X_1 \cap X_2 \cap X_3 \cap \cdots \cap X_n\} = P\{X_1\}P\{X_2\}P\{X_3\} \cdots P\{X_n\}. \quad (2.13)$$

To obtain the probability for the union of these events, we first note that the union may be related to the intersection of the nonevents $\tilde{X}_i$:

$$P\{X_1 \cup X_2 \cup X_3 \cup \cdots \cup X_n\} + P\{\tilde{X}_1 \cap \tilde{X}_2 \cap \tilde{X}_3 \cap \cdots \tilde{X}_n\} = 1, \quad (2.14)$$

which may be visualized by drawing a Venn diagram for three or four events. Now if we apply Eq. 2.13 to the independent $\tilde{X}_i$, we obtain, after rearranging terms

$$P\{X_1 \cup X_2 \cup X_3 \cup \cdots \cup X_n\} = 1 - P\{\tilde{X}_1\}P\{\tilde{X}_2\}P\{\tilde{X}_3\} \cdots P\{\tilde{X}_n\}. \quad (2.15)$$

Finally, from Eq. 2.3 we must have for each X_i,

$$P\{\tilde{X}_i\} = 1 - P\{X_i\}. \quad (2.16)$$

Thus we have,

$$P\{X_1 \cup X_2 \cup X_3 \cup \cdots \cup X_n\} = 1 - [1 - P\{X_1\}][1 - P\{X_2\}]$$
$$[1 - P\{X_3\}] \cdots [1 - P\{X_n\}], \quad (2.17)$$

or more compactly

$$P\{X_1 \cup X_2 \cup X_3 \cup \cdots \cup X_n\} = 1 - \prod_{i=1}^{n} [1 - P\{X_i\}]. \quad (2.18)$$

This expression may also be shown to hold for the $\tilde{X}_i$.

EXAMPLE 2.4

A critical seam in an aircraft wing must be reworked if any one of the 28 identical rivets is found to be defective. Quality control inspections find that 18% of the seams must be reworked. (*a*) Assuming that the defects are independent, what is the probability that a rivet will be defective? (*b*) To what value must this probability be reduced if the rework rate is to be reduced below 5%?

  Solution (*a*) Let X_i represent the failure of the *i*th rivet. Then, since

$$P\{X_1\} = P\{X_2\} = \cdots P\{X_{28}\},$$
$$0.18 = P\{X_1 \cup X_2 \cup \cdots \cup X_{28}\} = 1 - [1 - P\{X_1\}]^{28}$$
$$P\{X_1\} = 1 - (0.82)^{1/28} = 0.0071.$$

  (*b*) Since $0.05 = 1 - [1 - P\{X_1\}]^{28}$,
$$P\{X_1\} = 1 - (0.95)^{1/28} = 0.0018.$$

One other expression is very useful in the solution of certain reliability problems. It is sometimes referred to as the law of "total probability." Suppose we divide a Venn diagram into regions of X and $\tilde{X}$ as shown in Fig. 2.4 We can always decompose the probability of Y, denoted by the circle, into two mutually exclusive contributions:

$$P\{Y\} = P\{Y \cap X\} + P\{Y \cap \tilde{X}\}. \quad (2.19)$$

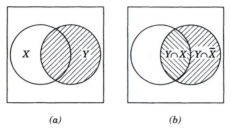

(a)   (b)

FIGURE 2.4 Venn diagram for total probability law.

Thus using Eq. 2.4, we have

$$P\{Y\} = P\{Y|X\}P\{X\} + P\{Y|\tilde{X}\}P\{\tilde{X}\}. \tag{2.20}$$

EXAMPLE 2.5

A motor operated relief valve opens and closes intermittently on demand to control the coolant level in an industrial process. An auxiliary battery pack is used to provide power for the approximately 1/2 percent of the time when there are plant power outages. The demand failure probability of the valve is found to be 3×10^{-5} when operated from the plant power and 9×10^{-5} when operated from the battery pack. Calculate the demand failure probability assuming that the number of demands is independent of the power source. Is the increase due to the battery pack operation significant?

Solution Let X signify a power outage. Then $P\{X\} = 0.005$ and $P\{\tilde{X}\} = 0.995$. Let Y signify valve failure. Then $P\{Y|\tilde{X}\} = 3 \times 10^{-5}$ and $P\{Y|X\} = 9 \times 10^{-5}$. From Eq. 2.20, the valve failure per demand is,

$$P\{Y\} = 9 \times 10^{-5} \times 0.005 + 3 \times 10^{-5} \times 0.095 = 3.03 \times 10^{-5}.$$

The net increase in the failure probability over operation entirely with plant power is only three percent.

2.3 DISCRETE RANDOM VARIABLES

Frequently in reliability considerations, we need to know the probability that a specific number of events will occur, or we need to determine the average number of events that are likely to take place. For example, suppose that we have a computer with N memory chips and we need to know the probability that none of them, that one of them, that two of them, and so on, will fail during the first year of service. Or suppose that there is a probability p that a Christmas tree light bulb will fail during the first 100 hours of service. Then, on a string of 25 lights, what is the probability that there will be n $(0 \leq n \leq 25)$ failures during this 100-hr period? To answer such reliability questions, we need to introduce the properties of discrete random variables. We do this first in general terms, before treating two of the most important discrete probability distributions.

Properties of Discrete Variables

A discrete random variable is a quantity that can be equal to any one of a number of discrete values $x_0, x_1, x_2, \ldots, x_n, \ldots, x_N$. We refer to such a variable with the bold-faced character $\mathbf{x}$, and denote by x_n the values to which it may be equal. In many cases these values are integers so that $x_n = n$. By *random* variables we mean that there is associated with each x_n a probability $f(x_n)$ that $\mathbf{x} = x_n$. We denote this probability as

$$f(x_n) = P\{\mathbf{x} = x_n\}. \tag{2.21}$$

We shall, for example, often be concerned with counting numbers of failures (or of successes). Thus we may let $\mathbf{x}$ signify the number n of failures in N tests. Then $f(0)$ is the probability that there will be no failure, $f(1)$ the probability of one failure, and so on. The probabilities of all the possible outcomes must add to one

$$\sum_n f(x_n) = 1, \tag{2.22}$$

where the sum is taken over all possible values of x_n.

The function $f(x_n)$ is referred to as the *probability mass function* (PMF) of the discrete random variable $\mathbf{x}$. A second important function of the random variable is the *cumulative distribution function* (CDF) defined by

$$F(x_n) = P\{\mathbf{x} \leq x_n\}, \tag{2.23}$$

the probability that the value of $\mathbf{x}$ will be less than or equal to the value x_n. Clearly, it is just the sum of probabilities:

$$F(x_n) = \sum_{n'=0}^{n} f(x_{n'}). \tag{2.24}$$

Closely related is the *complementary cumulative distribution function* (CCDF), defined by the probability that $\mathbf{x} > x_n$:

$$\tilde{F}(x_n) = P\{\mathbf{x} > x_n\}. \tag{2.25}$$

It is related to the PMF by

$$\tilde{F}(x_n) = 1 - F(x_n) = \sum_{n'=n+1}^{N} f(x_{n'}), \tag{2.26}$$

where x_N is the largest value for which $f(x_n) > 0$.

It is often convenient to display discrete random variables as bar graphs of the PMF. Thus, if we have, for example,

$$f(0) = 0, \quad f(1) = \tfrac{1}{16}, \quad f(2) = \tfrac{1}{4}, \quad f(3) = \tfrac{3}{8}, \quad f(4) = \tfrac{1}{4}, \quad f(5) = \tfrac{1}{16},$$

the PMF may be plotted as in Fig. 2.5a. Similarly, from Eq. 2.24 the bar graph for the CDF appears as in Fig. 2.5b.

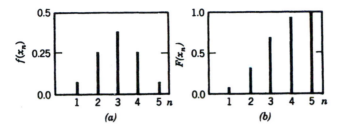

FIGURE 2.5 Discrete probability distribution: (*a*) probability mass function (PMF), (*b*) corresponding cumulative distribution function (CDF).

Several important properties of the random variable **x** are defined in terms of the probability mass function $f(x_n)$. The *mean* value, μ, of **x** is

$$\mu = \sum_n x_n f(x_n), \qquad (2.27)$$

and the *variance* of **x** is

$$\sigma^2 = \sum_n (x_n - \mu)^2 f(x_n), \qquad (2.28)$$

which may be reduced to

$$\sigma^2 = \sum_n x_n^2 f(x_n) - \mu^2. \qquad (2.29)$$

The mean is a measure of the expected value or central tendency of **x** when a very large sampling is made of the random variable, whereas the variance is a measure of the scatter or dispersion of the individual values of x_n about μ. It is also sometimes useful to talk about the most probable value of **x**: the value of x_n for which the largest value of $f(x_n)$ occurs, assuming that there is only one largest value. Finally, the median value is defined as that value **x** $= x_n$ for which the probability of obtaining a smaller value is $1/2$:

$$\sum_{n' \geq n} f(x_{n'}) = \tfrac{1}{2}, \qquad (2.30)$$

and consequently,

$$\sum_{n' \leq n} f(x_{n'}) = \tfrac{1}{2}. \qquad (2.31)$$

EXAMPLE 2.6

A discrete probability distribution is given by

$$f(x_n) = An \qquad n = 0, 1, 2, 3, 4, 5$$

(*a*) Determine A.

(*b*) What is the probability that **x** ≤ 3?

(*c*) What is μ?

(*d*) What is σ?

   Solution   (*a*) From Eq. 2.22

$$1 = \sum_{n=0}^{5} An = A(0 + 1 + 2 + 3 + 4 + 5) = 15A$$

$$A = \frac{1}{15}.$$

(*b*) From Eq. 2.23 and 2.24,

$$P\{\mathbf{x} \leq 3\} = F(3) = \sum_{n=0}^{3} \frac{n}{15} = \frac{1}{15} (0 + 1 + 2 + 3) = \frac{2}{5}.$$

(*c*) From Eq. 2.27

$$\mu = \sum_{n=0}^{5} n \frac{n}{15} = \frac{1}{15} (0 + 1 + 4 + 9 + 16 + 25) = \frac{11}{3}.$$

(*d*) Using Eq. 2.29, we first calculate

$$\sum_{n=0}^{5} x_n^2 f(x_n) = \sum_{n=0}^{5} \frac{1}{15} n^3 = \frac{1}{15} (0 + 1 + 8 + 27 + 64 + 125) = 15,$$

to obtain for the variance

$$\sigma^2 = 15 - \mu^2 = 15 - \left(\frac{11}{3}\right)^2 = 1.555$$

$$\sigma = 1.247.$$

The idea of the expected value is an important one. In general, if there is a function $g(x_n)$ of the random variable $\mathbf{x}$, the *expected value* $E\{g\}$ is defined for a discrete random variable as

$$E\{g\} = \sum_{n} g(x_n) f(x_n). \tag{2.32}$$

Thus the mean and variance given by Eqs. 2.27 and 2.28 may be written as

$$\mu = E\{x\} \tag{2.33}$$

$$\sigma^2 = E\{(x - \mu)^2\} \tag{2.34}$$

or as in Eq. 2.29,

$$\sigma^2 = E\{x^2\} - \mu^2. \tag{2.35}$$

The quantity $\sigma = \sqrt{\sigma^2}$ is referred to as the standard error or *standard deviation* of the distribution. The notion of expected value is also applicable to the continuous random variables discussed in the following chapter.

The Binomial Distribution

The binomial distribution is the most widely used discrete distribution in reliability considerations. To derive it, suppose that p is the probability of failure for some piece of equipment in a specified test and

$$q = 1 - p \tag{2.36}$$

is the corresponding success (i.e., nonfailure) probability. If such tests are truly independent of one another, they are referred to as Bernoulli trials.

We wish to derive the probability

$$f(n) = P\{\mathbf{n} = n | N, p\} \tag{2.37}$$

that in N independent tests there are n failures. To arrive at this probability, we first consider the example of the test of two units of identical design and construction. The tests must be independent in the sense that success or failure in one test does not depend on the result of the other. There are four possible outcomes, each with an associated probability: qq is the probability that neither unit fails, pq the probability that only the first unit fails, qp the probability that only the second unit fails, and pp the probability that both units fail. Since these are the only possible outcomes of the test, the sum of the probabilities must equal one. Indeed,

$$p^2 + 2pq + q^2 = (p + q)^2 = 1, \tag{2.38}$$

and by the definition of Eq. 2.37

$$f(0) = q^2, \quad f(1) = 2qp, \quad f(2) = p^2. \tag{2.39}$$

In a similar manner the probability of n independent failures may also be covered for situations in which a larger number of units undergo testing. For example, with $N = 3$ the probability that all three units fail independently is obtained by multiplying the failure probabilities of the individual units together. Since the units are identical, the probability that none of the three fails is qqq. There are now three ways in which the test can result in one unit failing: the first fails, pqq; the second fails, qpq; or the third fails, qqp. There are also three combinations that lead to two units failing: units 1 and 2 fail, ppq; units 1 and 3 fail, pqp; or units 2 and 3 fail, qpp. Finally, the probability of all three units failing is ppp.

In the three-unit test the probabilities for the eight possible outcomes must again add to one. This is indeed the case, for by combining the eight terms into four we have

$$q^3 + 3q^2p + 3qp^2 + p^3 = (q + p)^3 = 1. \tag{2.40}$$

The probabilities of the test resulting in 0, 1, 2, or 3 failures are just the successive terms on the left:

$$f(0) = q^3, \quad f(1) = 3q^2p, \quad f(2) = 3qp^2, \quad f(3) = p^3. \tag{2.41}$$

The foregoing process may be systematized for tests of any number of units. For N units Eq. 2.41 generalizes to

$$C_0^N q^N + C_1^N p q^{N-1} + C_2^N p^2 q^{N-2} + \cdots + C_{N-1}^N p^{N-1} q$$
$$+ C_N^N p^N = (q + p)^N = 1, \quad (2.42)$$

since $q = 1 - p$. For this expression to hold, it may be shown that the C_n^N must be the binomial coefficients. These are given by

$$C_n^N = \frac{N!}{(N - n)! n!}. \quad (2.43)$$

A convenient way to tabulate these coefficients is in the form of Pascal's triangle; this is shown in Table 2.2. Just as in the case of $N = 2$ or 3, the $N + 1$ terms on the left-hand side of Eq. 2.42 are the probabilities that there will be 0, 1, 2, ... , N failures. Thus the PMF for the binomial distribution is

$$f(n) = C_n^N p^n (1 - p)^{N-n}, \quad n = 0, 1, \ldots, N. \quad (2.44)$$

That the condition Eq. 2.22 is satisfied follows from Eq. 2.42. The CDF corresponding to $f(n)$ is

$$F(n) = \sum_{n'=0}^{n} C_{n'}^N p^{n'} (1 - p)^{N-n'}, \quad (2.45)$$

and of course if we sum over all possible values of n' as indicated in Eq. 2.22 we must have

$$\sum_{n=0}^{N} C_n^N p^n (1 - p)^{N-n} = 1. \quad (2.46)$$

The mean of the binomial distribution is

$$\mu = Np, \quad (2.47)$$

and the variance is

$$\sigma^2 = Np(1 - p). \quad (2.48)$$

TABLE 2.2 Pascal's Triangle

				1					$N = 0$
			1		1				$N = 1$
		1		2		1			$N = 2$
	1		3		3		1		$N = 3$
1		4		6		4		1	$N = 4$
1	5		10		10		5		1	$N = 5$
1	6	15		20		15		6	1	$N = 6$
1	7	21	35		35		21	7	1	$N = 7$
1	8	28	56	70		56	28	8	1	$N = 8$

EXAMPLE 2.7

Ten compressors with a failure probability $p = 0.1$ are tested. (*a*) What is the expected number of failures $E\{n\}$? (*b*) What is σ^2? (*c*) What is the probability that none will fail? (*d*) What is the probability that two or more will fail?

Solution (*a*) $E\{n\} = \mu = Np = 10 \times 0.1 = 1$.

(*b*) $\sigma^2 = Np(1 - p) = 10 \times 0.1(1 - 0.1) = 0.9$.

(*c*) $P\{\mathbf{n} = 0 | 10, p\} = f(0) = C_0^{10} p^0 (1 - p)^{10} = 1 \times 1 \times (1 - 0.1)^{10} = 0.349$.

(*d*) $P\{\mathbf{n} \geqslant 2 | 10, p\} = 1 - f(0) - f(1) = 1 - C_0^{10} p^0 (1 - p)^{10} - C_1^{10} p^1 (1 - p)^9$
$= 1 - (1 - 0.1)^{10} - 10 \times 0.1 \times (1 - 0.1)^9 = 0.264$.

The proof of Eqs. 2.47 and 2.48 requires some manipulation of the binomial terms. From Eqs. 2.27 and 2.44 we see that

$$\mu = \sum_{n=1}^{N} n C_n^N p^n (1 - p)^{N-n}, \tag{2.49}$$

where the $n = 0$ term vanishes and therefore is eliminated. Making the substitutions $M = N - 1$ and $m = n - 1$ we may rewrite the series as

$$\mu = p \sum_{m=0}^{M} (m + 1) C_{m+1}^{M+1} p^m (1 - p)^{M-m}. \tag{2.50}$$

Since it is easily shown that

$$(m + 1) C_{m+1}^{M+1} = (M + 1) C_m^M, \tag{2.51}$$

we may write

$$\mu = (M + 1) p \sum_{m=0}^{M} C_m^M p^m (1 - p)^{M-m}. \tag{2.52}$$

However, Eq. 2.46 indicates that the sum on the right is equal to one. Therefore, noting that $M + 1 = N$, we obtain the value of the mean given by Eq. 2.47.

To obtain the variance we begin by combining Eqs. 2.29, 2.44 and 2.47

$$\sigma^2 = \sum_{n=1}^{N} n^2 C_n^N p^n (1 - p)^{N-n} - N^2 p^2. \tag{2.53}$$

Employing the same substitutions for N and n, and utilizing Eq. 2.51, we obtain

$$\sigma^2 = (M + 1) p \left\{ \sum_{m=0}^{M} m C_m^M p^m (1 - p)^{M-m} + \sum_{m=0}^{M} C_m^M p^m (1 - p)^{M-m} \right\} - N^2 p^2. \tag{2.54}$$

But from Eqs. 2.46 and 2.49 we see that the first of the two sums is just equal to Mp and the second is equal to one. Hence

$$\sigma^2 = (M + 1) p(Mp + 1) - N^2 p^2. \tag{2.55}$$

Finally, since $M = N - 1$, this expression reduces to Eq. 2.48.

The Poisson Distribution

Situations in which the probability of failure p becomes very small, but the number of units tested N is large, are frequently encountered. It then becomes cumbersome to evaluate the large factorials appearing in the binomial distribution. For this, as well as for a variety of situations discussed in later chapters, the Poisson distribution is employed.

The Poisson distribution may be shown to result from taking the limit of the binomial distribution as $p \to 0$ and $N \to \infty$, with the product Np remaining constant. To obtain the distribution we first multiply the binomial PDF given by Eq. 2.44 by N^n / N^n and rearrange the factors to yield

$$f(n) = \left\{ \frac{N!}{(N-n)!N^n} \right\} (1-p)^{-n} \frac{(Np)^n}{n!} (1-p)^N. \qquad (2.56)$$

Now assume that $p \ll 1$ so that we may write $\ln (1-p) \approx -p$ and hence the last factor becomes

$$(1-p)^N = \exp[N \ln (1-p)] \approx e^{-Np}. \qquad (2.57)$$

Likewise as p becomes vanishingly small $(1-p)^{-n} \to 1$ for finite n, and as $N \to \infty$, we have

$$\frac{N!}{(N-n)!N^n} = \left(1 - \frac{n-1}{N}\right)\left(1 - \frac{n-2}{N}\right) \cdots \left(1 - \frac{1}{N}\right) 1 \to 1 \qquad (2.58)$$

Hence as $p \to 0$ and $N \to \infty$, with $Np = \mu$, Eq. 2.56 reduces to

$$f(n) = \frac{\mu^n}{n!} e^{-\mu}, \qquad (2.59)$$

which is the probability mass function for the Poisson distribution.

Unlike the binomial distribution, the Poisson distribution can be expressed in terms of a single parameter, μ. Thus $f(n)$ may be written as the probability

$$P\{\mathbf{n} = n | \mu\} = \frac{\mu^n}{n!} e^{-\mu}, \qquad n = 0, 1, 2, 3, \ldots. \qquad (2.60)$$

The normalization condition, Eq. 2.22, must, of course, be satisfied. This may be verified by first recalling the power series expansion for the exponential function

$$e^\mu = \sum_{n=0}^{\infty} \frac{\mu^n}{n!}. \qquad (2.61)$$

Thus we have

$$\sum_{n=0}^{\infty} f(n) = \sum_{n=0}^{\infty} \frac{\mu^n}{n!} e^{-\mu} = e^\mu e^{-\mu} = 1. \qquad (2.62)$$

In the foregoing equations we have chosen $Np = \mu$ because it may be shown to be the mean of the Poisson distribution. From Eqs. 2.59 and 2.61 we have

$$\sum_{n=0}^{\infty} nf(n) = \sum_{n=0}^{\infty} n\frac{\mu^n}{n!} e^{-\mu} = \mu. \tag{2.63}$$

Likewise, since it may be shown that

$$\sum_{n=0}^{\infty} n^2 f(n) = \sum_{n=0}^{\infty} n^2 \frac{\mu^n}{n!} e^{-\mu} = \mu(\mu + 1), \tag{2.64}$$

we may use Eq. 2.35 to show that the variance is equal to the mean,

$$\sigma^2 = \mu. \tag{2.65}$$

EXAMPLE 2.8
___

Do the preceding 10-compressor example approximating the binomial distribution by a Poisson distribution. Compare the results.

   Solution   (a) $\mu = Np = 1$.
(b) $\sigma^2 = \mu = 1$ (0.9 for binomial).
(c) $P\{n = 0 | \mu = 1\} = e^{-\mu} = 0.3678$ (0.3874 for binomial).
(d) $P\{n \geqslant 2 | \mu = 1\} = 1 - f(0) - f(1) = 1 - 2e^{-\mu} = 0.2642$ (0.2639 for binomial).

___

2.4   ATTRIBUTE SAMPLING

The discussions in the preceding section illustrate how the binomial and Poisson distributions can be determined, given the parameter p, which we often use to denote a failure probability. In reliability engineering and the associated discipline of quality assurance, however, one rarely has the luxury of knowing the value of p, *a priori*. More often, the problem is to estimate a failure probability, mean number of failures, or other related quantity from test data. Moreover, the amount of test data is often quite restricted, for normally one cannot test large numbers of products to failure. For the number of such destructive tests that may be performed is severely restricted both by cost and the completion time, which may be equal to the product design life or longer.

   Probability estimation is a fundamental task of statistical inference, which may be stated as follows. Given a very large—perhaps infinite—population of items of identical design and manufacture, how does one estimate the failure probability by testing a sample of size N drawn from this large population? In what follows we examine the most elementary case, that of attribute testing in which the data consists simply of a pass or fail for each item tested. We approach this by first introducing the point estimator and sampling distribution, and then discussing interval estimates and confidence levels. More extensive treatments are found in standard statistics texts; we shall return to the treatment of statistical estimates for random variables in Chapter 5.

Sampling Distribution

Suppose we want to estimate the failure probability p of a system and also gain some idea of the precision of the estimate. Our experiment consists of testing N units for failure, with the assumption that the N units are drawn randomly from a much larger population. If there are n failures, the failure probability, defined by Eq. 2.1, may be estimated by

$$\hat{p} = n/N \tag{2.66}$$

We use the caret to indicate that $\hat{p}$ is an estimate, rather than the true value p. It is referred to as a point estimate of p, since there is no indication of how close it may be to the true value.

The difficulty, of course, is that if the test is repeated, a different value of n, and therefore of $\hat{p}$, is likely to result. The number of failures is a random variable that obeys the binomial distribution discussed in the preceding section. Thus $\hat{p}$ is also a random variable. We may define a probability mass function (PMF) as

$$P\{\hat{\mathbf{p}} = \hat{p}_n | N, p\} = f(\hat{p}_n), \qquad n = 0, 1, 2, \dots N, \tag{2.67}$$

where $\hat{p}_n = n/N$ is just the value taken on by $\hat{\mathbf{p}}$ when there are n failures in N trials. The PMF is just the binomial distribution given by Eq. 2.44

$$f(\hat{p}_n) = C_n^N p^n (1 - p)^{N-n}. \tag{2.68}$$

This probability mass function is called the sampling distribution. It indicates that the probability for obtaining a particular value $\hat{p}_n$ from our test is just $f(\hat{p}_n)$, given that the true value is p.

For a specified value of p, we may gain some idea of the precision of the estimate for a given sample size N by plotting the $f(\hat{p}_n)$. Such plots are shown in Fig. 2.6 for $p = 0.25$ with several different values of N. We see—not surprisingly—that with larger sample sizes the distribution bunches increasingly about p, and the probability of obtaining a value of $\hat{p}$ with a large error becomes smaller. With $p = 0.25$ the probability that $\hat{p}$ will be in error by more than 0.10 is about 50% when $N = 10$, about 20% when $N = 20$, and only about 10% when $N = 40$.

We may show that Eq. 2.66 is an unbiased estimator: If many samples of size N are obtained, the mean value of the estimator (i.e., the mean taken over all the samples) converges to the true value of p. Equivalently, we must show that the expected value of $\hat{p}$ is equal to p. Thus for $\hat{p}$ to be unbiased we must have $E\{\hat{p}\} = p$. To demonstrate this we first note by comparing Eqs. 2.44 and 2.68 that $f(\hat{p}_n) = f(n)$. Thus with $\hat{p} = n/N$ we have

$$\mu_{\hat{p}} \equiv E\{\hat{p}\} = \sum_n \hat{p}_n f(\hat{p}_n) = \frac{1}{N} \sum_n n f(n). \tag{2.69}$$

The sum on the right, however is just Np, the mean value of n. Thus we have

$$\mu_{\hat{p}} = p. \tag{2.70}$$

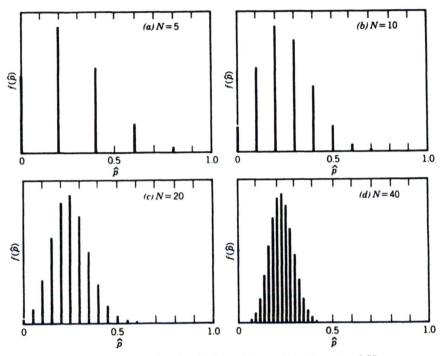

FIGURE 2.6 Probability mass function for binomial sampling where $p = 0.25$.

The increased precision of the estimator with increased N is demonstrated by observing that the variance of the sampling distribution decreases with increased N. From Eq. 2.29 we have

$$\sigma_{\hat{p}}^2 = \sum_n \hat{p}_n^2 f(\hat{p}_n) - \hat{\mu}_{\hat{p}}^2. \tag{2.71}$$

Inserting $\hat{\mu} = Np$, $\hat{p} = n/N$, and $f(\hat{p}_n) = f(n)$, we have

$$\sigma_{\hat{p}}^2 = \frac{1}{N^2} \left\{ \sum_n n^2 f(n) - \mu^2 \right\}, \tag{2.72}$$

but since the bracketed term is just $Np(1 - p)$, the variance of the binomial distribution, we have

$$\sigma_{\hat{p}}^2 = \frac{1}{N} p(1 - p), \tag{2.73}$$

or equivalently

$$\sigma_{\hat{p}} = \frac{1}{\sqrt{N}} \sqrt{p(1 - p)}. \tag{2.74}$$

Unfortunately, we do not know the value of p beforehand. If we did, we would not be interested in using the estimator to obtain an approximate value. Therefore, we would like to estimate the precision of $\hat{p}$ without knowing the exact value of p. For this we must introduce the somewhat more subtle notion of the confidence interval.

Confidence Intervals

The confidence interval is the primary means by which the precision of a point estimator can be determined. It provides lower and upper confidence limits to indicate how tightly the sampling distribution is compressed around the true value of the estimated quantity. We shall treat confidence interval more extensively in Chapter 5. Here we confine our attention to determining the values of

$$p^- = \hat{p} - A \tag{2.75}$$

and

$$p^+ = \hat{p} + B, \tag{2.76}$$

where these lower and upper confidence limits are associated with the point estimator $\hat{p}$.

To determine A and B, and therefore the limits, we first choose a risk level designated by α: $\alpha = 0.05$, which, for example, would be a 5% risk. Suppose we are willing to accept a risk of $\alpha/2$ in which the estimated lower confidence limit p^- will turn out to be larger than p, the true value of the failure probability. This may be stated as the probability

$$P\{p^- > p\} = \alpha/2, \tag{2.77}$$

which means we are $1 - \alpha/2$ confident that the calculated lower confidence limit will be less or equal to the true value:

$$P\{p^- \leq p\} = 1 - \alpha/2. \tag{2.78}$$

To determine the lower confidence limit we first insert Eq. 2.75 and rearrange the inequality to obtain

$$P\{\hat{p} \leq p + A\} = 1 - \alpha/2. \tag{2.79}$$

But this is just the CDF for the sampling distribution evaluated at $p + A$. Thus from the definition of the Cumulative Distribution Function given in Eq. 2.24 we may write

$$\sum_{\hat{p}_n \leq p+A} f(\hat{p}_n) = 1 - \alpha/2. \tag{2.80}$$

Recalling that $\hat{p}_n = n/N$ and copying the Probability Mass Function explicitly from Eq. 2.68, we have

$$\sum_{n=0}^{N(p+A)} C_n^N p^n (1 - p)^{N-n} = 1 - \alpha/2. \tag{2.81}$$

Thus to find the lower confidence limit we must determine the value of A for which this condition is most closely satisfied for specified α, N and p.

Similarly, to obtain the upper limit at the same confidence we require

$$P\{p \leq p^+\} = 1 - \alpha/2, \qquad (2.82)$$

which upon insertion of Eq. 2.76 yields

$$P\{\hat{p} \geq p - B\} = 1 - \alpha/2 \qquad (2.83)$$

and leads to the analogous condition on B,

$$\sum_{n=N(p-B)}^{N} C_n^N p^n (1 - p)^{N-n} = 1 - \alpha/2. \qquad (2.84)$$

To express the confidence interval more succinctly, the combined results of the foregoing equations are frequently expressed as the probability

$$P\{p^- \leq p \leq p^+\} = 1 - \alpha. \qquad (2.85)$$

Solutions for Eqs. 2.81 and 2.84 have been presented in convenient graphical form for obtaining p^+ and p^- from the point estimator $\hat{p} = n/N$. These are shown for a 95% confidence interval, corresponding to $\alpha/2 = 0.025$, in Fig. 2.7 for values of N ranging from 10 to 1000. The corresponding graphs for other confidence intervals are given in Appendix B.

The results in Fig. 2.7 indicate the limitations of classical sampling methods if highly accurate estimates are required, particularly when small failure probabilities are under considerations. Suppose, for example, that 10 items are tested with only one failure; our 95% confidence interval is then $0.001 < p < 0.47$. Much larger samples are needed to obtain reasonable error bounds on the parameter p. For sufficiently large values of N, typically $Np > 5$ and $N(1 - p) > 5$, the confidence interval may be expressed as

$$p^{\pm} = p \pm z_{\alpha/2} \frac{1}{\sqrt{N}} \sqrt{p(1 - p)} \qquad (2.86)$$

with $z_{0.1} = 1.28$, $z_{0.05} = 1.54$, $z_{0.025} = 1.96$ and $z_{0.005} = 2.58$. The origin of this expression is discussed in Chapter 5. Note that in all binomial sampling the true value of p is unknown. Thus $\hat{p}$, the unbiased point estimator, must be utilized to evaluate this expression.

EXAMPLE 2.9

Fourteen of a batch of 500 computer chips fail the final screening test. Estimate the failure probability and the 80% confidence interval.

Solution $\hat{p} = 14/500 = 0.028$. Since $pN = 14$ (>5), Eq. 2.86 can be used.

With $z_{0.1} = 1.28$, $p^{\pm} = 0.028 \pm 1.28 \dfrac{1}{\sqrt{500}} \sqrt{0.028(1 - 0.028)}$

$$p^{\pm} = 0.028 \pm 0.009 \text{ or } p^- = 0.019, \ p^+ = 0.037$$

We must take care in interpreting the probability statements related to confidence limits and intervals. Equation 2.85 is best understood as follows.

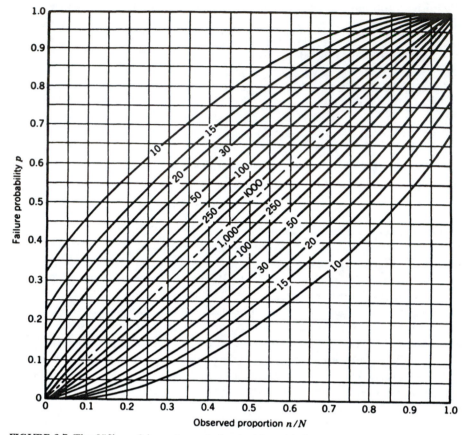

FIGURE 2.7 The 95% confidence intervals for the binomial distribution. [From E. S. Pearson and C. J. Clopper, "The Use of Confidence or Fiducial Limits Illustrated in the Case of the Binomial," *Biometrica*, **26**, 204 (1934). With permission of Biometrica.]

Suppose that a large number of samples each of size N are taken and that the values of p^- and p^+ are tabulated. Note that p^- and p^+, along with $\hat{p}$, are random variables and thus are expected to take on different values for each sample. The 90% confidence interval simply signifies that for 90% of the samples, the true value of p will lie between the calculated confidence limits.

2.5 ACCEPTANCE TESTING

Binomial sampling of the type we have discussed has long been associated with acceptance testing. Such sampling is carried out to provide an adequate degree of assurance to the buyer that no more than some specified fraction of a batch of products is defective. Central to the idea of acceptance sampling is that there be a unique pass-fail criterion.

 The question naturally arises why all the units are not inspected if it is important that p be small. The most obvious answer is expense. In many cases it may simply be too expensive to inspect every item of large-size batches of mass-produced items. Moreover, for a given budget, much better quality assurance is often achieved if the funds are expended on carrying out thorough inspections, tests, or both on a randomly selected sample instead of carrying out more cursory tests on the entire batch.

When the tests involve reliability-related characteristics, the necessity for performing them on a sample becomes more apparent, for the tests may be destructive or at least damaging to the sample units. Consider two examples. If safety margins on strength or capacity are to be verified, the tests may involve stress levels far above those anticipated in normal use: large torques may be applied to sample bolts to ensure that failure is by excessive deformation and not fracture; electric insulation may be subjected to a specified but abnormally high voltage to verify the safety factor on the breakdown voltage. If reliability is to be tested directly, each unit of the sample must be operated for a specified time to determine the fraction of failures. This time may be shortened by operating the sample units at higher stress levels, but in either case some sample units will be destroyed, and those that survive the test may exhibit sufficient damage or wear to make them unsuitable for further use.

Binomial Sampling

Typically, an acceptance testing procedure is set up to provide protection for both the producer and the buyer in the following way. Suppose that the buyer's acceptance criteria requires that no more than a fraction p_1 of the total batch fail the test. That is, for the large (theoretically infinite) batch the failure probability must be less than p_1. Since only a finite sample size N is to be tested, there will be some risk that the population will be accepted even though $p > p_1$. Let this risk be denoted by β, the probability of accepting a batch even though $p > p_1$. This is referred to as the buyer's risk; typically, we might take $\beta \approx 10\%$.

The producers of the product may be convinced that their product exceeds the buyer's criteria with a failure fraction of only $p_0 (p_0 < p_1)$. In taking only a finite sample, however, they run the risk that a poor sample will result in the batch being rejected. This is referred to as the producer's risk and it is denoted by α, the probability that a sample will be rejected even though $p < p_0$. Typically, an acceptable risk might be $\alpha \approx 5\%$.

Our object is to construct a binomial sampling scheme in which p_0 and p_1 result in predetermined values of α and β. To do this, we assume that the sample size is much less than the batch size. Let **n** be the random variable denoting the number of defective items, and n_d be the maximum number of defective items allowable in the sample. The buyer's risk β is then the probability that there will be no more than n_d defective items, given a failure probability of p_1:

$$\beta = P\{\mathbf{n} \leq n_d | N, p_1\}. \tag{2.87}$$

Using the binomial distribution, we obtain

$$\beta = \sum_{n=0}^{n_d} C_n^N p_1^n (1 - p_1)^{N-n}. \tag{2.88}$$

Similarly, the producer's risk α is the probability that there will be more than n_d defective items in the batch, even though $p = p_0$:

$$\alpha = P\{n > n_d | N, p_0\} \tag{2.89}$$

or

$$\alpha = \sum_{n=n_d+1}^{N} C_n^N p_0^n (1 - p_0)^{N-n}. \tag{2.90}$$

From Eqs. 2.88 and 2.90 the values of n_d and N for the sampling scheme can be determined. With n_d and N thus determined, the characteristics of the resulting sampling scheme can be presented graphically in the form of an operating curve. The operating curve is just the probability of acceptance versus the value p, the true value of the failure probability:

$$P\{n \le n_d | N, p\} = \sum_{n=0}^{n_d} C_n^N p^n (1 - p)^{N-n}. \tag{2.91}$$

In Fig. 2.8 is shown a typical operating curve, with β being the probability of acceptance when $p = p_1$, and α the probability of rejection when $p = p_0$.

The Poisson Limit

As in the preceding section, the binomial distribution may be replaced by the Poisson limit when the sample size is very large $N \gg 1$, and the failure probabilities are small $p_0, p_1 \ll 1$. This leads to considerable simplifications in carrying out numerical computations. Defining $m_0 = Np_0$ and $m_1 = Np_1$, we may replace Eqs. 2.88 and 2.90 by the corresponding Poisson distributions:

$$\beta = \sum_{n=0}^{n_d} \frac{m_1^n}{n!} e^{-m_1} \tag{2.92}$$

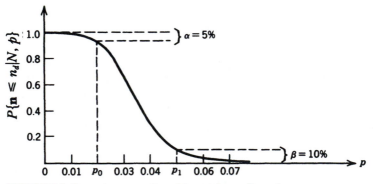

FIGURE 2.8 Operating curve for a binomial sampling scheme.

TABLE 2.3[a] Binomial Sampling Chart for $\alpha = 0.05$; $\beta = 0.10$

n_d	m_0	m_1	p_1/p_0	n_d	m_0	m_1	p_1/p_0
0	0.0513	2.303	44.9	13	8.463	18.96	2.24
1	0.3531	3.890	11.0	14	9.246	20.15	2.18
2	0.8167	5.323	6.52	15	10.04	21.32	2.12
3	1.365	6.681	4.89	16	10.83	22.49	2.08
4	1.969	7.994	4.06	17	11.63	23.64	2.03
5	2.613	9.275	3.55	18	12.44	24.78	1.99
6	3.285	10.53	3.21	19	13.25	25.91	1.96
7	3.980	11.77	2.96	20	14.07	27.05	1.92
8	4.695	12.99	2.77	21	14.89	28.20	1.89
9	5.425	14.21	2.62	22	15.68	29.35	1.87
10	6.168	15.45	2.50	23	16.50	30.48	1.85
11	6.924	16.64	2.40	24	17.34	31.61	1.82
12	7.689	17.81	2.32	25	18.19	32.73	1.80

[a] Adapted from E. Schindowski and O. Schürz, *Statistische Qualitätskontrolle*, VEB Verlag Technik, Berlin, 1972.

and

$$\alpha = 1 - \sum_{n=0}^{n_d} \frac{m_0^n}{n!} e^{-m_0}. \tag{2.93}$$

Given α and β, we may solve these equations numerically for m_0 and m_1 with $n_d = 0, 1, 2, \ldots$. The results of such a calculation for $\alpha = 5\%$ and $\beta = 10\%$ are tabulated in Table 2.3. One uses the table by first calculating p_1/p_0; n_d is then read from the first column, and N is determined from $N = (m_0/p_0)$ or $N = (m_1/p_1)$. This is best illustrated by an example.

EXAMPLE 2.10

Construct a sampling scheme for n_d and N, given

$$\alpha = 5\%, \beta = 10\%, p_0 = 0.02, \text{ and } p_1 = 0.05.$$

Solution We have $p_1/p_0 = 0.05/0.02 = 2.5$. Thus from Table 2.3 $n_d = 10$. Now $N = m_0/p_0 = 6.168/0.02 \cong 308$.

Multiple Sampling Methods

We have discussed in detail only situations in which a single sample of size N is used. Acceptance of the items is made, provided that the number of defective items does not exceed n_d, which is referred to as the acceptance number. Often more varied and sophisticated sampling schemes may be used to glean additional information without an inordinate increase in sampling effort.* Two such schemes are double sampling and sequential sampling.

* See, for example, A. V. Feigenbaum, *Total Quality Control*, 3rd ed., McGraw-Hill, New York, 1983, Chapter 15.

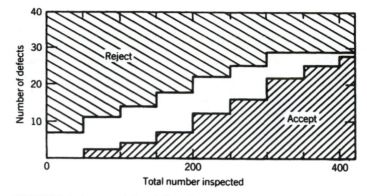

FIGURE 2.9 A sequential sampling chart.

In double sampling a sample size N_1 is drawn. The batch, however, need not be rejected or accepted as a result of the first sample if too much uncertainty remains about the quality of the batch. Instead, a second sample N_2 is drawn and a decision made on the combined sample size $N_1 + N_2$. Such schemes often allow costs to be reduced, for a very good batch will be accepted or a very bad batch rejected with the small sample size N_1. The larger sample size $N_1 + N_2$ is reserved for borderline cases.

In sequential sampling the principle of double sampling is further extended. The sample is built up item by item, and a decision is made after each observation to accept, reject, or take a larger sample. Such schemes can be expressed as sequential sampling charts, such as the one shown in Fig. 2.9. Sequential sampling has the advantage that very good (or bad) batches can be accepted (or rejected) based on very small sample sizes, with the larger samples being reserved for those situations in which there is more doubt about whether the number of defects will fall within the prescribed limits. Sequential sampling does have a disadvantage. If the test of each item takes a significant length of time, as usually happens in reliability testing, the total test time is likely to take too long. The limited time available then dictates that a single sample be taken and the items tested simultaneously.

Bibliography

Feigenbaum, A. V., *Total Quality Control*, 3rd ed., McGraw-Hill, NY, 1983.

Ireson W. G., (ed.) *Reliability Handbook*, McGraw-Hill, NY, 1966.

Lapin, L. L., *Probability and Statistics for Modern Engineering*, Brooks/Cole, Belmont, CA, 1983.

Montgomery, D. C., and G. C. Runger, *Applied Statistics and Probability for Engineers*, Wiley, NY, 1994.

Pieruschka, E., *Principles of Reliability*, Prentice-Hall, Englewood Cliffs, NJ, 1963.

Exercises

2.1 Suppose that $P\{X\} = 0.32$, $P\{Y\} = 0.44$, and $P\{XUY\} = 0.58$.

    (a) Are the events mutually exclusive?

    (b) Are they independent?

    (c) Calculate $P\{X|Y\}$.

    (d) Calculate $P\{Y|X\}$.

2.2 Suppose that X and Y are independent events with $P\{X\} = 0.28$ and $P\{Y\} = 0.41$. Find (a) $P\{\tilde{X}\}$, (b) $P\{X \cap Y\}$, (c) $P\{\tilde{Y}\}$, (d) $\{X \cap \tilde{Y}\}$, (e) $P\{X \cup Y\}$, (f) $P\{\tilde{X} \cap \tilde{Y}\}$.

2.3 Suppose that $P\{A\} = 1/2$, $P\{B\} = 1/4$, and $P\{A \cap B\} = 1/8$. Determine (a) $P\{A|B\}$, (b) $P\{B|A\}$, (c) $P\{A \cup B\}$, (d) $P\{\tilde{A}|\tilde{B}\}$.

2.4 Given: $P\{A\} = 0.4$, $P\{A \cup B\} = 0.8$, $P\{A \cap B\} = 0.2$. Determine (a) $P\{B\}$, (b) $P\{A|B\}$, (c) $P\{B|A\}$.

2.5 Two relays with demand failures of $p = 0.15$ are tested.

    (a) What is the probability that neither will fail?

    (b) What is the probability that both will fail?

2.6 For each of the following, draw a Venn diagram similar to Fig. 2.3 and shade the indicated areas: (a) $(X \cup Y) \cap \tilde{Z}$, (b) $\tilde{X} \cap \tilde{Y} \cap Z$, (c) $(\tilde{X} \cup \tilde{Y}) \cap Z$, (d) $(X \cap \tilde{Y}) \cup Z$.

2.7 An aircraft landing gear has a probability of 10^{-5} per landing of being damaged from excessive impact. What is the probability that the landing gear will survive a 10,000 landing design life without damage?

2.8 Consider events A, B and C. If $P\{A\} = 0.8$, $P\{B\} = 0.3$, $P\{C\} = 0.4$, $P\{A|B \cap C\} = 0.5$, $P\{B|C\} = 0.6$.

    (a) Determine whether events B and C are independent.

    (b) Determine whether events B and C are mutually exclusive.

    (c) Evaluate $P\{A \cap B \cap C\}$

    (d) Evaluate $P\{B \cap C|A\}$

2.9. A particulate monitor has a power supply consisting of two batteries in parallel. Either battery is adequate to operate the monitor. However, since the failure of one battery places an added strain on the other, the conditional probability that the second battery will fail, given the failure of the first, is greater than the probability that the first will fail. On the basis of testing it is known that 7% of the monitors in question will have at least one battery failed by the end of their design life, whereas in 1% of the monitors both batteries will fail during the design life.

    (a) Calculate the battery failure probability under normal operating conditions.

(b) Calculate the conditional probability that the battery will fail, given that the other has failed.

2.10 Two pumps operating in parallel supply secondary cooling water to a condenser. The cooling demand fluctuates, and it is known that each pump is capable of supplying the cooling requirements 80% of the time in case the other fails. The failure probability for each pump is 0.12; the probability of both failing is 0.02. If there is a pump malfunction, what is the probability that the cooling demand can still be met?

2.11 For the discrete PMF,

$$f(x_n) = Cx_n^2; \qquad x_n = 1, 2, 3.$$

(a) Find C.
(b) Find $F(x_n)$.
(c) Calculate μ and σ.

2.12 Repeat Exercise 2.11 for

$$f(x_n) = Cx_n(6 - x_n), \qquad x_n = 0, 1, 2, \ldots, 6.$$

2.13 Consider the discrete random variable defined by

x_n	0	1	2	3	4	5
$f(x_n)$	$\dfrac{11}{36}$	$\dfrac{9}{36}$	$\dfrac{7}{36}$	$\dfrac{5}{36}$	$\dfrac{3}{36}$	$\dfrac{1}{36}$

Compute the mean and the variance.

2.14 A discrete random variable x takes on the values 0, 1, 2, and 3 with probabilities 0.4, 0.3, 0.2, and 0.1, respectively. Compute the expected values of x, x^2, $2x + 1$, and e^{-x}.

2.15 Evaluate the following:
(a) C_3^5, (b) C_2^9, (c) C_7^{12}, (d) C_{19}^{20}.

2.16 A discrete probability mass function is given by $f(0) = 1/6$, $f(1) = 1/3$, $f(2) = 1/2$.

(a) Calculate the mean value μ.
(b) Calculate the standard deviation σ.

2.17 Ten engines undergo testing. If the failure probability for an individual engine is 0.10, what is the probability that more than two engines will fail the test?

2.18 A boiler has four identical relief valves. The probability that an individual relief valve will fail to open on demand is 0.06. If the failures are independent:

(a) What is the probability that at least one valve will fail to open?
(b) What is the probability that at least one valve will open?

2.19 If the four relief valves were to be replaced by two valves in the preceding problem, to what value must the probability of an individual valve's failing be reduced if the probability that no valve will open is not to increase?

2.20 The discrete uniform distribution is

$$f(n) = 1/N, \qquad n = 1, 2, 3, 4, \ldots N.$$

(a) Show that the mean is $(N + 1)/2$.

(b) Show that the variance is $(N^2 - 1)/12$.

2.21 The probability of an engine's failing during a 30-day acceptance test is 0.3 under adverse environmental conditions. Eight engines are included in such a test. What is the probability of the following? (a) None will fail. (b) All will fail. (c) More than half will fail.

2.22 The probability that a clutch assembly will fail an accelerated reliability test is known to be 0.15. If five such clutches are tested, what is the probability that the error in the resulting estimate will be more than 0.1?

2.23 A manufacturer produces 1000 ball bearings. The failure probability for each ball bearing is 0.002.

(a) What is the probability that more than 0.1% of the ball bearings will fail?

(b) What is the probability that more than 0.5% of the ball bearings will fail?

2.24 Verify Eqs. 2.63 and 2.64.

2.25 Suppose that the probability of a diode's failing an inspection is 0.006.

(a) What is the probability that in a batch of 500, more than 3 will fail?

(b) What is the mean number of failures per batch?

(*Note:* Use the Poisson distribution.)

2.26 The geometric distribution is given by

$$f(n) = p(1 - p)^{n-1}, \qquad n = 1, 2, 3, 4, \ldots \infty$$

(a) Show that Eq. 2.22 is satisfied.

(b) Find that the expected value of n is $1/p$.

(c) Show that the variance of $f(n)$ is $1/p^2$.

(*Note:* The summation formulas in Appendix A may be useful.)

2.27 One thousand capacitors undergo testing. If the failure probability for each capacitor is 0.0010, what is the probability that more than two capacitors will fail the test?

2.28 Let p equal the probability of failure and n be the trial upon which the first failure occurs. Then n is a random variable governed by the geometric distribution given in exercise 2.26. An engineer wanting to study the failure mode proof tests on a new chip. Since there is only one test setup she must run them one chip at a time. If the failure probability is $p = 0.2$.

(a) What is the probability that the first chip will not fail?

(b) What is the probability that the first three trials will produce no failures?

(c) How many trials will she need to run before the probability of obtaining a failure reaches $1/2$?

2.29 A manufacturer of 16K byte memory boards finds that the reliability of the manufactured boards is 0.98. Assume that the defects are independent.

(a) What is the probability of a single byte of memory being defective?

(b) If no changes are made in design or manufacture, what reliability may be expected from 128K byte boards?

(*Note:* 16K bytes $= 2^{14}$ bytes, 128K bytes $= 2^{17}$ bytes.)

2.30 The PMF for a discrete distribution is

$$f(n) = \frac{1}{2}\frac{\lambda^n}{n!}\exp(-\lambda) + \frac{1}{2}\frac{\eta^n}{n!}\exp(-\eta), \qquad n = 0, 1, 2, 3, 4, \ldots \infty$$

(a) Determine μ_n

(b) Determine σ_n^2

2.31 Diesel engines used for generating emergency power are required to have a high reliability of starting during an emergency. If the failure to start on demand probability of 1% or less is required, how many consecutive successful starts would be necessary to ensure this level of reliability with a 90% confidence?

2.32 An engineer feels confident that the failure probability on a new electromagnetic relay is less than 0.01. The specifications require, however, only that $p < 0.04$. How many units must be tested without failure to prove with 95% confidence that $p < 0.04$?

2.33 A quality control inspector examines a sample of 30 microcircuits from each purchased batch. The shipment is rejected if 4 or more fail. Find the probability of rejecting the batch where the fraction of defective circuits in the entire (large) batch is

(a) 0.01,

(b) 0.05,

(c) 0.15.

2.34 Suppose that a sample of 20 units passes an acceptance test if no more than 2 units fail. Suppose that the producer guarantees the units for a failure probability of 0.05. The buyer considers 0.15 to be the maximum acceptable failure probability.

(a) What is the producers risk?

(b) What is the buyer's risk?

2.35 Suppose that 100 pressure sensors are tested and 14 of them fail the calibration criteria. Make a point estimate of the failure probability, then use Eq. 2.86 to estimate the 90% and the 95% confidence interval.

2.36 Draw the operating curve for the 2 out of 20 sampling scheme of exercise 2.34.

(a) What must the failure probability be to obtain a producer's risk of no more than 10%?

(b) What must the failure probability be for the buyer to have a risk of no more than 10%?

2.37 Construct a binomial sampling scheme where the producer's risk is 5%, the buyer's risk 10%, $p_0 = 0.03$, and $p_1 = 0.06$. (Use Table 2.3)

2.38 A standard acceptance test is carried out on 20 battery packs. Two fail.

(a) What is the 95% confidence interval for the failure probability?

(b) Make a rough estimate of how many tests would be required if the 95% confidence interval were to be within ± 0.1 of the true failure probability. Assume the true value is $p = 0.2$.

2.39 A buyer specifies that no more than 10% of large batches of items should be defective. She tests 10 items from each batch and accepts the batch if none of the 10 is defective. What is the probability that she will accept a batch in which more than 10% are defective?

CHAPTER 3

Continuous Random Variables

"All business proceeds on beliefs or judgements of probabilities and not just on certainties."

Charles Eliot

3.1 INTRODUCTION

In Chapter 2 probabilities of discrete events, most frequently failures, were discussed. The discrete random variables associated with such events are used to estimate the number of events that are likely to take place. In order to proceed further with reliability analysis, however, it is necessary to consider how the probability of failure depends on a variety of other variables that are continuous: the duration of operation time, the strength of the system, the magnitudes of stresses, and so on. If the repeated measurement of such variables is carried out, however, the same value will not be obtained with each test. These values are referred to as continuous random variables for they cannot be described with certainty, but only with the probability that they will take on values within some range. In Section 3.2 we first introduce the mathematical apparatus required to describe random variables. In Section 3.3 the normal and related distributions are presented. In section 3.4 the Weibull and extreme-valve distributions are described.

3.2 PROPERTIES OF RANDOM VARIABLES

In this section we examine some of the important properties of continuous random variables. We first define the quantities that determine the behavior of a single random variable. We then examine how these properties are transformed when the variable is changed.

Probability Distribution Functions

We denote a continuous random variable with bold-faced type as **x** and the values that **x** may take on are specified by x, that is, in normal type. The properties of a random variable are specified in terms of probabilities. For example, $P\{\mathbf{x} < x\}$ is used to designate the probability that **x** has a value less than x. Similarly, $P\{a < \mathbf{x} < b\}$ is the probability that **x** has a value between a and b. Two particular probabilities are most often used to describe a random variable. The first one,

$$F(x) = P\{\mathbf{x} \le x\}, \tag{3.1}$$

the probability that **x** has a value less than or equal to x, is referred to as the *cumulative distribution function*, or CDF for short. Second, the probability that **x** lies between x and $x + \Delta x$ as Δx becomes infinitesimally small is denoted by

$$f(x)\, \Delta x = P\{x \le \mathbf{x} \le x + \Delta x\}, \tag{3.2}$$

where $f(x)$ is the *probability density function*, referred to hereafter as the PDF. Since both $f(x)$ and $F(x)$ are probabilities, they must be greater than or equal to zero for all values of x.

These two functions of x are related. Suppose that we allow **x** to take on any values $-\infty \le \mathbf{x} \le +\infty$. Then the CDF is just the integral of the PDF over all $\mathbf{x} \le x$:

$$F(x) = \int_{-\infty}^{x} f(x')\, dx'. \tag{3.3}$$

We also may invert this relationship by differentiating to obtain

$$f(x) = \frac{d}{dx} F(x). \tag{3.4}$$

The probability distributions $f(x)$ and $F(x)$ are normalized as follows: We first note that the probability that **x** lies between a and b may be obtained by integration

$$\int_{a}^{b} f(x)\, dx = P\{a \le \mathbf{x} \le b\}. \tag{3.5}$$

Now, **x** must have some value between $-\infty$ and $+\infty$. Thus

$$P\{-\infty \le \mathbf{x} \le \infty\} = 1. \tag{3.6}$$

The combination of this relationship with Eq. 3.5 with $a = -\infty$ and $b = +\infty$ then yields the normalization condition

$$\int_{-\infty}^{\infty} f(x)\, dx = 1. \tag{3.7}$$

Then, setting $\mathbf{x} = \infty$ in Eq. 3.3, we find the corresponding condition on the CDF to be

$$F(\infty) = 1. \tag{3.8}$$

One more function that is often used is the *complementary cumulative distribution function* or CCDF, which is defined as

$$\tilde{F}(x) = P\{\mathbf{x} > x\}, \tag{3.9}$$

where we use the tilde to designate the complementary distribution, since $\mathbf{x} > x$ is the same as $\mathbf{x}$ *not* $\leq x$. The definition of $f(x)$ and Eq. 3.7 allows us to write $\tilde{F}(x)$ as

$$\tilde{F}(x) = \int_x^\infty f(x') \, dx' = 1 - \int_{-\infty}^x f(x') \, dx', \tag{3.10}$$

or combining this expression with Eq. 3.3 yields

$$\tilde{F}(x) = 1 - F(x). \tag{3.11}$$

Thus far we have assumed that $\mathbf{x}$ can take on any value $-\infty \leq \mathbf{x} \leq +\infty$. In many situations we must deal with variables that are restricted to a smaller domain. For example, time is most often restricted to $0 \leq \mathbf{t} \leq \infty$. In such cases the foregoing relationships may be modified quite simply. For example, in considering only positive values of time we have

$$F(t) = 0, \qquad t < 0, \tag{3.12}$$

and therefore for time, Eq. 3.3 becomes

$$F(t) = \int_0^t f(t') \, dt'. \tag{3.13}$$

Similarly, the condition of Eq. 3.7 becomes

$$\int_0^\infty f(t) \, dt = 1. \tag{3.14}$$

In Fig. 3.1 the relation between $f(x)$ and $F(x)$ is illustrated for a typical random variable with the restriction that $0 \leq \mathbf{x} \leq \infty$. In what follows we retain the $\pm\infty$ limits on the random variables, with the understanding that these are to be appropriately reduced in situations in which the domain of the variable is restricted.

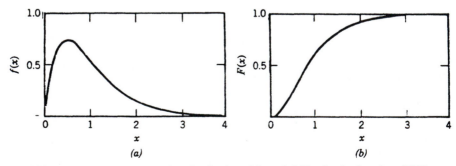

FIGURE 3.1 Continuous probability distribution: (*a*) probability density function (PDF), (*b*) corresponding cumulative distribution function (CDF).

EXAMPLE 3.1

The PDF of the lifetime of an appliance is given by

$$f(t) = 0.25te^{-0.5t}, \qquad t \geqslant 0,$$

where t is in years. (*a*) What is the probability of failure during the first year? (*b*) What is the probability of the appliance's lasting at least 5 years? (*c*) If no more than 5% of the appliances are to require warranty services, what is the maximum number of months for which the appliance can be warranted?

Solution   First calculate the CDF and CCDF:

$$F(t) = \int_0^t dt\, 0.25te^{-0.5t} = 1 - (1 + 0.5t)\,e^{-0.5t},$$

$$\tilde{F}(t) = (1 + 0.5t)\,e^{-0.5t}.$$

(*a*) $F(1) = 1 - (1 + 0.5 \times 1)\,e^{-0.5 \times 1} = 0.0902.$

(*b*) $\tilde{F}(5) = (1 + 0.5 \times 5)\,e^{-0.5 \times 5} = 0.2873.$

(*c*) We must have $\tilde{F}(t_0) \geqslant 0.95$, where t_0 is the warranty period in years. From part (*a*) it is clear that the warranty must be less than one year, since $\tilde{F}(1) = 1 - F(1) = 0.91.$

Try 6 months, $t_0 = \frac{6}{12}$; $\tilde{F}(\frac{6}{12}) = 0.973.$
Try 9 months, $t_0 = \frac{9}{12}$; $\tilde{F}(\frac{9}{12}) = 0.945.$
Try 8 months, $t_0 = \frac{8}{12}$; $\tilde{F}(\frac{8}{12}) = 0.955.$
The maximum warranty is 8 months.

Characteristics of a Probability Distribution

Often it is not necessary, or possible, to know the details of the probability density function of a random variable. In many instances it suffices to know certain integral properties. The two most important of these are the mean and the variance.

The mean or expectation value of **x** is defined by

$$\mu = \int_{-\infty}^{\infty} xf(x)\, dx. \tag{3.15}$$

The variance is given by

$$\sigma^2 = \int_{-\infty}^{\infty} (x - \mu)^2 f(x)\, dx. \tag{3.16}$$

The variance is a measure of the dispersion of values about the mean. Note that since the integrand on the right-hand side of Eq. 3.16 is always nonnegative, the variance is always nonnegative. In Fig. 3.2 examples are shown of probability density functions with different mean values and with different values of the variance, respectively.

More general functions of a random variable can be defined. Any function, say $g(x)$, that is to be averaged over the values of a random variable we

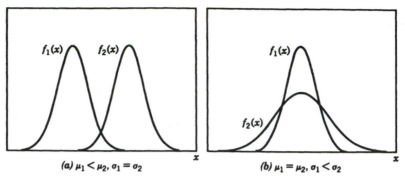

FIGURE 3.2 Probability density functions.

write as

$$E\{g(x)\} \equiv \int_{-\infty}^{\infty} g(x) f(x) \, dx. \tag{3.17}$$

The quantity $E\{g(x)\}$ is referred to as the expected value of $g(x)$. It may be interpreted more precisely as follows. If we sampled an infinitely large number of values of **x** from $f(x)$ and calculated $g(x)$ for each one of them, the average of these values would be $E\{g\}$. In particular, the nth moment of $f(x)$ is defined to be

$$E\{x^n\} = \int_{-\infty}^{\infty} x^n f(x) \, dx. \tag{3.18}$$

With these definitions we note that $E\{x^0\} = 1$, and the mean is just the first moment:

$$\mu = E\{x\} \tag{3.19}$$

Similarly, the variance may be expressed in terms of the first and second moments. To do this we write

$$\sigma^2 = E\{(x - \mu)^2\} = E\{x^2 - 2x\mu + \mu^2\}. \tag{3.20}$$

But since μ is independent of x, it can be brought outside of the integral to yield

$$\sigma^2 = E\{x^2\} - 2E\{x\}\mu + \mu^2. \tag{3.21}$$

Finally, using Eq. 3.19, we have

$$\sigma^2 = E\{x^2\} - E\{x\}^2. \tag{3.22}$$

In addition to the mean and variance, two additional properties are sometimes used to characterize the PDF of a random variable; these are the skewness and the kurtosis. The skewness is defined by

$$sk = \frac{1}{\sigma^3} \int_{-\infty}^{\infty} (x - \mu)^3 f(x) \, dx. \tag{3.23}$$

It is a measure of the asymmetry of a PDF about the mean. In Fig. 3.3 are shown two PDFs with identical values of μ and σ^2, but with values of the

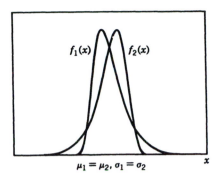

FIGURE 3.3 Probability density functions with skewness of opposite signs.

skewness that are opposite in sign but of the same magnitude. The kurtosis, like the variance, is a measure of the spread of $f(x)$ about the mean. It is given by

$$ku = \frac{1}{\sigma^4} \int_{-\infty}^{\infty} (x - \mu)^4 f(x) \, dx. \tag{3.24}$$

EXAMPLE 3.2

A lifetime distribution has the form

$$f(t) = \beta t e^{-\alpha t},$$

where t is in years. Find β, μ, and σ in terms of α.

Solution We shall use the fact that (see Appendix A)

$$\int_0^{\infty} d\xi \, \xi^n e^{-\xi} = n!.$$

From Eq. 3.14,

$$\int_0^{\infty} dt \, \beta t e^{-\alpha t} = 1.$$

With $\zeta = \alpha t$, we therefore have

$$\frac{\beta}{\alpha^2} \int_0^{\infty} d\zeta \, \zeta e^{-\zeta} = \frac{\beta}{\alpha^2} \times 1 = 1.$$

Thus $\beta = \alpha^2$ and we have $f(t) = \alpha^2 t e^{-\alpha t}$.
The mean is determined from Eq. 3.15:

$$\mu \equiv \int_0^{\infty} dt \, t f(t) = \alpha^2 \int_0^{\infty} dt \, t^2 e^{-\alpha t} = \frac{1}{\alpha} \int_0^{\infty} d\zeta \, \zeta^2 e^{-\zeta} = \frac{2!}{\alpha}.$$

Therefore, $\mu = 2/\alpha$.
The variance is found from Eq. 3.22, which reduces to

$$\sigma^2 \equiv \int_0^{\infty} dt \, t^2 f(t) - \mu^2,$$

but

$$\int_0^\infty dt\, t^2 f(t) = \alpha^2 \int_0^\infty dt\, t^3 e^{-\alpha t} = \frac{1}{\alpha^2} \int_0^\infty d\zeta\, \zeta^3 e^{-\zeta} = \frac{3!}{\alpha^2} = \frac{6}{\alpha^2}$$

and therefore,

$$\sigma^2 = \frac{6}{\alpha^2} - \left(\frac{2}{\alpha}\right)^2 = \frac{2}{\alpha^2}.$$

Thus $\sigma = \sqrt{2}/\alpha$.

EXAMPLE 3.3

Calculate μ and σ in Example 3.1.

Solution Note that the distribution in Examples 3.1 and 3.2 are identical if $\alpha = 0.5$. Therefore $\mu = 4$ years, and $\sigma = 2\sqrt{2}$ years.

Transformations of Variables

Frequently, in reliability considerations, the random variable for which data are available is not the one that can be used directly in the reliability estimates. Suppose, for example, that the distribution of speeds of impact $f(v)$ is known for a mechanical snubber. If the wear on the snubber, however, is proportional to the kinetic energy, $e = \frac{1}{2}mv^2$, the energy is also a random variable and it is the distribution of energies $f_e(e)$ that is needed. Such problems are ubiquitous, for much of engineering analysis is concerned with functional relationships that allow us to predict the value of one variable (the dependent variable) in terms of another (the independent variable).

To deal with situations such as the change from speed to energy in the foregoing example, we need a means for transforming one random variable to another. The problem may be stated more generally as follows. Given a distribution $f_x(x)$ or $F_x(x)$ of the random variable $\mathbf{x}$, find the distribution $f_y(y)$ of the random variable $\mathbf{y}$ that is defined by

$$y = y(x). \tag{3.25}$$

We then refer to $f_y(y)$ as the derived distribution. Hereafter, we use subscripts $\mathbf{x}$ and $\mathbf{y}$ to distinguish between the distributions whenever there is a possibility of confusion. First, consider the case where the relation between y and x has the characteristics shown in Fig. 3.4; that is, if $x_1 < x_2$, then $y(x_1) < y(x_2)$. Then $y(x)$ is a monotonically increasing function of x; that is, $dy/dx > 0$. To carry out the transformation, we first observe that

$$P\{\mathbf{x} \leq x\} = P\{\mathbf{y} \leq y(x)\}, \tag{3.26}$$

or simply

$$F_x(x) = F_y(y) \tag{3.27}$$

FIGURE 3.4 Function of a random variable x.

To obtain the PDF $f_y(y)$ in terms of $f_x(x)$, we first write the preceding equation as

$$\int_{-\infty}^{x} f_x(x') \, dx' = \int_{-\infty}^{y(x)} f_y(y') \, dy'. \tag{3.28}$$

Differentiating with respect to x, we obtain

$$f_x(x) = f_y(y) \frac{dy}{dx} \tag{3.29}$$

or

$$f_y(y) = f_x(x) \left| \frac{dx}{dy} \right|. \tag{3.30}$$

Here we have placed an absolute value about the derivative. With the absolute value, the result can be shown to be valid for either monotonically increasing or monotonically decreasing functions.

The most common transforms are of the linear form

$$y = ax + b, \tag{3.31}$$

and the foregoing equation becomes simply

$$f_y(y) = \frac{1}{|a|} f_x \left(\frac{y-b}{a} \right). \tag{3.32}$$

Note that once a transformation has been made, new values of the mean and variance must be calculated, since in general

$$\int g(x) f_x(x) \, dx \neq \int g(y) f_y(y) \, dy. \tag{3.33}$$

EXAMPLE 3.4

Consider the distribution $f_x(x) = \alpha e^{-\alpha x}$, $0 \leqslant x \leqslant \infty$, $\alpha > 1$.
(*a*) Transform to the distribution $f_y(y)$, where $y = e^x$.
(*b*) Calculate μ_x and μ_y.

Solution   (*a*) $dy/dx = e^x$; therefore, Eq. 3.30 becomes $f_y(y) = e^{-x}f_x(x)$. We also have $x = \ln y$. Therefore,

$$f_y(y) = e^{-\ln y}\alpha e^{-\alpha \ln y} = \frac{\alpha}{y^{\alpha+1}}, \qquad 1 \le y \le \infty.$$

(*b*) $\mu_x = \int_0^\infty x\alpha e^{-\alpha x}\, dx = \frac{1}{\alpha}$,

$\mu_y = \int_1^\infty y\alpha y^{-(\alpha+1)}\, dy = \frac{\alpha}{\alpha - 1}$.

3.3 NORMAL AND RELATED DISTRIBUTIONS

Continuous random variables find extensive use in reliability analysis for the description of survival times, system loads and capacities, repair rates, and a variety of other phenomena. Moreover, a substantial number of standardized probability distributions are employed to model the behavior of these variables. For the most part we shall introduce these distributions as they are needed for model reliability phenomena in the following chapters. We introduce here the normal distribution and the related lognormal and Dirac delta distributions, for they appear in a variety of different contexts throughout the book. Moreover, they provide convenient vehicles for applying the concepts of the foregoing discussion.

The Normal Distribution

Unquestionably, the normal distribution is the most widely applied in statistics. It is frequently referred to as the Gaussian distribution. To introduce the normal distribution, we first consider the following function of the random variable **x**,

$$f(x) = \frac{1}{\sqrt{2\pi}\, b} \exp\left[-\frac{1}{2}\left(\frac{x-a}{b}\right)^2\right], \qquad -\infty \le x \le \infty, \tag{3.34}$$

where *a* and *b* are parameters that we have yet to specify. It may be shown that $f(x)$ meets the conditions for a probability density function. First, it is clear that $f(x) \ge 0$ for all *x*. Second, by performing the integral

$$\int_{-\infty}^{\infty} \frac{1}{\sqrt{2\pi}\, b} \exp\left[-\frac{1}{2}\left(\frac{x-a}{b}\right)^2\right] dx = 1 \tag{3.35}$$

it may be shown that the condition on the PDF given by Eq. 3.7 is met. The evaluation of Eq. 3.35 cannot be carried out by rudimentary means. Rather, the method of residues from the theory of complex variables must be employed. For convenience, some of the more common integrals involving the normal distribution are included in Appendix A.

A unique feature of the normal distribution is that the mean and variance appear explicitly as the two parameters *a* and *b*. To demonstrate this, we insert

Eq. 3.34 into the definitions of the mean and variance, Eqs. 3.15 and 3.16. Using the evaluated integrals in Appendix A, we find

$$\mu \equiv \int_{-\infty}^{\infty} dx \frac{x}{\sqrt{2\pi}\,b} \exp\left[-\frac{1}{2}\left(\frac{x-a}{b}\right)^2\right] = a, \tag{3.36}$$

$$\sigma^2 \equiv \int_{-\infty}^{\infty} dx\,(x-\mu)^2 \frac{1}{\sqrt{2\pi}\,b} \exp\left[-\frac{1}{2}\left(\frac{x-a}{b}\right)^2\right] = b^2. \tag{3.37}$$

Consequently, we may write the normal PDF directly in terms of the mean and variance as

$$f(x) = \frac{1}{\sqrt{2\pi}\,\sigma} \exp\left[-\frac{1}{2}\left(\frac{x-\mu}{\sigma}\right)^2\right], \qquad -\infty \leqslant x \leqslant \infty. \tag{3.38}$$

Similarly, the CDF corresponding to Eq. 3.34 is

$$F(x) = \int_{-\infty}^{x} \frac{1}{\sqrt{2\pi}\,\sigma} \exp\left[-\frac{1}{2}\left(\frac{x'-\mu}{\sigma}\right)^2\right] dx'. \tag{3.39}$$

When we use the normal distribution, it is often beneficial to make a change of variables first in order to express $F(x)$ in a standardized form. To this end, we define the random variable **z** in terms of **x** by

$$z \equiv (x-\mu)/\sigma. \tag{3.40}$$

Recalling that PDFs transform according to Eq. 3.30, we have

$$f_z(z) = f(x)\left|\frac{dx}{dz}\right| = \frac{1}{\sqrt{2\pi}\,\sigma} \exp\left[-\frac{1}{2}\left(\frac{x-\mu}{\sigma}\right)^2\right] |\sigma|, \tag{3.41}$$

which for $x = \mu + \sigma z$

$$f_z(z) = \frac{1}{\sqrt{2\pi}} \exp(-\tfrac{1}{2}z^2). \tag{3.42}$$

This implies that for the reduced variate z, $\mu_z = 0$ and $\sigma_z^2 = 1$.

The PDF is plotted in Fig. 3.5. Its appearance causes it to be referred to frequently as the bell-shaped curve. The standardized form of the CDF may also be found by applying Eq. 3.40 to $F(x)$,

$$F(x) \equiv \Phi[(x-\mu)/\sigma], \tag{3.43}$$

where the standardized error function on the right is defined as

$$\Phi(z) = \frac{1}{\sqrt{2\pi}} \int_{-\infty}^{z} \exp(-\tfrac{1}{2}\zeta^2)\,d\zeta. \tag{3.44}$$

The integrand of this expression is just the standardized normal PDF. A graph of $\Phi(z)$ is given in Fig. 3.6; note that each unit on the horizontal axis corresponds to one standard deviation σ, and that the mean value is now at the origin. A tabulation of $\Phi(z)$ is included in Appendix C. Although values

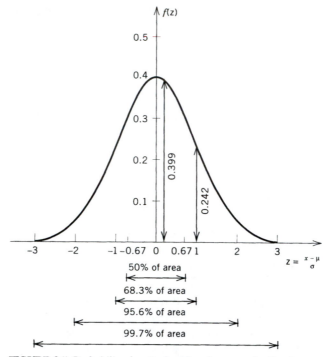

FIGURE 3.5 Probability density function for a standardized normal distribution.

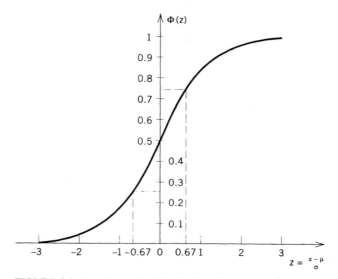

FIGURE 3.6 Cumulative distribution function for a standardized normal distribution.

for $z < 0$ are included in Appendix C, this is only for convenience, since for the normal distribution we may use the property $f(-z) = f(z)$ to obtain $\Phi(-z)$ from

$$\Phi(-z) = 1 - \Phi(z). \tag{3.45}$$

EXAMPLE 3.5

The time to wear out of a cutting tool edge is distributed normally with $\mu = 2.8$ hr and $\sigma = 0.6$ hr.
(a) What is the probability that the tool will wear out in less than 1.5 hr?
(b) How often should the cutting edges be replaced to keep the failure rate less than 10% of the tools?

Solution (a) $P\{t < 1.5\} = F_t(1.5) = \Phi(z)$, where

$$z = (t - \mu)/\sigma, \qquad z = (1.5 - 2.8)/0.6 = -2.1667$$

From Appendix C: $\Phi(-2.1667) = 0.0151$.
(b) $P\{t < t\} = 0.10$; $\Phi(z) = 0.10$. Then from Appendix C, $z \approx -1.28$. Therefore, we have

$$-t + \mu = 1.28\sigma, \qquad t = \mu - 1.28\sigma = 2.8 - 1.28 \times 0.6 = 2.03 \text{ hr.}$$

The normal distribution arises in many contexts. It may be expected to occur whenever the random variable **x** arises from the sum of a number of random effects, no one of which dominates the total. It is widely used to represent measurement errors, dimensional variability in manufactured goods, material properties, and a host of other phenomena.

A specific illustration might be as follows. Suppose that an elevator cable consists of strands of wire. The strength of the cable is then

$$x = x_1 + x_2 + \cdots x_i + \cdots x_N, \tag{3.46}$$

where x_i is the strength of the ith strand. Even though the PDF of the individual strands x_i is not a normal distribution, the strength of the cable will be given by a normal distribution, provided that N, the number of strands, is sufficiently large.

The normal distribution also has the following property. If **x** and **y** are random variables that are normally distributed, then

$$u = ax + by, \tag{3.47}$$

where a and b are constants, is also distributed normally. Moreover, it may be shown that the mean and variance of u are related to those of x and y by

$$\mu_u = a\mu_x + b\mu_y \tag{3.48}$$

and

$$\sigma_u^2 = a^2\sigma_x^2 + b^2\sigma_y^2. \tag{3.49}$$

The same relationships may be extended to linear combinations of three or more random variables.

Often the normal distribution is adopted as a convenient approximation, even though there may be no sound physical basis for assuming that the previously stated conditions are met. In some situations this may be justified on the basis that it is the limiting form of several other distributions, the binomial and the Poisson, to name two. More important, if one is concerned only with very general characteristics and not the details of the shape, the normal distribution may sometimes serve as a widely tabulated, if rough, approximation to empirical data. One must take care, however, not to pursue too far the idea that the normal distribution is generally a reasonable representation for empirical data. If the data exhibit a significant skewness, the normal distribution is not likely to be a good choice. Moreover, if one is interested in the "tails" of the distribution, where $|(x - \mu)/\sigma| \gg 1$, improper use of the normal distribution is likely to lead to large errors. Extreme values of distribution must often be considered when determining safety factors and related phenomena. Distributions appropriate to such extreme-value problems are taken up in section 3.4.

The Dirac Delta Distribution

If the normal distribution is used to describe a random variable **x**, the mean μ is the measure of the average value of x and the standard deviation σ is a measure of the dispersion of x about μ. Suppose that we consider a series of measurements of a quantity μ with increasing precision. The PDF for the measurements might look similar to Fig. 3.7. As the precision is increased—decreasing the uncertainty—the value of σ decreases. In the limit where there is no uncertainty $\sigma \to 0$, **x** is no longer a random variable, for we know that $\mathbf{x} = \mu$.

The Dirac delta function is used to treat this situation. It may be defined as

$$\delta(x - \mu) = \lim_{\sigma \to 0} \frac{1}{\sqrt{2\pi}\,\sigma} \exp\left[-\frac{1}{2\sigma^2}(x - \mu)^2 \right]. \qquad (3.50)$$

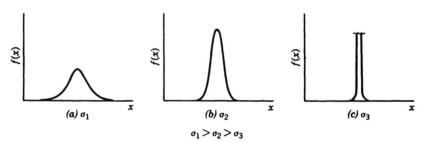

$$\sigma_1 > \sigma_2 > \sigma_3$$

FIGURE 3.7 Normal distributions with different values of the variance.

Two extremely important properties immediately follow from this definition:

$$\delta(x - \mu) = \begin{cases} \infty, & x = \mu \\ 0, & x \neq \mu \end{cases}, \tag{3.51}$$

and

$$\int_{\mu-\varepsilon}^{\mu+\varepsilon} \delta(x - \mu)\, dx = 1, \qquad \varepsilon > 0. \tag{3.52}$$

Specifically, even though $\delta(0)$ is infinite, the area under the curve is equal to one.

The primary use of the Dirac delta function in this book is to simplify integrals in which one of the variables has a fixed value. This appears, for example, in the treatment of expected values.

Suppose that we want to calculate the expected value of $g(x)$, as given by Eq. 3.17 when $f(x) = \delta(x - x_0)$; then

$$E\{g(x)\} = \int_{-\infty}^{\infty} g(x)\,\delta(x - x_0)\, dx \tag{3.53}$$

may be written as

$$E\{g(x)\} = \int_{x_0-\varepsilon}^{x_0+\varepsilon} g(x)\,\delta(x - x_0)\, dx, \qquad \varepsilon > 0, \tag{3.54}$$

since $\delta(x - x_0) = 0$ away from $x = x_0$. If $g(x)$ is continuous, we may pull it outside the integral for very small ε to yield

$$E\{g(x)\} \simeq g(x_0) \int_{x_0-\varepsilon}^{x_0+\varepsilon} \delta(x - x_0)\, dx. \tag{3.55}$$

Therefore, for arbitrarily small ε, we obtain

$$E\{g(x)\} \equiv \int_{x_0-\varepsilon}^{x_0+\varepsilon} g(x)\,\delta(x - x_0)\, dx = g(x_0). \tag{3.56}$$

A more rigorous proof may be provided by using Eq. 3.50 in Eq. 3.53 and expanding $g(x)$ in a power series about x_0.

The Lognormal Distribution

As indicated earlier, if a random variable $\mathbf{x}$ can be expressed as a sum of the random variables, x_i, $i = 1, 2, \ldots, N$ where no one of them is dominant, then $\mathbf{x}$ can be described as a normal distribution, even though the $\mathbf{x}_i$ are described by nonnormal distributions that may not even be the same for different values of i. A second frequently arising situation consists of a random variable $\mathbf{y}$ that is a product of the random variables y_i:

$$y = y_1 y_2 \cdots y_N. \tag{3.57}$$

For example, the wear on a system may be proportional to the product of the magnitudes of the demands that have been made on it. Suppose that we take the natural logarithm of Eq. 3.57:

$$\ln y = \ln y_1 + \ln y_2 + \cdots + \ln y_N. \tag{3.58}$$

The analogy to the normal distribution is clear. If no one of the terms on the right-hand side has a dominant effect, then $\ln y$ should be distributed normally. Thus, if we define

$$x \equiv \ln y, \tag{3.59}$$

then x is distributed normally and y is said to be distributed lognormally.

To obtain the lognormal distribution for y, we first write the normal distribution for x,

$$f_x(x) = \frac{1}{\sqrt{2\pi}\,\sigma_x} \exp\left[-\frac{1}{2\sigma_x^2}(x - \mu_x)^2 \right], \tag{3.60}$$

where μ_x is the mean value of $\mathbf{x}$, and σ_x^2 is the variance of the distribution in $\mathbf{x}$. Now suppose that we let x be the natural logarithm of the variable y. In order to find the PDF in y, we must transform the distribution according to Eq. 3.30:

$$f_y(y) = f_x(x) \left| \frac{dx}{dy} \right|. \tag{3.61}$$

Noting that

$$\frac{dx}{dy} = \frac{d}{dy}\ln y = \frac{1}{y}, \tag{3.62}$$

and using $x = \ln y$ to eliminate x from Eqs. 3.60 and 3.61, we obtain

$$f_y(y) = \frac{1}{\sqrt{2\pi}\,\omega y} \exp\left\{ -\frac{1}{2\omega^2}\left[\ln\left(\frac{y}{y_0}\right) \right]^2 \right\}, \tag{3.63}$$

where we have made the replacements

$$\mu_x \equiv \ln y_0; \qquad \sigma_x = \omega. \tag{3.64}$$

The corresponding CDF is obtained by integrating over y with a lower limit of $y = 0$. The results can be expressed in terms of the standardized normal integral as

$$F_y(y) = \Phi\left[\frac{1}{\omega}\ln\left(\frac{y}{y_0}\right) \right]. \tag{3.65}$$

The PDF and the CDF for the lognormal distribution are plotted as a function of y in Fig. 3.8. Note that for small values of ω, the lognormal and normal distributions have very similar appearances.

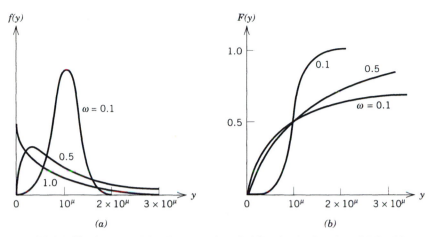

FIGURE 3.8 The lognormal distribution (a) probability density function (PDF), (b) cumulative distribution function (CDF).

The mean of the lognormal distribution may be obtained by applying Eq. 3.15 to Eq. 3.63:

$$\mu_y = y_0 \exp(\omega^2/2). \tag{3.66}$$

Note that it is not equal to the parameter y_0 for which the distribution is a maximum. On the contrary, y_0 may be shown to be the median value of **y**. Similarly, the variance in **y** is not equal to ω but rather is

$$\sigma_y^2 = y_0^2 \, exp(\omega^2)[\exp(\omega^2) - 1]. \tag{3.67}$$

Lognormal distributions are widely applied in reliability engineering to describe failure caused by fatigue, uncertainties in failure rates, and a variety of other phenomena. It has the property that if variables x and y have lognormal distributions, the product random variable $z = xy$ is also lognormally distributed.

The lognormal distribution also finds use in the following manner. Suppose that the best estimate of a variable is y_0 and there is a 90% certainty that y_0 is known within a factor of n. That is, there is a probability of 0.9 that it lies between y_0/n and $y_0 n$, where $n > 1$. We then have

$$0.05 = \int_0^{y_0/n} \frac{1}{\sqrt{2\pi}\,\omega y} \exp\left\{ -\frac{1}{2\omega^2}\left[\ln\left(\frac{y}{y_0}\right)\right]^2 \right\} dy. \tag{3.68}$$

With the change of variables $\zeta = (1/\omega) \ln(y/y_0)$ Eq. 3.68 may be written as

$$0.05 = \int_{-\infty}^{-(1/\omega)\ln n} \frac{1}{\sqrt{2\pi}} \exp(-\tfrac{1}{2}\zeta^2) \, d\zeta. \tag{3.69}$$

This integral is the CDF for the standardized normal distribution, given by Eq. 3.44. Thus we have

$$0.05 = \Phi\left(-\frac{1}{\omega}\ln n\right), \tag{3.70}$$

where Φ is the standardized normal CDF. Similarly, it may be shown that

$$0.95 = \Phi\left(+ \frac{1}{\omega} \ln n \right). \tag{3.71}$$

From the table in Appendix C it is seen that the argument for which $\Phi = 0.05$ or 0.95 is ∓ 1.645. Thus we have

$$\frac{1}{\omega} \ln n = 1.645. \tag{3.72}$$

Therefore, the parameter ω is given by

$$\omega = \frac{1}{1.645} \ln n. \tag{3.73}$$

With y_0 and ω determined, the μ_y can be determined from Eq. 3.66.

EXAMPLE 3.6

Fatigue life data for an industrial rocker arm is fit to a lognormal distribution. The following parameters are obtained: $y_0 = 2 \times 10^7$ cycles, $\omega = 2.3$. (*a*) To what value should the design life be set if the probability of failure is not to exceed 1.0%? (*b*) If the design life is set to 1.0×10^6 cycles, what will the failure probability be?

Solution   (*a*) Let y be the number of cycles for which the failure probability is 1%. Then, from Eq. 3.65, we have

$$0.01 = F_y(y) = \Phi\left[\frac{1}{2.3} \ln\left(\frac{y}{2 \times 10^7} \right) \right].$$

From Appendix C we find

$$\Phi(-2.32) \approx 0.01.$$

Thus

$$-2.32 = \frac{1}{2.3} \ln\left(\frac{y}{2 \times 10^7} \right)$$

and

$$y = 2 \times 10^7 \exp(-2.32 \times 2.3)$$

$$= 9.63 \times 10^4 \text{ cycles.}$$

(*b*) In Eq. 3.65 we have

$$z \equiv \frac{1}{\omega} \ln\left(\frac{y}{y_0} \right) = \frac{1}{2.3} \ln\left(\frac{10^6}{2.0 \times 10^7} \right)$$

$$= -1.302.$$

From Appendix C, $\Phi(-1.302) \approx 0.096$ so that

$$F_y(y) = 0.096 \text{ probability of failure.}$$

3.4 WEIBULL AND EXTREME VALUE DISTRIBUTIONS

The Weibull and extreme value distributions are widely employed for reliability related problems. Their relationship to one another is analogous to that between the lognormal and the normal distribution. The Weibull distribution, like the log normal, ranges $0 \leq x < \infty$, while extreme value like normal distributions have the range $-\infty < x < \infty$. Moreover, the distributions are related through a logarithmic transformation.

Weibull Distribution

The Weibull distribution is widely used in reliability analysis for describing the distribution of times to failure and of strengths of brittle materials, such as ceramics. It is quite flexible in matching a wide range of phenomena. It is particularly justified for situations where a "worst link" or the largest of many competing flaws is responsible for failure. The Weibull CDF is given by

$$F(x) = 1 - \exp[-(x/\theta)^m], \qquad 0 \leq x \leq \infty \qquad (3.74)$$

where θ is the scale and m is the shape parameter. The derivative may be performed as indicated in Eq. 3.4 to obtain the PDF

$$f(x) = \frac{m}{\theta}\left(\frac{x}{\theta}\right)^{m-1} \exp[-(x/\theta)^m], \qquad 0 \leq x \leq \infty. \qquad (3.75)$$

The PDF for the Weibull distribution is shown in Fig. 3.9 for several different values of m.

The mean and the variance of the distribution are obtained from Eqs. 3.15 and 3.16, respectively. They are rather complicated functions of the scale and shape parameters:

$$\mu = \theta\Gamma(1 + 1/m) \qquad (3.76)$$

and

$$\sigma^2 = \theta^2[\Gamma(1 + 2/m) - \Gamma(1 + 1/m)^2]. \qquad (3.77)$$

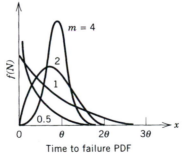

FIGURE 3.9 The Weibull distribution.

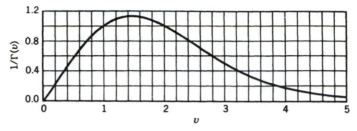

FIGURE 3.10 The gamma function.

In these expressions the complete gamma function $\Gamma(\nu)$ is defined by the integral

$$\Gamma(\nu) = \int_0^\infty \zeta^{\nu-1} e^{-\zeta} \, d\zeta. \tag{3.78}$$

Figure 3.10 shows the dependence of $1/\Gamma(\nu)$ for the values $0 < \nu < 1$, since $\nu > 1$, can be obtained from the identity:

$$\Gamma(\nu) = (\nu - 1)\Gamma(\nu - 1). \tag{3.79}$$

A wide spread use of the Weibull distribution is in describing weakest link phenomena. This may be illustrated by considering a proverbial chain, where the strengths of the N link are described by the random variables $\mathbf{x}_1$, $\mathbf{x}_2$, $\mathbf{x}_3 \ldots \mathbf{x}_N$. The strength of the chain is then also a random variable, say $\mathbf{y}$, which takes on the value of the weakest link. Thus

$$P\{\mathbf{y} > y\} = P\{\mathbf{x}_1 > y \cap \mathbf{x}_2 > y \cap \mathbf{x}_3 > y \cap \cdots \cap \mathbf{x}_N > y\}. \tag{3.80}$$

If the link strengths are independent,

$$P\{\mathbf{y} > y\} = P\{\mathbf{x}_1 > y\} P\{\mathbf{x}_2 > y\} P\{\mathbf{x}_3 > y\} \cdots P\{\mathbf{x}_N > y\}. \tag{3.81}$$

If all of the links are governed by identical strength distributions we can express the probabilities on the right in terms of a single CDF, $F_x(x)$:

$$P\{\mathbf{x}_i > y\} = 1 - P\{\mathbf{x}_i \le y\} = 1 - F_x(y). \tag{3.82}$$

Likewise, since the CDF for $\mathbf{y}$ may be written as $F_y(y) = 1 - P\{\mathbf{y} > y\}$, Eq. 3.81 becomes

$$F_y(y) = 1 - [1 - F_x(y)]^N. \tag{3.83}$$

Now, suppose the link strengths are governed by a Weibull distribution,

$$F_x(x) = 1 - \exp[-(x/\theta)^m]; \tag{3.84}$$

then combining these two equations, we have

$$F_y(y) = 1 - [e^{-(y/\theta)^m}]^N = 1 - e^{-N(y/\theta)^m}. \tag{3.85}$$

Thus the chain strength may also be expressed as a Weibull distribution

$$F_y(y) = 1 - \exp[-(y/\theta')^m] \tag{3.86}$$

with the same shape parameter, and a scale parameter of

$$\theta' = N^{-1/m}\theta. \tag{3.87}$$

Even in situations where the underlying distribution is not explicitly known, but the failure mechanism arises from many competing flaws, the Weibull distribution often provides a good empirical fit to the data.

EXAMPLE 3.7

A chain is made of links whose strengths are Weibull distributed with m = 5 and θ = 1,000 lbs. (*a*) What is the mean strength of one link.? (*b*) What is the mean strength of a chain of 100 links? (*c*) At what load is there a 5% probability that the 100 link chain will fail?

Solution (*a*) From Eq. 3.76: $\mu_x = 1,000 \; \Gamma(1.20) = 1,000 \cdot 0.918 = 918$ lbs.
(*b*) From Eq. 3.87: $\theta' = 100^{-1/5} \cdot 1000 = 398$ lbs.
    Thus $\mu_y = 398 \; \Gamma(1.20) = 398 \cdot 0.918 = 365$ lbs.
(*c*) $0.05 = 1 - \exp[-(y/\theta')^m]$ or $y = \theta'[\ln(1/0.95)]^{1/5} = 398 \cdot 0.552 = 220$ lbs.

A special case of the Weibull distribution is probably the most widely used in reliability engineering. Taking $m = 1$ results in the single-parameter exponential distribution. The CDF is

$$F(x) = 1 - e^{-x/\theta}, \qquad 0 \leq x \leq \infty \tag{3.88}$$

and the PDF is

$$f(x) = \frac{1}{\theta} e^{-x/\theta}, \qquad 0 \leq x \leq \infty. \tag{3.89}$$

The mean and the variance are both given in terms of the single parameter as $\mu = \theta$ and $\sigma^2 = \theta^2$ respectively.

Extreme Value Distributions

Extreme value distributions, or more precisely asymptotic extreme value distributions, frequently arise in situations where the number of variables—flaws, acceleration, etc.—from which the data is gathered is very large. Both maximum and minimum extreme value distributions are applied in reliability engineering. There are a number of different types of extreme value distributions. We will confine our attention here to the type I or Gumbel distributions. The PDF for the maximum and minimum Gumbel distributions are plotted in Fig. 3.11. Note that they have long tails on the right and left respectively.
    The CDF for the maximum extreme value distribution is given by

$$F(x) = \exp[-e^{-(x-u)/\theta}], \qquad -\infty < x < \infty. \tag{3.90}$$

Differentiating according to Eq. 3.4 then produces the PDF:

$$f(x) = \frac{1}{\theta} e^{-(x-u)/\theta} \exp[-e^{-(x-u)/\theta}], \qquad -\infty < x < \infty. \tag{3.91}$$

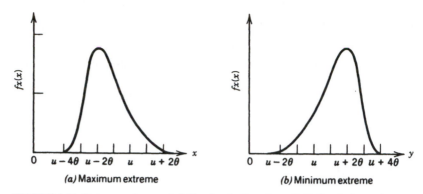

FIGURE 3.11 Extreme-value probability density functions. E. J. Gumbel op. cit.

The PDF is plotted in Fig. 3.11a. The mean and the variance are given by

$$\mu = u + \gamma\Theta, \tag{3.92}$$

where $\gamma = 0.5772157 \ldots$, and

$$\sigma^2 = \frac{\pi^2}{6}\Theta^2. \tag{3.93}$$

Like the normal and lognormal distribution, a reduced variant can be defined which simplifies the CDF. If we take $w = (x - u)/\Theta$, then the CDF becomes

$$F_w(w) = e^{-e^{-w}}, \tag{3.94}$$

which explains why type I extreme value distributions are frequently referred to as double exponential distributions.

The maximum extreme value distribution often works well in combining loads on a system when it is the maximum load that determines whether the system will fail. Suppose that x_1, x_2, $x_3 \ldots x_N$ are the magnitudes of the individual loads, and let y denote the maximum of these loads. To determine the probability that y will not exceed some specified value y, we may write

$$P\{y \le y\} = P\{x_1 \le y \cap x_2 \le y \cap x_3 \le y \cap \cdots \cap x_N \le y\}. \tag{3.95}$$

If the magnitudes of the successive loads are independent of one another, this expression simplifies to

$$P\{y \le y\} = P\{x_1 \le y\}P\{x_2 \le y\}P\{x_3 \le y\} \cdots P\{x_N \le y\}. \tag{3.96}$$

We also note, from Eq. 3.1, that each of these probabilities is just a CDF. Thus if the loads are identically distributed we may rewrite this equation as

$$F_y(y) = F_x(y)^N. \tag{3.97}$$

Now, assume that the CDF for each loading is the maximum extreme value distribution, given by Eq. 3.90. We then have

$$F_y(y) = \{\exp[-e^{-(y-u)/\Theta}]\}^N = \exp[-Ne^{-(y-u)/\Theta}], \tag{3.98}$$

and the CDF for y can be written as a single extreme-value distribution

$$F_y(y) = \exp[-e^{-(y-u')/\Theta}], \tag{3.99}$$

where the displacement parameter has been increased to a value of

$$u' = u + \Theta \ln(N), \tag{3.100}$$

and Θ remains unchanged.

EXAMPLE 3.8

The stress on a landing gear fastener is governed during landing by a maximum extreme-value distribution with a displacement parameter of $u = 8.0$ kips (kilopounds) and $\Theta = 1.5$ kips. (*a*) What is the mean value for individual loading. (*b*) What is the mean value of the maximum load over the 10,000 landing design life of the fastener? (*c*) What strength should the fastener be designed to if there is to be no more than a 1% chance of overloading during the 10,000 landing design life?

   Solution   (*a*) From Eq. 3.92, $\mu = 8.0 + 0.5772 \cdot 1.5 = 8.87$ kips.

(*b*) From Eq. 3.100 we have $u' = 8.0 + 1.5 \ln(10,000) = 21.8$ kips.
    Again from Eq. 3.92 we have $\mu = 21.8 + 0.5772 \cdot 1.5 = 22.7$ kips.

(*c*) Solve Eq. 3.99 for y: $y = u' - \Theta \ln[\ln(1/F)]$.
    With $F = 0.99$, we have $y = 21.8 - 1.5 \ln[\ln(1/099)] = 21.8 - 1.5(-4.60)$
    or $y = 28.7$ kips.

   The minimum extreme-value distribution is frequently used as an alternative to the Weibull in describing strength distributions and related phenomena. The CDF for the corresponding minimum extreme-value distribution is

$$F(x) = 1 - \exp[-e^{(x-u)/\Theta}], \qquad -\infty < x < \infty, \tag{3.101}$$

and the corresponding PDF is

$$f(x) = \frac{1}{\Theta} e^{(x-u)/\Theta} \exp[-e^{(x-u)/\Theta}], \qquad -\infty < x < \infty. \tag{3.102}$$

The PDF is plotted in Fig. 3.11b. The mean and variance are given by

$$\mu = u - \gamma\Theta \tag{3.103}$$

and

$$\sigma^2 = \frac{\pi^2}{6} \Theta^2. \tag{3.104}$$

If we define a reduced variate by $w = (u - x)/\Theta$, we again obtain Eq. 3.94 as the CDF of the reduced variate w.

   It is noteworthy that the minimum extreme value distribution is closely related to the Weibull distribution and as a result is often used for similar purposes, such as representing distributions of times to failure. If we let

$$x = \ln(y),\tag{3.105}$$

then the foregoing equations in x for the minimum extreme-value distribution reduce to a Weibull distribution in y; the Weibull parameters are given in terms of those for the extreme-value distribution by

$$\theta = e^u\tag{3.106}$$

and

$$m = 1/\Theta.\tag{3.107}$$

Thus the Weibull distribution has the same relationship to the minimum extreme-value distribution as the lognormal has to the normal: In both cases they are related by Eq. 3.105, and in the first, the domain of the random variable is $-\infty < x < \infty$, while in the second it is $0 < y < \infty$.

Bibliography

Ang, A. H-S., and W. H. Tang, *Probability Concepts in Engineering Planning and Design,* Vol. 1, Wiley, NY, 1975.

Gumbel, E. J., *Statistics of Extremes,* Columbia Univ. Press, NY, 1958.

Lapin, L. L., *Probability and Statistics for Modern Engineering,* Brooks/Cole, Belmont, CA, 1983.

Montgomery, D. C., and G. C. Runger, *Applied Statistics and Probability for Engineers,* Wiley, NY, 1994.

Olkin, I., Z. J. Gleser, and G. Derman, *Probability Models and Applications,* Macmillan Co., NY, 1980.

Pieruschka, E., *Principles of Reliability,* Prentice-Hall, Englewood Cliffs, NJ, 1963.

Exercises

3.1 For the PDF

$$f(x) = \begin{cases} bx(1-x), & 0 \leqslant x \leqslant 1, \\ 0, & \text{otherwise} \end{cases}$$

determine b, μ, and σ.

3.2 Consider the following PDF:

$$f(x) = 1/2 \qquad 0 < x < 2,$$
$$= 0 \qquad \text{otherwise}$$

Determine the mean and variance.

3.3 A motor is known to have an operating life (in hours) that fits the distribution

$$f(t) = \frac{a}{(t+b)^3}, \qquad t \geq 0.$$

The mean life of the motor has been estimated to be 3000 hr.

(a) Find a and b.

(b) What is the probability that the motor will fail in less than 2000 hr?

(c) If the manufacturer wants no more than 5% of the motors returned for warranty service, how long should the warranty be?

3.4 For a random variable for which the PDF is

$$f(x) = \begin{cases} 0, & x < -1 \\ A, & -1 < x < 1 \\ 0, & x > 1 \end{cases}$$

Determine (a) A, (b) μ, (c) σ^2, (d) sk, (e) ku.

3.5 Suppose that

$$F(x) = 1 - e^{-0.2x} - 0.2xe^{-0.2x}, \qquad 0 \leq x \leq \infty.$$

(a) Find $f(x)$.

(b) Determine μ and σ^2.

(c) Find the expected value of e^{-x}.

3.6 Repeat Exercise 3.4 for $f(x) = A \exp(-|x|)$, $-\infty \leq x \leq \infty$.

3.7 Suppose that the maximum flaw size in steel bars is given by

$$f(x) = 4xe^{-2x}, \qquad 0 \leq x \leq \infty,$$

where x is in microns.

(a) What is the mean value of the maximum flaw size?

(b) If flaws of lengths greater than 1.5 microns are detected and the bars rejected, what fraction of the bars will be accepted?

(c) What is the mean value of the maximum flaw size for the bars that are accepted?

3.8 The following PDF has been proposed for the distribution of pit depths in a tailpipe of thickness x_0:

$$f(x) = A \sinh[\alpha(x_0 - x)], \qquad 0 \leq x \leq x_0.$$

(a) Determine A in terms of α.

(b) Determine $F(x)$: the CDF.

(c) Determine the mean pit depth. What is the probability that there will be a pit of more than twice the mean depth?

3.9 The PDF for the maximum depths of undetected cracks in steel piping is

$$f(x) = \frac{1}{\gamma} \frac{e^{-x/\gamma}}{(1 - e^{-\tau/\gamma})},$$

where τ is the pipe thickness and $\gamma = 6.25$ mm.

(a) What is the CDF?

(b) For a 20-mm-thick pipe, what is the probability that a crack will penetrate more than half of the pipe thickness?

3.10 For a random variable for which the PDF is $f(x)$, $-\infty \leqslant x \leqslant \infty$ find the following in terms of the moments $\overline{x^n} \equiv \int_{-\infty}^{+\infty} x^n f(x)\ dx$:
(a) μ, (b) σ^2, (c) sk, (d) ku.

3.11 Under design pressure the minimum unflawed thickness of a pipe required to prevent failure is τ_0.

(a) Using the maximum crack depth PDF from Exercise 3.9, show that if the probability of failure is to be less than ε, the total pipe thickness must be at least

$$\tau = \gamma \ln\left[1 + \frac{1}{\varepsilon}(e^{\tau_0/\gamma} - 1) \right].$$

(b) For $\gamma = 6.25$ mm and a minimum unflawed thickness of $\tau_0 = 4$ cm, what must the total thickness be if the probability of failure is 0.1%?

(c) Repeat part b for a probability of failure of 0.01%.

(d) Show that for $\tau_0 \gg \gamma$ and $\varepsilon \ll 1$, τ is approximately $\tau_0 + \gamma \ln(1/\varepsilon)$.

3.12 Suppose

$$f_x(x) = \begin{cases} 0, & x < 0 \\ 1, & 0 < x < 1 \\ 0, & x > 1 \end{cases}$$

(a) If $y = x^2$, find $f_y(y)$. (b) If $z = 3x$, find $f_z(z)$.

3.13 Express the skewness in terms of the moments $E\{x^n\}$.

3.14 The beta distribution is defined by

$$f(x) = \frac{1}{B}x^{r-1}(1 - x)^{t-r-1}, \qquad 0 \leqslant x \leqslant 1.$$

Show

(a) that if t and r are integers,

$$B = \frac{(r-1)!(t-r-1)!}{(t-1)!},$$

(b) that $\mu = r/t$,

(c) that

$$\sigma^2 = \frac{\mu(1-\mu)}{t+1} = \frac{r(t-r)}{t^2(t+1)},$$

(d) that if t and r are integers, $f(x)$ may be written in terms of the binomial distribution:

$$f(x) = (t-1)\, C_{r-1}^{t-2} x^{r-1} (1-x)^{t-r-1}.$$

3.15 Transform the beta distribution given in the Exercise 3.14 by

$$y = a + (b-a)x,\ a \leqslant y \leqslant b.$$

(a) Find $f_y(y)$. (b) Find μ_y.

3.16 A PDF of impact velocities is given by $\alpha e^{-\alpha v}$. Find the PDF for impact kinetic energies E, where $E = \frac{1}{2} mv^2$.

3.17 The tensile strength of a group of shock absorbers is normally distributed with a mean value of 1,000 lb. and a standard deviation of 40 lb. The shock absorbers are proof tested at 950 lb.
(a) What fraction will survive the proof test?
(b) If it is decided to increase the strength of the shock absorbers (i.e., to increase the mean strength while leaving the standard deviation unchanged) so that 99% pass the test, what must the new value of the mean strength be?
(c) If it is decided to improve quality control (i.e., to decrease the variance while leaving the mean strength unchanged) so that 99% pass the test, what must the new value of the standard deviation be?

3.18 An elastic bar is subjected to a force l. The resulting strain energy is given by

$$\varepsilon = cl^2,$$

where c is $d/2AE$, with d the length of the bar, A the area, and E the modulus of elasticity. Suppose that the PDF of the force can be represented by standardized normal form $f_l(l)$. Find the PDF $f_\varepsilon(\varepsilon)$ for the strain energy.

3.19 The life of a tool bit is normally distributed with

$$\text{mean: } \bar{t} = 10\ \text{hr} \qquad \text{variance: } \sigma^2 = 4\ \text{hr}^2.$$

What is the L_{10} of the tool?

(L_{10} = time at which 10% of the tools have failed.)

3.20 Suppose

$$f_x(x) = \begin{cases} 0, & x < 1 \\ 1, & 1 < x < 2 \\ 0, & x > 2 \end{cases}$$

(a) if $y = \ln(x)$ find the PDF for y. (b) if $z = \exp(x)$ find the PDF for z.

3.21 The total load on a building may often be represented as the sum of three contributions: the dead load **d**, from the weight of the structure; the live load **1**, from human beings, furniture, and other movable weights; and the wind load **w**. Suppose that the loads from each of the sources on a support column are represented as normal distributions with the following properties:

$$\mu_d = 6.0 \text{ kips} \qquad \sigma_d = 0.4 \text{ kips,}$$
$$\mu_1 = 9.2 \text{ kips} \qquad \sigma_1 = 1.2 \text{ kips,}$$
$$\mu_w = 4.6 \text{ kips} \qquad \sigma_w = 1.1 \text{ kips.}$$

Determine the mean and standard deviation of the total load.

3.22 Verify that μ and σ^2 appearing in Eq. 3.38 are indeed the mean and variance of $f(x)$; that is, verify Eqs. 3.36 and 3.37.

3.23 If the strength of a structural member is known with 90% confidence to a factor of 3, to what factor is it known with (a) 99% confidence, (b) with 50% confidence? Assume a lognormal distribution.

3.24 Verify Eqs. 3.66 through 3.67.

3.25 The L_{10} of a bearing is the life of the bearing at which 10% failures may be expected. A new bearing design follows a Weibull distribution with $m = 2$, and a L_{10} of one year. (a) What fraction of the bearings would you expect to fail in six months? (b) If you had to guarantee no more than 1% failures, to what length of time would you limit the design life?

3.26 One-inch long ceramic fibers are known to have a strength given by a Weibull distribution with a scale parameter of 8 lb and a shape parameter of 7.0. Assume weakest link theory.

(a) What will the scale and shape parameters be for fibers that are two inches long?

(c) If 1.0% of the one inch fiber breaks under the stress of a particular application, what fraction of the two-inch fibers would you expect to break under the same stress?

   (d) If two, two-inch fibers are used in parallel to increase the strength, what fraction would you expect to break?

   (e) How many lb of force were the fibers under?

3.27 The distribution of detectable flaw sizes in tubing is given by Eq. 3.88 with $\theta = 1/17$ cm. There are an average of three detectable flaws per centimeter of tubing.

   (a) What fraction of the flaws will have a size larger than 0.8 cm?

   (b) What is the probability of finding a flaw larger than 0.8 cm in a 100-m length of tubing?

   (c) In 1000 meters of tubing?

3.28 Suppose a system contains 12 of the bearings from exercise 3.25 and the system fails with the failure of the first bearing failure. Estimate the system L_{10}.

CHAPTER 4

Quality and Its Measures

"The first step of the engineer in trying to satisfy these wants is, therefore, that of translating as nearly as possible these wants into the physical characteristics of the thing manufactured to satisfy these wants. In taking this step intuition and judgment play an important role as well as the broad knowledge of the human element involved in the wants of the individuals. The second step of the engineer is to set up ways and means of obtaining a product which will differ from the arbitrarily set standards of these quality characteristics by no more than may be left to chance."

Walter A. Shewhart,
Economic Control of Quality of Manufactured Products, 1931.

4.1 QUALITY AND RELIABILITY

Quality and reliability are intertwined in the design and manufacture of products and in their usage. With the mathematical apparatus set forth in the two preceding chapters we can become more quantitative in examining the relationships that were introduced in Chapter 1. Our objective is to provide an outline to those quality considerations that provides the broad framework useful for the more focused treatment of reliability contained in the chapters to come.

Recall from the discussion in Chapter 1 that the definition of quality leads to two related considerations. First, quality is associated with the ability to design products that incorporate characteristics and features that are highly optimized to meet the customer's needs and desires. Whereas some of these characteristics may be esthetic, and therefore inherently qualitative in nature, the majority can be specified as quantitative performance characteristics. Second, quality is associated with the reduction of variability in these performance characteristics. It is the control and reduction of performance variability with which we shall be most concerned.

Quality is diminished as the result of three broad causes of performance variability:

1. variability in the manufacturing processes
2. variability in the operating environment
3. product deterioration.

Quality improvement measures that reduce or counteract these three causes of performance variability result in large positive impacts on product reliability, for failures usually may be traced to these causes and their interactions. Generally, the product variabilities arising from lack of precision or deficiencies in manufacturing processes lead to failures concentrated early in the product life. These are referred to as early failures or infant mortality. Variability caused by extremes in the operating environment is associated with failures that are equally likely to occur randomly throughout product life; their occurrence probability is independent of the product age. Finally, deterioration most frequently leads to wear or aging failures concentrated toward the end of product life.

To further pursue the improvement of quality—and therefore of reliability—it is instructive to relate the sources of variability and failure to the stages of the product development cycle. Product development falls roughly into three categories:

1. product design
2. process design
3. manufacturing.

Product design encompasses both conceptual and detailed stages. In conceptual design the customer's wants are translated into performance specifications and both the functional principles and physical configuration of the product are synthesized. In detailed design the detailed configuration of the components and parts is set forth and part parameters and tolerances are specified. Process design also includes conceptual and detailed phases in which the manufacturing processes to be employed are first chosen and then the detailed tooling specifications are made. Finally, after the processes are designated and the factory is organized, manufacturing begins and is monitored. To obtain high quality products it is necessary to effectively connect the customer's wants to the design process, and to consider concurrently the manufacturing processes that are to be employed as the product is designed. Only with strong efforts to integrate the product design with the selection of the manufacturing processes can the desirable performance characteristics be produced with a minimum of variability and cost.

In Table 4.1 the three product development activities are related to the three sources of variability and failure. On reflection, it becomes clear that

TABLE 4.1 Stages at which Product Performance Variability can be Reduced

Development Stage	Source of Variability		
	Manufacturing Processes	Operating Environment	Product Deterioration
Product Design	O	O	O
Process Design	O	X	X
Manufacture	O	X	X

O - variability reduction possible
X - variability reduction impossible

much quality and reliability must be designed into a product. Once the design is completely specified, nothing more can be accomplished in process design or manufacturing to reduce the product's susceptibility to failures that are brought about primarily by environmental stresses or product deterioration. Only the product variability leading to infant mortality failures can be substantially reduced through process design and manufacturing quality control.

While the highest importance may be placed on product design, process design is arguably a close second. The conceptual process design—the choice of what processes are to be used and the possible development of new processes—and the detailed determination of process parameters and variability largely determine the conformance to the target values that can be maintained in the manufacturing process. Process design has a large impact on manufacturing variability.

The reduction of variability through the design of product and process is termed off-line quality control, to contrast it with the on-line control that is exercised while production is in progress. The name of Dr. Genichi Taguchi is strongly associated with off-line quality control, for he has lead in developing quantitative methodologies for quality improvement. In the following section, we examine the rationale behind off-line quality control and discuss the techniques through which it is implemented. In Section 4.3 we examine the minimization of variability in the manufacturing process, employing the Six Sigma methodology for relating process quality control to design specifications.

4.2   THE TAGUCHI METHODOLOGY

To gain an understanding of off-line quality control we first formulate quality in terms of the Taguchi loss function. We then examine his approach to robust design: design that decreases performance sensitivity to the variabilities introduced by manufacturing, operating environment, or deterioration. Finally, we briefly outline the experimental design formalism through which the designs of both products and manufacturing processes may be optimized.

Quality Loss Measures

To access the quality of a product the optimized target values of the performance characteristics are compared with the distribution of values that has actually been achieved in the production process. The characteristic variability is represented by a probability density function, say $f(x)$, where x, the characteristic, is a continuous random variable. Since the variability most often results from many small causes in the manufacturing processes, no one of which is dominant, $f(x)$ is frequently represented by a normal distribution,

$$f(x) = \frac{1}{\sqrt{2\pi}\,\sigma} \exp\left[-\frac{1}{2}\left(\frac{x-\mu}{\sigma}\right)^2 \right] \tag{4.1}$$

with a mean μ and a standard deviation σ.

This probability distribution must be compared to a target value and to the specification limits to assess the quality achieved. Suppose that τ is the characteristic target value, and the specification is that x has a value within the interval $\tau \pm \Delta$. The upper and lower specification limits are then defined by

$$LSL = \tau - \Delta \quad \text{and} \quad USL = \tau + \Delta.$$

Often the distribution mean is assumed to be on target (i.e., $\mu = \tau$), and the tolerance limits are taken to be roughly three standard deviations above and below the target. This situation is shown in Fig. 4.1a. Using the CDF for the standard normal distribution, we can see that the fraction of product for which the characteristic is out of specification is $2\,\Phi(-\Delta/\sigma)$. According to the classical interpretation of the specification limits, any product with a characteristic falling between the *LSL* and *USL* is equally acceptable. This implies that no quality loss is incurred so long as x lies between these limits. Conversely if the characteristic falls outside the limits, it is unacceptable. If

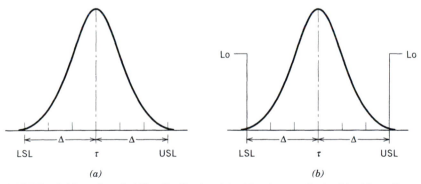

(a)   *(b)*

FIGURE 4.1 Normal probability distribution (a) with tolerance limits (b) with traditional quality loss.

L_0 is the loss in dollars associated with failure to meet the tolerance per product, then we may define a quality loss function according to

$$L(x) = \begin{cases} L_o, & x < LSL \\ 0, & LSL \le x \le USL \\ L_o, & x > USL \end{cases}, \tag{4.2}$$

which is shown graphically in Fig. 4.1b. Note that the expected quality loss per product is defined by

$$\overline{L} = \int L(x) f(x) \ dx. \tag{4.3}$$

Thus using Eq. 4.2 and the centered normal distribution, we obtain

$$\overline{L} = 2L_o\Phi(-\Delta/\sigma). \tag{4.4}$$

The loss function pictured in Fig. 4.1b is sometimes characterized as the goal-post philosophy: If you kick the ball anywhere between the goal posts the quality reward is the same, i.e., zero quality loss. Taguchi argues that this is not realistic. Any deviation from the design target is undesirable, and the loss in quality grows continuously with the deviation from the target value.

Some illustrations demonstrate the weakness of the goal-post loss function. Consider the three distributions shown in Fig. 4.2, all of which have roughly the same expected value of the goal-post loss function (i.e., L_o multiplied by the area under the curve outside of the specification limits). They have, however, very different quality implications. Case a is what one would normally expect: a normal distribution with $\mu = \tau$. In case b the mean is on target, but the variance has increased significantly as a result of the change in the distribution's shape. The distribution for case c is normal, and the variance has decreased significantly from case a. Now, however, the mean is shifted downward substantially from the target value. Taguchi illustrates the quality losses incurred in cases b and c through two frequently-quoted case studies.*

Color TV tubes were produced at two locations under a single set of specifications. It was determined, however, that at the second location many

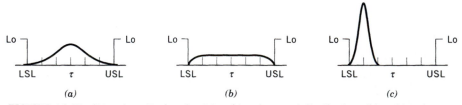

FIGURE 4.2 Traditional quality loss for (a) unbiased normal distribution, (b) unbiased non-normal distribution (c) biased normal distribution.

* G. Taguchi and Y. Wu, *Introduction to Off-Line Quality Control*, Central Japan Quality Control Association, Nagaya, 1979.

more customer complaints were recorded about the picture being dim or about premature tube burnout caused by too bright a picture. A detailed study of the tube brightness revealed the problem. The first plant's brightness distribution was normally distributed about the target values as shown in Fig. 4.2a. The second plant's distribution was nearly uniform as shown in Fig. 4.2b. Thus, even though the tubes from the second plant were within the goal-post specifications, large numbers of sets were produced near the upper or the lower specification limits, and it was these sets that were causing complaints. The consumer did not view the sets in terms of go/no-go specifications. For even within the specified limits, increased deviations from the optimum brightness caused increased numbers of dissatisfied customers.

Figure 4.2c illustrates a quality problem associated with Polyethylene film produced in Japan for use as greenhouse coverings. The film needed to be thick enough to resist wind damage but not so thick as to prevent the passage of light. To satisfy these competing needs, the specification stated that the thickness should be 1.0 mm ± 0.2 mm. The producer made the film thinner in order to manufacture additional square meters of the film at the same materials cost. Since the film thickness could be controlled to ±.02 mm consistently, the nominal thickness was reduced from 1.0 mm to 0.82 mm. The ability to produce the film within 0.02 mm of the nominal assured that the product would still meet specifications while at the same time yielding a significant savings in the required amount of polyethylene feed stock.

Strong typhoon winds, however, destroyed a large number of greenhouses in which the film was used. The replacement cost of the film had to be paid by the customer, and these costs were much higher than expected. The film producer had failed to consider that the customer's cost would rise while the producer's fell. The film was of poor quality and reliability. For even though there was a small variability in the production process, the decrease in the nominal thickness caused the film to be more susceptible to failure under the extreme environmental stress caused by the typhoon.

Experiences such as these prompted Taguchi to formulate a continuous loss function that more closely represents the quality degradation associated with increased deviation from the performance characteristic target value:

$$L(x) = k(x - \tau)^2, \tag{4.5}$$

where the coefficient is determined by setting the loss equal to L_o at both lower and upper specification limits as indicated in Fig. 4.3a. $Lo = k\Delta^2$ so that

$$k = L_o/\Delta^2. \tag{4.6}$$

With this loss function the expected loss accounts for both deviations of the mean from the target value and variability about the mean. Moreover, the expected loss evaluation does not require $f(x)$ to be normally distributed. To demonstrate, we substitute the Taguchi loss $L(x)$ into Eq. 4.3:

$$\overline{L} = \int k(x - \tau)^2 f(x) \, dx. \tag{4.7}$$

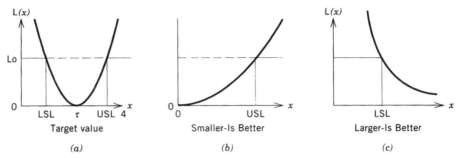

FIGURE 4.3 Taguchi loss functions.

If we write $x - \tau = (x - \mu) + (\mu - \tau)$, the expected loss may be recast as

$$\overline{L} = \int (x - \mu)^2 f(x) \, dx + 2(\mu - \tau) \int (x - \mu) f(x) \, dx$$
$$+ (\mu - \tau)^2 \int f(x) \, dx. \tag{4.8}$$

With the definitions of μ and σ and the normalization of the probability density function defined in Chapter 3, the first term becomes the variance, the second vanishes, and the third is referred to as the bias. We obtain

$$\overline{L} = k[\sigma^2 + (\mu - \tau)^2]. \tag{4.9}$$

Hence, only the mean and variance of the characteristic distribution $f(x)$ are required to evaluate the expected value of the loss function.

EXAMPLE 4.1

The specification for a shaft diameter is 10 ± 0.01 cm. The diameter distribution of manufactured shafts is known to be normal, but it is found that 1.5% of the shaft diameters are greater than the upper specification limit and 0.04% are smaller than the lower specification limit. If the cost of producing an out-of-tolerance shaft is $3.50, what is the expected value of the Taguchi loss function?

Solution $\Phi[(10.01 - \mu)/\sigma] = 1.0 - 0.015 = 0.985$, $\Phi[(9.99 - \mu)/\sigma] = 0.0004$ Thus from Appendix C: $(10.01 - \mu)/\sigma = 2.17$, $(9.99 - \mu)/\sigma = -3.35$, Hence, $\mu + 2.17\sigma = 10.01$ and $\mu - 3.35\sigma = 9.99$. Solve for $\mu = 10.002$, and $\sigma = 0.0036$. Since the specification half width is $\Delta = 0.01$ we may combine Eqs. 4.6 and 4.9 to obtain:

$$\overline{L} = \frac{\$3.50}{0.01^2} [(0.0036)^2 + (10.00 - 10.002)^2] = \$0.60$$

For the many situations where the performance characteristic should be minimized, such as in fuel consumption, emissions, or engine noise, only an upper specification limit, *USL* is set. For these situations, Taguchi defines the smaller-is-better loss function as

$$L(x) = kx^2, \tag{4.10}$$

where k is determined by equating the loss function to the quality loss at the *USL*, as indicated in Fig. 4.3b. Thus

$$k = (USL)^{-2} L_0. \tag{4.11}$$

The expected loss is obtained by combining Eqs. 4.3 and 4.10

$$\overline{L} = k \int_0^\infty x^2 f(x)\, dx. \tag{4.12}$$

EXAMPLE 4.2

The distribution of a contaminant in an industrial solvent is known to be approximated by an exponential distribution. If 0.5% of the solvent containers are found to exceed the upper specification limit and must be discarded at a cost of $12.00 per container, what is the expected value of the Taguchi loss function?

Solution From Eq. 3.88 we have $F(USL) = 1 - e^{-USL/\theta} = 0.995$ or $e^{-USL/\theta} = 0.005$. Thus $USL/\theta = ln(1/0.005) = 5.298$. Then from Eqs. 4.11 and 4.12:

$$\overline{L} = \frac{\$12.00}{USL^2} \int_0^\infty \frac{x^2}{\theta} e^{-x/\theta}\, dx = \frac{\$12.00\,\theta^2}{USL^2} \int_0^\infty \xi^2 e^{-\xi}\, d\xi = \$12.00(USL/\theta)^{-2} \cdot 2$$

Thus $\overline{L} = \$12.00(5.298)^{-2} \cdot 2 = \$0.95$.

For performance characteristics where larger-is-better, such as strength, impact resistance, computing speed, or carrying capacity, only the lower specification limit, *LSL*, is designated. The Taguchi loss function is then

$$L(x) = kx^{-2}, \tag{4.13}$$

with k determined by setting the loss function equal to L_0 at the *LSL*, as indicated in Fig. 4.3c. Hence,

$$k = (LSL)^2 L_0, \tag{4.14}$$

and the expected loss is

$$\overline{L} = k \int_0^\infty x^{-2} f(x)\, dx. \tag{4.15}$$

EXAMPLE 4.3

The strength of components made of a new ceramic are found to be Weibull distributed with a shape factor of $m = 4$ and a scale parameter of $\theta = 500$lb. The lower specification limit on strength is 100 lb. What is the expected Taguchi loss if each failed specimen costs $30.00?

Solution Inserting the Weibull distribution from Eq. 3.75 into Eq. 4.15, we have, for $m = 4$, $\overline{L} = L_0 LSL^2 \theta^{-4} \int_0^\infty xe^{-(x/\theta)^4}\, dx$. Changing variables, $z = \sqrt{2}(x/\theta)^2$ and multiplying numerator and denominator by $\sqrt{2\pi}$, we can express the integral in term of the CDF of the standard normal distribution. Hence:

$$\overline{L} = L_0 LSL^2 \theta^{-2} 2\sqrt{2\pi} \int_0^\infty \frac{1}{\sqrt{2\pi}} e^{-\frac{1}{2}z^2}\, dz = L_0 \sqrt{2\pi}\, LSL^2 \theta^{-2} \Phi(\infty) = L_0 \sqrt{2\pi}\, LSL^2 \theta^{-2}.$$

Therefore:

$$\overline{L} = \$30.00 \cdot \sqrt{2\pi} \cdot 100^2 \cdot 500^{-2} = \$3.01.$$

In the quest for high conformance, reducing quality loss for smaller-is-better and larger-is-better performance characteristics is equivalent to characteristic minimization and maximization, respectively. Many performance characteristics fall into one of these two classes. The situation is more complex for target characteristics, for as indicated in Eq. 4.9, one must reduce the quality loss which arises both from the variance and bias terms, σ^2 and $(\mu - \tau)^2$, respectively. Target characteristics appear frequently in product design, but they are more prevalent in the design of manufacturing processes. In order to obtain product characteristics that are maximized or minimized, it is necessary for the process parameters to be on target. For example, to maximize engine power or minimize fuel consumption, a plethora of dimensional and materials design parameters must be produced with precision. But to accomplish this, manufacturing processes must be designed such that their performance characteristics (i.e., their ability to produce precision dimensions, coating thicknesses, alloy compositions, etc.) are on target, with very little variability.

A basic premise of Taguchi methodology is that it is much easier to eliminate bias from the target characteristics than to reduce the variance. Thus quality improvement is achieved most effectively by first concentrating on variance reduction, even if a side effect is to increase the bias. Once the variance is reduced, the removal of the bias is more straightforward. The plastic sheet problem discussed earlier provides a transparent example. Achieving a small variance in the thickness requires precision sheet-forming machinery and careful control of the composition of the polymer feed stock and of the temperature, pressure, and other process variables. Changing the mean thickness of the sheet, however, required only a single change of process parameter for the forming machinery. This two-step approach for reducing variability in performance characteristics serves as a basis for the robust design methodology that we treat next.

Robust Design

A robust design may be defined as one for which the performance characteristics are very insensitive to variations in the manufacturing process, variability in environmental operating conditions, and deterioration with age. Taguchi designates these factors as product noise, outer noise, and inner noise respectively.* Likewise, in his writings he frequently refers to performance characteristics as functional or product characteristics. In attempting to develop highly robust products it is useful to distinguish between the techniques that may be employed during the conceptual and detailed design phases.

---

* G. Taguchi, *Introduction to Quality Engineering,* Asian Productivity Organization, 1986 (Distributed by American Supplier Institute, Inc., Dearborn, MI).

In conceptual design the specifications of customer needs and desires are translated into a product concept. The physical principles to be employed, the geometrical configuration, and the materials of construction are determined in this stage. In a conceptual engine design, for example, the fuel to be burned, the number of cylinders, the configuration (opposed or V) the coolant (water or air) and the engine block material would be included among the host of issues to be settled. Each decision made in the conceptual design process has quality and reliability implications that are fixed once the product concept has been delineated. Concepts requiring fewer and simpler parts may reduce susceptibility to manufacturing variability. Configurations conducive to natural convection may reduce sensitivity to environmental temperature changes. And judicial materials selection may stave off deterioration from corrosion, warpage, or fatigue. Even with the conceptual design complete, however, much remains to be done to make a product more robust.

The conceptual product design, often existing as a set of sketches, configuration drawings, models, and notes is transformed through detailed design to a set of working drawings and specifications that are sufficiently complete so that the product—or at least a prototype—can be built. Within detailed design a distinction is frequently made between parameter and tolerance design, since each dimension, material composition, or other design parameter must have tolerance limits associated with it before the task is complete.

The Taguchi robust design methodology focuses on choosing mean values of the design parameters such that the product performance characteristics are made less sensitive to parameter variance. If this is accomplished, the performance sensitivity to manufacturing variability will be reduced. Likewise, since the design parameters tend to vary with temperature and other environmental conditions as well as with wear, sensitivity to environmental and aging effects also will be reduced. The product quality is thus increased and a concomitant increase in reliability may be expected. This is a more intelligent approach than reducing performance variability simply by specifying tighter design parameter tolerances. Tighter tolerances will increase manufacturing costs and they are not likely to decrease performance sensitivity to environmental or aging effects.

The two-step robust design methodology is illustrated schematically in Fig. 4.4*a*, *b* and *c*. Initially, as indicated in Fig. 4.4*a*, the mean value of the performance characteristic is on target, but the variance is too large. First, optimize the value of one or more design parameters to minimize the perfor-

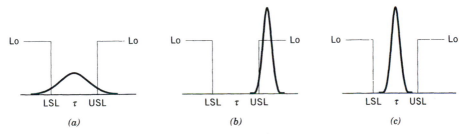

FIGURE 4.4 Distribution of performance characteristic *x*.

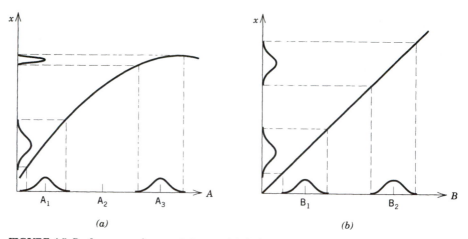

FIGURE 4.5 Performance characteristic *x* vs. (a) design parameter *A*, (b) design parameter *B*.

mance sensitivity to the value of that parameter, regardless of the effect on the performance mean. To achieve this transformation a design parameter must be identified for which the performance characteristic displays a nonlinear response. Such a situation is shown in Fig. 4.5*a* where increasing the value of the design parameter *A*, increases the mean value of the performance characteristic *x*, but decreases the variance in *x*. Success in this effort leads to a performance distribution such as that shown in Fig. 4.4*b*, were the variance is greatly reduced, though a large positive bias from the target value has been introduced. Second, identify an adjustment parameter to bring the mean back on target without increasing the variance. The result is shown in Fig. 4.4*c*. Such a parameter must have a linear effect on the performance characteristic. As indicated in Fig. 4.5*b*, increasing the parameter *B* will increase the mean value of the performance characteristic *x*, while leaving its variance unaffected.

Two examples—one electrical and the other mechanical—illustrate the foregoing procedure.* Consider first a circuit that is required to provide a specified output voltage. This voltage is determined primarily by the gain of a transistor and the value of a resistor. The transistor is a nonlinear devise. As a result, graphs of output voltage versus transistor gain appear as the two curved lines shown in Fig. 4.6 for resistor values R_1 and R_2. Suppose the prototype design achieves the target voltage, indicated by the arrow, with resistance R_1 and transistor gain G_1 as shown. The inherent variability in the transistor gain depicted by the bell-shaped curve about G_1, however, causes an unacceptably wide distribution of output voltages as indicated by curve a.

Improving performance quality directly through tolerance reduction is difficult, because a substantially higher quality component—the transistor—would be required to reduce the width of the curve centered about G_1, thus increasing costs. In robust design, parameter values are used to improve

* P. J. Ross, *Taguchi Techniques for Quality Engineering,* McGraw-Hill, New York, 1988.

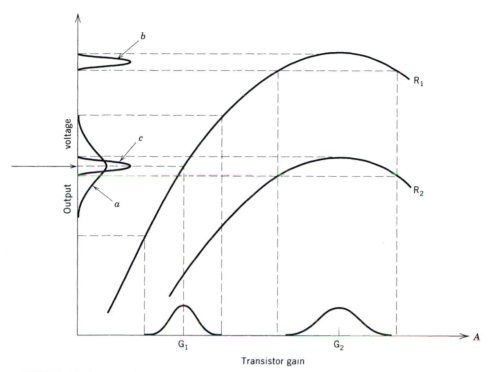

FIGURE 4.6 Output voltage vs transistor gain. (From Ross, P. *Taguchi Techniques for Quality Engineering*, pgs. 176, 178, 258, McGraw-Hill, New York, 1988. Reprinted by permission.)

performance quality before the tightening of tolerances is considered. To accomplish this we again follow the two-step procedure of decreasing variance and then removing bias. If we operate the transistor at a higher gain, at point G_2, the gain variance will also increase as indicated by the normal distribution about G_2. Nevertheless, the nonlinear relationship between gain and output voltage causes the output voltage distribution, given by curve b, to have a much narrower distribution.

Increasing the gain in going from case a to case b introduces a large positive bias in the output voltage. We must now proceed with the second step to eliminate this bias. After examining several possible values of the resistance, we choose the value R_2 that results in the lower voltage versus gain curve plotted in Fig. 4.6. The resistance R_2 brings the output voltage back on target, and as indicated by curve c, the narrow spread in the output voltage is maintained. Thus we have achieved a smaller quality loss in the performance characteristic without resorting to the use of a higher quality—and therefore more expensive—transistor.

Finally, note that in addition to allowing a lower quality component to be used, the forgoing parameter optimization reduced the effects of operating environment and transistor aging on the output voltage. Since the transistor gain is likely to be somewhat effected by the ambient temperature, reducing the output voltage sensitivity to the gain also reduces its sensitivity to ambient temperature. Likewise, the output voltage in the improved design is less sensitive to the drifts in transistor gain, which are likely to be a result of aging.

The engine, metal, oil-fill tube and associated rubber cap pictured in Fig. 4.7 provides a second instructive example. The cap must be easy to remove or install. It must also seal the tube against the engine crankcase pressure. Consequently, the force required to release the cap must be small enough for any owner to remove and insert the cap easily, but large enough that the crankcase pressure will not be capable of blowing the cap off under foreseeable operating conditions. Thus, the required release force is a performance characteristic. The vertical axis of Fig. 4.8 shows the upper force limit determined by minimum user strength and the lower force limit determined by maximum crankcase pressure; the target is centered between the limits.

The force resisting installation or removal results from the crimped ridge in the metal tube over which the rubber cap must deflect. The cap can be removed or inserted only when it deflects sufficiently for its outside diameter (OD) to become less than the inside diameter (ID) of the crimp in the tube. Roughly speaking, the force required is proportional to the product of the required deflection and the cap stiffness. The resisting force can thus be increased by increasing the difference between the cap OD of the tube crimp ID. The required force can also be made larger by observing that the cap stiffness increases with wall thickness.

The deflection is much more difficult to control than the stiffness. The stiffness predominantly depends on the wall thickness, which is easily controlled within a small percentage variation. The required deflection is determined by a small difference in diameters that is likely to be very sensitive to variability in the manufacturing process. It will also be sensitive to environmental conditions since different coefficients of thermal expansion are likely to change the necessary deflection with temperature.

Two force versus deflection curves are shown in Fig. 4.8 for different wall thicknesses and therefore for different cap stiffness. The initial design

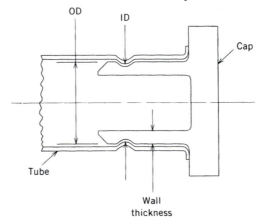

FIGURE 4.7 Engine oil fill tube and cap. (From Ross, P. *Taguchi Techniques for Quality Engineering*, pgs. 176, 178, 258, McGraw-Hill, New York, 1988. Reprinted by permission.)

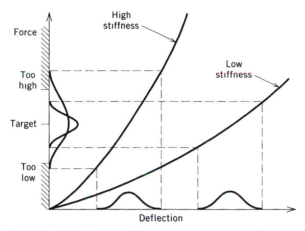

FIGURE 4.8 Cap removal force using parameter design.
(From Ross, P. *Taguchi Techniques for Quality Engineering*, pgs.
176, 178, 258, McGraw-Hill, New York, 1988. Reprinted by
permission.)

corresponds to the high stiffness curve, which results in an unacceptably
wide force distribution spread about the target characteristic. The stiffness is
decreased by making the wall thickness of the cap smaller. This reduces the
spread in the force distribution significantly. However, if the same ID and
OD are retained, the result is a mean force that is too small to resist the
crankcase pressure. If the required deflection is then increased by increasing
the ID–OD difference, the mean force is brought back on target. As indicated
in Fig. 4.8, a design is then achieved in which the variability in the performance
characteristic is decreased by changing parameters, but without tightening
manufacturing tolerances.

Manufacturing processes as well as the products themselves can be improved greatly through the use of the robust design methodology. By setting
the process parameters to minimize the variability in the process output,
higher quality parts and components are obtained without a commensurate
increase in cost for manufacturing equipment. Moreover, in process optimization, it is often clear from the beginning what factor can be used for the
adjustment; it is often the length of time that the process is applied. To
illustrate, consider a spray coating operation. The thickness of the coating is
specified within a very narrow tolerance interval, i.e., a very smooth finish is
required. Suppose that the variability in the coating thickness is sensitive to
the temperature at which it is applied to the surface. The process engineer
first varies the application temperature and determines the temperature at
which the variance in the thickness is minimized. She then adjusts the spray
time until the mean thickness coincides with the target value.

The Design of Experiments

The robust design examples considered thus far could be illustrated graphically because in each case two identifiable design parameters are manipulated
to reduce the variance of the performance characteristic and return the mean

to the target value. More often, however, many parameters interact in determining the behavior of each performance characteristic. It is often unclear which of these are important, and which are not. This situation arises frequently regardless of whether the performance characteristic is of the larger-is-better, smaller-is-better, or target value variety.

In some situations the relationships between parameters and the performance characteristics may be studied through computer modeling. This is often the case, for example, in circuit analysis and in the many mechanical stress problems that are amenable to solution by finite element analysis. In other situations, however, understanding of the process has not reached the point where computer simulation can be utilized effectively. Then, experiments must be performed on product or process prototypes, and the performance evaluated with different sets of parameters. In either event—whether the experiments are computational or physical—efficacy demands that the optimal parameter combination be found with the fewest experiments possible, because the cost of the optimization effort tends to rise in direct proportion to the number of experiments that must be performed.

Picking parameters by trial and error would be an exceedingly wasteful effort and would not likely come close to the optimal conditions within a reasonable number of trials. Varying one parameter at a time is more systematic, but is still relatively inefficient. Moreover, false conclusions may be reached if the factors interact with one another. This can be illustrated with a simple two-parameter case. Suppose we represent a performance characteristic as the elevation in the contour plots shown in Fig. 4.9. The design parameters, x and y, are to be selected to maximize the characteristic. Thus, the object of the experimentation is to locate the point marked by a #. The fundamental difference between Fig. 4.9a and b is that the contour ellipses in Fig. 4.9b appear to be rotated with respect to the axes, while those in Fig. 4.9a are not. In statistical terms the parameters are said to interact in Fig. 4.9b, while those in Fig 4.9a do not.

Changing a single variable at a time will successfully find the optimum in Fig. 4.9a, where there are no interactions. Starting at (x_0, y_0), we first vary x by performing a number of experiments while holding y constant at a value $y = y_0$. Assume a maximum at x_1 is found. Then y is varied by doing an

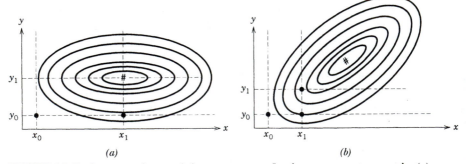

FIGURE 4.9 Performance characteristic contour maps for design parameters x and y (a) no interaction between x and y (b) interaction between x and y.

additional set of experiments while holding x constant at $x = x_1$. The maximum found at $y = y_1$, and indeed the optimal value, is at (x_1, y_1).

This procedure will give a false result in Fig. 4.9b, however, where an interaction is present. Starting at (x_0, y_0) we again vary x, holding y constant at $y = y_0$, and find a maximum at x_1. But now varying y with $x = x_1$ yields a maximum at y_1, but (x_1, y_1) is far from the optimal point marked with a #. In this situation one would need to iterate several times, next holding $y = y_1$ and searching for the maximum x_2, then holding $x = x_2$ and searching for the maximum $y = y_2$, and so on. The number of experiments required and therefore the cost of the exercise could soon become prohibitive.

This simple two parameter problem indicates experiments in which only one variable changes at a time are ineffective when statistical interactions exist between parameters. The weakness becomes more pronounced as the number of design parameters increases. As a result, more powerful strategies have been developed in which all of the parameters are changed simultaneously in order to reduce the total number of experiments needed to locate the optimum. These strategies are collectively referred to as designed experiments.

The most complete of the designed experiments is the full factorial experiment in which m values, called levels, of each parameter are used in all possible combinations. Consequently, if there are n parameters, a full factorial experimental design requires that m^n experiments be performed. For the two-parameter example above, 4 experiments would be required with 2 levels, 9 with 3 levels and so on. When several parameters must be examined, the number of required experiments rises very rapidly. A two-level experiment with ten parameters, for example, requires $2^{10} = 1024$ experiments. Even if the experiments consist of computer simulations, the numbers can soon become excessive. One strategy for reducing the number of experiments without commensurate loss of information is the fractional factorial experiment.

The difference between full factorial, fractional factorial, and single parameter at a time experiments is illuminated by examining three parameters, with two possible values (or levels) for each. The three strategies are shown schematically in Fig. 4.10 where the dimensions correspond to the parameters. Experiments are run for the (x, y, z) combinations indicated by solid circles

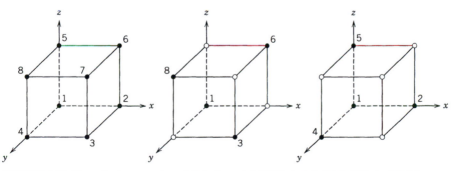

FIGURE 4.10 Three factor experimental designs: (a) full factorial, (b) half factorial, (c) one-factor at a time.

and are omitted where the open circles are shown. Thus, Fig. 4.10*a* is a full factorial design, with the 2^3 or eight experiments corresponding to all possible combinations of the low and high level of each parameter. Only four experiments are run using either the half-factorial design in Fig. 4.10*b* or the single parameter at a time variation in Fig. 4.10*c*. Note that in the fractional factorial design there are two experiments done at the high and at the low level of each parameter, whereas in the single-parameter-at-a-time design two experiments are performed at the low level of *x, y* and *z,* but only one at the high level of each of these parameters.

Comparisons of Fig. 4.10*b* and *c* allow us to examine how more effective use is made of a given number of experiments in the half factorial designed experiment than by changing a single parameter at a time. Assume we want to maximize the value of a performance characteristic η. To determine the effect of the parameter *x* using the single parameter at a time experiment in Fig. 4.10*c*, we calculate the difference between the two experiments for which *y* and *z* are held constant:

$$\Delta \eta_x = \eta_2 - \eta_1. \tag{4.16}$$

Consequently, only two experiments are utilized. In contrast, the partial factorial design of Fig. 4.10*b* utilizes all four experimental results; we compute the effect as an average difference between experiments in which *x* is at level 2 and at level 1,

$$\Delta \eta_x = (\eta_6 + \eta_3 - \eta_1 - \eta_8)/2. \tag{4.17}$$

The use of more experiments reduces the effects of the noise due to random errors in individual measurements. It also tends to average out effects due to changes with respect to *y* and *z,* since both high and low level values of *y* and *z* are included. The same argument applies to determining the effects of the *y* and *z* parameters. The fractional factorial design also allows one to estimate the effects of selected statistical interactions between variables.

Fractional factorial experiments become more valuable as the number of parameters increases and the number of levels per parameter is increased to three or possibly more. They eliminate many of the difficulties of single-parameter-at-a-time experiments but require many fewer trials than a full-factorial experiment. Taguchi has packaged techniques for performing fractional factorial experiments in a particularly useful form called orthogonal arrays. Moreover, he has coupled the parameter selection with techniques for including the noise arising from temperature, vibration, humidity, or other environmental effects.

Figure 4.11*a* is an example from the collection of orthogonal arrays provided by Taguchi for dealing with different numbers of parameters and levels. For this three-level experiment the effects of four design parameters are to be studied. A full-factorial experiment would require $3^4 = 81$ trials. The array shown in Fig. 4.11*a* reduces the number of trials to nine, each represented by a row of the array. The columns represent the four design parameters, with the entries in each column representing the test level for that parameter in each of the nine experiments. Observe that each level for

	Design Parameters					Noise Factors		
Run #	θ_A	θ_B	θ_C	θ_D		W_1'	W_2	W_3
1	1	1	1	1		1	1	1
2	1	2	2	2		1	2	2
3	1	3	3	3		2	1	2
4	2	1	2	3		2	2	1
5	2	2	3	1				
6	2	3	1	2				
7	3	1	3	2				
8	3	2	1	3				
9	3	3	2	1				

FIGURE 4.11 Orthogonal arrays: (a) three-level design parameter array, (b) two-level noise array.

each parameter appears in the same number of experiments: level 1 of θ_B for example appears in trials 1, 4 and 7; level 2 in trials 2, 5 and 8; and level 3 in trials 3, 6 and 9.

The balance between parameter levels in the orthogonal array allows averages to be computed that isolate the effect of each parameter by averaging over the levels of the remaining parameters. Procedures for estimating the effects of each of the parameters on the performance characteristic η are sometimes referred to as analysis of means (or ANOM). Suppose that η_1, η_2, η_3, ... η_9 are the results of the nine experiments. Let $\overline{\eta}_{A1}$ be the performance characteristic averaged over those experiments for which θ_A is at level one, $\overline{\eta}_{A2}$ over those experiments for which θ_A is at level two, and so on. We then have

$$\overline{\eta}_{A1} = (\eta_1 + \eta_2 + \eta_3)/3,$$

$$\overline{\eta}_{A2} = (\eta_4 + \eta_5 + \eta_6)/3, \tag{4.18}$$

$$\overline{\eta}_{A3} = (\eta_7 + \eta_8 + \eta_9)/3.$$

Similarly we would have

$$\overline{\eta}_{B1} = (\eta_1 + \eta_4 + \eta_7)/3 \tag{4.19}$$

and so on.

Plots are instructive in determining the main effect of each parameter on the performance characteristic. To determine the effect of θ_A, we plot $\overline{\eta}_{A1}$, $\overline{\eta}_{A2}$ and $\overline{\eta}_{A3}$ versus the value of θ_A at each of the three levels. If the result appears as in Fig. 4.12a, there is no effect on the performance characteristic, and the value of θ_A may be chosen on the basis of cost. If the plot appears as in Fig. 4.12b or c, however, there is a significant effect. Then, since the object of this particular exercise is to maximize η, the value of θ_A that corresponds to the largest value of η should be chosen. The procedure is illustrated with the following example.

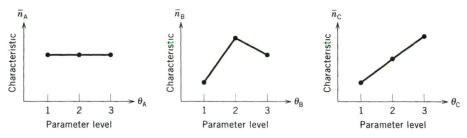

FIGURE 4.12 Performance characteristic vs. design parameters.

EXAMPLE 4.4

A manufacturer of filaments for incandescent lamps wants to determine the effect of the concentration of two alloy metals and of the speed and temperature at which the filaments are extruded on the filament life. A three-level experiment is to be used. The three levels of parameters θ_A and θ_B are the concentrations of alloy metals A and B, parameter θ_C is the extrusion speed, and parameter θ_D the extrusion temperature. Levels 1, 2, and 3 correspond to low, intermediate, and high values of each parameter. Nine sets of specimens are prepared according to the parameter levels given in Fig. 4.11 a. Each experiment consists of testing the thirty specimens to failure and recording the mean time to failure (MTTF) for that set. The resulting MTTFs for the nine experiments are: 105, 106, 109, 119, 119, 115, 129, 122, 125 hr.

Determine which parameters are most significant and estimate the optimal factor levels to maximize filament life.

Solution   Calculate the three level averages for parameter θ_A from Eq. 4.18, and the averages for θ_B, θ_C, and θ_D can be obtained analogously:

$$\eta_{A1} = (105 + 106 + 109)/3 = 106.7 \qquad \eta_{B1} = (105 + 119 + 129)/3 = 117.7$$
$$\eta_{A2} = (119 + 119 + 115)/3 = 117.7 \qquad \eta_{B2} = (106 + 119 + 122)/3 = 115.7$$
$$\eta_{A3} = (129 + 122 + 125)/3 = 125.3 \qquad \eta_{B3} = (109 + 115 + 125)/3 = 116.3$$

$$\eta_{C1} = (105 + 115 + 122)/3 = 114.0 \qquad \eta_{D1} = (105 + 119 + 125)/3 = 116.3$$
$$\eta_{C2} = (106 + 119 + 125)/3 = 116.7 \qquad \eta_{D2} = (106 + 115 + 129)/3 = 116.7$$
$$\eta_{C3} = (109 + 119 + 129)/3 = 119.0 \qquad \eta_{D3} = (109 + 119 + 122)/3 = 116.7$$

Graphs showing the main effects of the four parameters are shown in Fig. 4.13. Clearly parameter θ_A is most significant, and whereas θ_B, and θ_C have significantly less effect, θ_D has virtually no effect on the results. To maximize the MTTF, θ_A should be set at

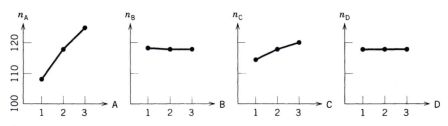

FIGURE 4.13 Performance characteristic vs. design parameters for example 4.4.

level 3, θ_B, and θ_C at levels 1 and 3, respectively; θ_D can be determined strictly on the basis of cost.

The foregoing procedure provides a means of determining which factors have the largest effects on performance. It also allows the optimum settings for the various parameters to be determined. Thus far we have implicitly assumed, however, that all the factors are significant. No quantitative method has been provided for determining whether the changes in parameter level are significant or are just the result of random effects or measurement errors. In the foregoing example, for instance, repeated measurements of the MTTF for a given set of the four parameters would not be expected to yield identical results, since the time-to-failure is an inherently random variable. By averaging over many measurements this randomness is reduced, but it still may be significant. Thus the following question must be addressed: Are the changes that occur with different parameter levels significant, or would changes of comparable magnitude occur if the experiments were repeated with a single set of parameters?

Such questions, related to the determination of which effects are significant and which are not, can be addressed with a powerful statistical technique referred to as the analysis of the variance or ANOVA. The step-by-step procedures of applying ANOVA to the results of partial-factorial experiments may be found in a number of texts, but are too lengthy to be treated here. Suffice it to say that the techniques are extremely valuable in the early stages of designed experiments, where many design parameters must be screened to determine which have a significant impact on performance, and which can safely be ignored in optimization studies.

Arrays such as that shown in Fig. 4.11*a* are often called design arrays, and the design parameters θ_A, θ_B, θ_C, θ_D are referred to as control factors in the Taguchi literature, since they can be prescribed by the designer. Frequently, it is desirable also to understand the sensitivity of the performance characteristic to those environmental factors that cannot easily be controlled under field conditions: ambient temperature, humidity, and vibration, for example. For such situations a second orthogonal array, referred to as a noise array, is added to the experimental procedure. Standard nomenclature is then to designate design and noise arrays as inner and outer arrays, since they deal with what Taguchi defines as product and outer noise: noise due to parameter and environmental variability, respectively.

An example of a noise array—this one being a two-level array for three environmental noise factors—is shown in Fig. 4.11*b*. In order to do the parameter optimization with this noise array included, each of the nine experiments with different parameter combinations must be repeated four times with the noise levels specified in the outer array. Thus 36 trials must be carried out. If w_2 is temperature and levels one and two are 50°F and 100°F, then for each of the nine parameter combinations the first and third runs would be at 50°F and the second and fourth at 100°F. The analysis would then be the same as with Eq. 4.18, but now each of the values of η_i on the right of these equations would be averaged over the four runs corresponding to the rows of the noise array.

Carefully designed experiments typically take place in three-phase proto-cal. In the first, several design parameters—perhaps ten or more—are screened using a two-level orthogonal array. The ANOVA then identifies the two to four design parameters and their interactions that are most important in determining the performance characteristic η. The second phase then involves performing experiments with a three-level array only for the design parameters that are found to be most significant. The ANOM of the second phase experiments then estimates of value of the performance characteristic and the optimal combination of design parameters. The third and final phase consists of a confirmation experiment to assure that the predicted value of η is achieved with the design parameters that have been selected.

Taguchi, adopting terminology common in electrical engineering, speci-fies η, the quantity to be maximized, not as the performance characteristic itself, but as the signal-to-noise or S/N ratio. For larger-is-better or smaller-is-better performance characteristics, η is expressed in terms of the expected quality loss $\overline{L}$ given by Eq. 4.15 or 4.12 respectively, as the logarithmic rela-tionship

$$\eta = -10 \ log_{10}(\overline{L}^2). \tag{4.20}$$

In the discussion of robust design emphasis is placed on the two step procedure in which design parameters are first selected to reduce the variance of the performance characteristic about the mean, even if a shift in the mean results. In using designed experiments based on orthogonal arrays for this purpose, Taguchi recommends that the ratio μ/σ, the inverse of the coeffi-cient of variation for the characteristic distribution $f(x)$, be used as a basis for the signal to noise ratio

$$\eta = -10 \ log_{10}(\sigma^2/\mu^2) \tag{4.21}$$

Once design parameters have been chosen to maximize this signal-to-noise ratio, an adjustment factor is employed to bring μ back on target. A number of other signal-to-noise ratio's are also defined in Taguchi's writing for the analysis of differing forms of the loss function.

4.3   THE SIX SIGMA METHODOLOGY

Thus far we have discussed the measurement of quality loss. We have also examined robust design methods for minimizing the effects of variability in parts fabrication and assembly on the performance characteristics. The achievement of a robust design allows the specification limits on parts dimen-sions, materials composition, and the myriad of other parameters that appear on shop drawing and specifications to be less stringent without a commensu-rate loss of reliability. Nonetheless, while good design will reduce the cost of the manufacturing processes, those processes still must be implemented to reduce the number of parts that do not meet specifications to very small numbers. For as products become more complex, the number of parameters that must fall within specification limits increases rapidly. To deal with this challenge, process capability concepts and the stringent requirements associated with them must be understood.

After providing some basic definitions, we examine the six sigma criteria which are increasingly coming into use for the improvement of product quality. Although the terminology and notation is somewhat different than that used in defining Taguchi loss function concepts, the approaches have much in common, for they take into account the related problems of reducing process variability and maintaining the process mean on target. Taguchi analysis is aimed primarily at off-line quality control; it targets the design of products and manufacturing processes to make performance as insensitive to part variability as possible. The six sigma methodology is focused primarily on controlling manufacturing processes such that the production of an out-of-tolerance part is an exceedingly rare event. In the analysis the normal distribution is a widely assumed model for parameter variability. This is justifiable, since variability in such parameters tends to arise from many small causes, no one of which is dominant.

Process Capability Indices

The basic quantity about which much of the analysis is centered is the capability index, C_p. It is the ratio of the specification interval,

$$USL - LSL = 2\Delta \tag{4.22}$$

to the process variability. The process parameter is assumed to be distributed normally, with the variability represented by 6σ, six times the standard deviation. Thus

$$C_p = (USL - LSL)/6\sigma. \tag{4.23}$$

The factor 6 is employed since traditionally specification limits have been most often taken to be three standard deviations above and below the target value. Equation 4.22 may be used to eliminate the USL and LSL and express the capability index in terms of the specification half-width Δ. We then have

$$C_p = \Delta/3\sigma. \tag{4.24}$$

The definition of the capability index assumes that the mean value of the parameter x is the target value, causing the distribution to be centered between the tolerance limits as indicated in Fig. 4.14. Since x is assumed to be normally distributed, the fraction of out-of-specification parts can be determined from Φ, which is the CDF of the standardized normal distribution defined in Chapter 3. Of the parts that don't meet specifications, half will have values of $x < \tau - \Delta$ and the other half will have values of $x > \tau + \Delta$.

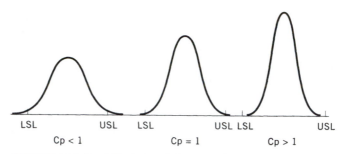

FIGURE 4.14 Capability index C_p for normal distributions.

Thus introducing the reduced variant

$$z = (x - \mu)/\sigma, \tag{4.25}$$

and taking $x = \tau - \Delta$ at the lower specification limit, we obtain $z = -\Delta/\sigma$. If we use Eq. 4.24, we may write z in terms of C_p: $z = -3C_p$. The fraction of rejected parts is then twice the area under the normal CDF to the left of the *LSL*. Hence

$$p = 2\Phi(-z) = 2\Phi(-3C_p). \tag{4.26}$$

The corresponding yield is defined as the fraction of parts accepted:

$$Y = 1 - 2\Phi(-3C_p). \tag{4.27}$$

From the definition of the capability index and the assumption of a centered normal distribution, a value of $C_p = 1.0$ corresponds to 0.27% out-of-tolerance parts, or a yield of $Y = 99.73\%$. As indicated in Fig. 4.14, a larger capability index implies that the fraction of items out of specification is smaller, while a smaller index corresponds to a larger fraction being outside the specification interval.

The capability index C_p is used as a measure of the short term or part-to-part variation of parameters against the specification interval. For example, if metal parts are being machined, no two successive parts will have exactly the same dimension. Machine vibrations, variability in the local material properties, and other random causes result in the part-to-part spread that gives rise to the normal distribution. If these short term variations are completely random, however, the distribution mean should remain equal to the target value.

Over longer periods of time more systematic variations in the manufacturing process are likely to cause the distribution mean to drift away from the target value. Possible causes for such drift are tool wear, changes in ambient temperature, operator change, and differing properties in batches of materials. To take these effects into account a second index, often referred to as the location index, is defined as

$$C_{pk} = C_p(1 - k), \tag{4.28}$$

where k is defined as the ratio of the mean drift to the specification half-width:

$$k = |\tau - \mu|/\Delta. \tag{4.29}$$

Thus if either the part-to-part variability increases or the process mean drifts from the target value, the index C_{pk} will decrease.

EXAMPLE 4.5

Calculate C_p, k, and C_{pk} for the distribution of shaft diameters in Example 4.1

Solution   From Example 4.1 we know that $\mu = 10.002$, $\sigma = 0.0036$, and $\Delta = 0.01$. From Eq. 4.24 $C_p = 0.01/(3 \times 0.0036) = 1.02$. Since $\tau = 10.00$, from Eq. 4.29 $k = |10.00 - 10.002|/0.01 = 0.2$ and from Eq. 2.28 $C_{pk} = (1 - 0.2) \times 1.02 = 0.816$.

The quantities C_p and C_{pk} are often referred to as the short- and long-term process capability, respectively. If the long-term drifts tend also to be of a random nature, it is useful to picture C_{pk} in terms of a normal distribution with an enlarged standard deviation. This is illustrated in Fig. 4.15 where the part-to-part variation at a number of different times is indicated by normal distributions. With mean shifts which are randomly distributed over long periods of time, we obtain the normal distribution indicated in Fig. 4.15 by

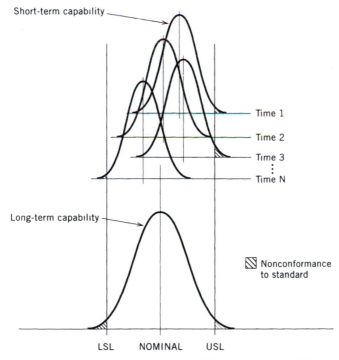

FIGURE 4.15 Effect of long term variability on process capability. (From Harry M. L. and Lawson, J. R., *Six Sigma Producility Analysis and Process Characterization*, pgs. 3–5 and 6–9, Addison-Wesley Publishing Co. Inc. and Motorola, Inc. 1992. Reprinted by permission.)

time averaging. The capability index may be written in this form as

$$C_{pk} = \Delta/3\sigma_k, \tag{4.30}$$

where σ_k is a measure of the increased spread of the distribution. We may view the standard deviation appearing in C_{pk} as

$$\sigma_k = c\sigma, \tag{4.31}$$

where the σ on the right is again the contribution of the part-to-part variability that appears in C_p, whereas c is a multiplier greater than one that arises from the variability induced over longer periods of time by the movement of the mean away from the target value τ. Clearly, we may also combine Eqs. 4.28, 4.30 and 4.31 to obtain $k = 1 - 1/c$, where k is referred to as the equivalent shift in the mean.

Since Eqs. 4.30 and 4.31 are equivalent to assuming that the time-averaged, long-term variability is also normally distributed about the target value, the long-term yield can be calculated simply by replacing C_p by C_{pk} in Eq. 4.27:

$$Y = 1 - 2\Phi(-3C_{pk}). \tag{4.32}$$

A third, and final, capability index, C_{pm}, is finding increased use. Like C_{pk} it measures both the variation about the mean and the bias of the mean from the target value. This index is closely related to the Taguchi loss function and thereby does not implicitly assume that the PDF is normally distributed. We define

$$C_{pm} = \Delta/3\sigma_m, \tag{4.33}$$

where the newly defined variance

$$\sigma_m^2 = \sigma^2 + (\mu - \tau)^2 \tag{4.34}$$

is the sum of contribution of the variance about the mean and the bias. We see from Eq. 4.9 that C_{pm} is closely related to the expected value $\overline{L}$, of the Taguchi loss function. Combining Eqs 4.6, 4.9 and 4.34, we have $\sigma_m^2 = \Delta^2 \overline{L}/L_o$, or equivalently

$$C_{pm} = \frac{1}{3}\sqrt{L_0/\overline{L}}. \tag{4.35}$$

Yield and System Complexity

Historically, the target in manufacturing processes has been to yield a short-term capability index of $C_{pk} = 1$. Consequently, the process was considered satisfactory if the specification limits were three standard deviations from the process mean. This resulted in 0.27% out-of-specification parts. Over a wide range of processes, it was found that the long-term variability tended to be considerably larger,[*] with values of c commonly in the range $1.4 < c < 1.6$.

[*] M. J. Harry and J. R. Lawson, *Six Sigma Producibility Analysis and Process Characterization*, Addison-Wesley Publishing Company, Reading, MA, 1992.

For example, if we take $c = 1.5$, for which $k = 1/3$, we find that with $C_p = 1$ the long-term capability index is only $C_{pk} = 2/3$. Thus Eq. 4.32 indicates that over time the yield is reduced to $1 - 2\Phi(-2)$ or 95.55%.

Yields computed in this way, however, apply only to a single part, and then only to a part with one specification. Real parts typically have a number of specifications that must be met. As products or systems grow more complex, having many parts, the total number of specifications grows very rapidly. Computer memory chips, for example, have many identical diodes, each of which must meet a performance specification. Conversely, an engine may have fewer parts, but each part may have a substantial number of specifications on critical dimensions, materials properties, and so on. In each case a large number of specifications must be satisfied if the product is to meet performance requirements. Indeed the complexity of the system may be measured roughly by the number of such specifications.

To better understand the relationship between complexity and yield, consider a device with M specifications, and let X_i signify the event that the ith specification is met. If all of the specifications must be met for the device to be satisfactory, then the yield will be

$$Y = P\{X_1 \cap X_2 \cap X_3 \cdots \cap X_M\}. \tag{4.36}$$

If we consider the specifications to be independent, then

$$Y = P\{X_1\}P\{X_2\}P\{X_3\} \cdots P\{X_M\}. \tag{4.37}$$

For simplicity, assume that the probability of each specification *not* being met is p, or equivalently $P\{X_i\} = 1 - p$. Hence

$$Y = (1 - p)^M. \tag{4.38}$$

Since the natural logarithm and exponential are inverse operations we may rewrite this equation as

$$Y = \exp[\ln(1 - p)^M]. \tag{4.39}$$

However, $\ln(1 - p)^M = M\ln(1 - p)$. Furthermore, for any reasonable values of the capability indices we can assume that $p \ll 1$, and for small values of p the approximation $\ln(1 - p) \approx -p$ is adequate. Hence the yield equation reduces to

$$Y = e^{-pM}. \tag{4.40}$$

The importance of small rejection probabilities per specification is obvious. The yield decays exponentially as the number of specifications increases, unless the probability p of violating each specification is reduced. To maintain the same yield, the value of p must be halved for each doubling in the number of specifications.

EXAMPLE 4.6

A manufacturer of circuits knows that 5 percent of the circuit boards fail in proof testing due to independent diode failures. The failure of any diode causes board failure. (*a*) If there are 100 diodes on the board, what is the probability of any one diode's failing? (*b*) If the size of the boards is increased to contain 500 diodes, what percent of the new boards will fail the proof testing? (*c*) What must the failure probability per diode be if the 5% failure rate is to be maintained for the 500 diode boards?

Solution (*a*) $Y_{100} = 1 - 0.05 = e^{-100p}$, thus $p = -\frac{1}{100} \ln{(0.95)} = 0.5 \times 10^{-3}$

(*b*) $1 - Y_{500} = 1 - e^{-500p} = 1 - \exp[-500 \times 0.5 \times 10^{-3}] = 0.22 = 22\%$

(*c*) $Y_{500} = 0.95 = e^{-500p'}$, thus $p' = -\frac{1}{500} \ln{(0.95)} = 0.1 \times 10^{-3}$.

Six Sigma Criteria

The exponential decay of yield with the number of required components or specifications has given rise to the demand to decrease the variability in manufacturing processes relative to the specification width. As indicated by our example, although it only leads to a 0.27 percent rejection rate on a single specification, the traditional three sigma criteria will quickly tend to 100 percent rejection as the number of specifications is increased: in a 100 specification system, for example, 76 percent will be found acceptable. If the long-term variability is also taken into account, using the multiplier of $c = 1.5$, then only 1.1 percent are acceptable.

This dilemma has appeared in many industries. It is perhaps most pronounced in microelectronics where integrated circuits may require millions of individual diodes to function properly. In order to produce highly complex systems that are also reliable, the probability of any one specification not being met must be measured in parts per million or ppm (where 1 ppm = 0.0001 percent). As a result, the Motorola Corporation formulated a strict set of criteria, and a methodology for implementing them that has seen increasingly wide spread use in recent years. The methodology is referred to as six sigma since the basic requirement is that the tolerance half-width be at least six standard deviations of the process distribution for short-term variation. This implies that $C_p > 2.0$. The fraction rejected on a short-term basis is then reduced to

$$p < 2\Phi(-6) = 0.002 \text{ ppm} \qquad (4.41)$$

The improvement in yield when going from the traditional three sigma criteria to four, five, and finally six sigma is illustrated in Fig. 4.16*a* as a function of the number of specifications that must be met.

The six sigma methodology also places a tighter criterion on the long-term multiplier *c*. Under the six sigma methodology it is required that long-term variability be reduced to $c < 1.333$. Thus from Eq. 4.28 we have $C_{pk} > 1.5$, and from Eq. 4.32 we see that the rejection rate will be less than 6.8 ppm. The relationship between C_{pk}, yield and complexity is shown in Fig. 4.16*b*.

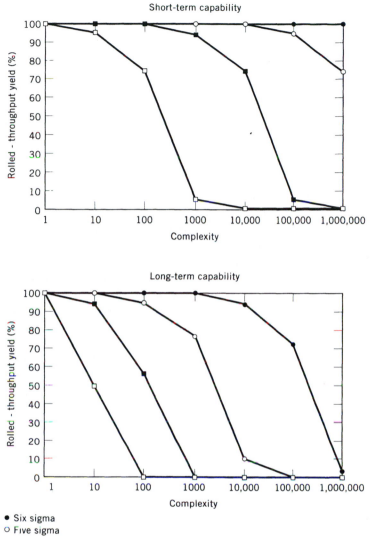

- ● Six sigma
- ○ Five sigma
- ■ Four sigma
- □ Three sigma

Note. Long-term capability based on an equivalent, one-sided 1.5σ mean shift

FIGURE 4.16 Yield vs. system complexity. (From Harry M. L. and Lawson, J. R., *Six Sigma Producility Analysis and Process Characterization*, pgs. 3–5 and 6–9, Addison-Wesley Publishing Co. Inc. and Motorola, Inc. 1992. Reprinted by permission.)

Implementation

The implementation of the six sigma criteria requires close interaction between the design and manufacturing processes. Assume a manufacturing process is to be implemented with the requirement that specified values of C_p and C_{pk} must be obtained. Since the specification limits have been set by the designer, these requirements can be met only by achieving sufficiently small σ and σ_k in the manufacturing process. Success requires first bringing the process into control. This entails making the process stable so that over the short term there is a well-defined σ. Then the systematic causes of long-term variation must be eliminated to reduce the value of c, and therefore of σ_k, to specified levels.

The techniques for bringing a process into control and then reducing and maintaining the smallest possible levels of short- and long-term variability require two engineering talents. An intimate knowledge of the manufacturing process and its physical basis is needed to identify and eliminate the causes of variability. The tools of statistical process control(SPC) must be mastered in order to identify the sources of long-term variation in the presence of background noise, to measure the reductions in variability, and to gain early warning of disturbing influences. The methods of SPC are discussed briefly in the concluding section of Chapter 5.

Reducing the causes of long-term variation may require a number of systematic changes to the manufacturing process. These may include better operator training, improved control over batch to batch variability of stock materials, more frequent tool changes, and better control over ambient temperature, dust or other environmental conditions, to name a few. Once the process has been brought into control, and the identifiable causes of long-term variation are reduced to a minimum, process capability, and therefore yield, cannot be further improved without decreasing σ^2, the short term process variance, or increasing Δ, the specification half interval.

To decrease σ^2, one must return to the process design and make it more robust. That is, one must perform designed experiments to find combinations of process parameters, which will yield a smaller part-to-part variance in the production output. Similar experiments may by performed to optimize the compositions of the feed stock materials. If the process parameter improvements achieved by robust design efforts are inadequate, then either of two alternatives may be considered, each of which is likely to add substantially to the production costs. Higher purity materials or better quality machinery of the same type may be specified to reduce the short-term variability. Alternately, a totally different process that is inherently more expensive may be required.

Alternately, Δ may be increased. To permit such an increase, however, one must retreat to earlier in the product development cycle in order to make the product performance characteristics less sensitive to the particular component or part parameter. Only then can an increase in the specification interval be justified. If this is inadequate, then features of the conceptual design or of the performance requirements may require reexamination. This iterative procedure for improving process and product design makes clear the necessity for concurrent engineering—the simultaneous design of the product and manufacturing processes. Costly delays or diminished quality and reliability are avoided only if the proposed manufacturing processes and their inherent limitations are considered concurrently while design concepts are worked out and product parameters and tolerances set.

Bibliography

Feigenbaum, A. V., *Total Quality Control*, 3rd ed., McGraw-Hill, NY, 1983.

Harry, M. J., and J. R. Lawson, *Six Sigma Producibility Analysis and Process Characterization*, Addison-Wesley, Reading, MA, 1992.

Mitra, A., *Fundamentals of Quality Control and Improvement*, Macmillan, NY, 1993.

Peace, G. S., *Taguchi Methods*, Addison-Wesley, Reading, MA, 1993.

Phadke, M. S., *Quality Engineering Using Robust Design*, Prentice-Hall, Englewood Cliffs, NJ, 1989.

Ross, P. J., *Taguchi Techniques for Quality Engineering*, McGraw-Hill, NY, 1988.

Taguchi, G., and Y. Wu, *Introduction to Off-Line Quality Control*, Central Japan Quality Control Association, Nagaya, 1979.

Taguchi, G., *Introduction to Quality Engineering*, Asian Productivity Organization, 1986. (Distributed by American Supplier Institute, Inc., Dearborn, MI.)

Taguchi, G., *Taguchi on Robust Technology Development*, ASME Press, NY, 1993.

Exercises

4.1 The allowable drift on a voltage regulator has a specification of 0.0 ± 0.8 volts. Each time a regulator does not satisfy this specification, there is an $80.00 cost for rework.

   a. Write the expression for the Taguchi loss function and evaluate the coefficient.

   b. If the PDF for the drift in volts is

$$f(x) = (3/4)(1 - x^2) \qquad |x| < 1$$

$$f(x) = 0 \qquad |x| > 1$$

   what is the expected value of the Taguchi loss?

   c. With the PDF given in b, what fraction of the regulators do not meet the specifications?

4.2 Widgets are manufactured with an impurity probability density function of

$$f(x) = \begin{cases} x, & 0 \le x \le 1 \\ 2 - x, & 1 \le x \le 2 \\ 0, & otherwise \end{cases}$$

  (a) Sketch the PDF.
  (b) Determine the mean.
  (c) Determine the variance.
  (d) The Taguchi smaller-is-better loss function for the widgets is given by $L(x) = 10x^2$.

Determine the expected value $\overline{L}$ of the loss function.

4.3 The probability density function for impurities is given by

$$f(x) = \begin{cases} 0, & x < 0 \\ 1/USL, & 0 < x < USL \\ 0, & x > USL \end{cases}$$

where *USL* is the upper specification limit. Evaluate the expected smaller-is-better quality loss, assuming that L_o is the penalty for exceeding the *USL*.

4.4 The target value for release pressure on a safety valve is p_o with a tolerance of $\pm \Delta p$. The manufacturer barely manages to meet this criterion with a PDF of

$$f(p) = \begin{cases} \dfrac{1}{2\Delta p}, & |p - p_o| < \Delta p \\ 0, & |p - p_o| \geq \Delta p \end{cases}$$

If L_o is the valve replacement cost, what is the average Taguchi loss $\overline{L}$ for the valves?

4.5 The luminescence of a surface is described by a PDF of

$$f(x) = 4xe^{-2x}, \qquad 0 \leq x \leq \infty.$$

The specifications are 1.0 ± 0.5.

  (a) What is the probability that the specification will not be met?
  (b) What is the expected value of the Taguchi loss function if the cost of being out of specification is $5.00?
  (c) Calculate the signal-to-noise ratio.
  {Note: see useful integrals in Appendix A.}

4.6 Suppose four parameters are to be chosen to maximize a toughness parameter. Nine experiments are to be analyzed using the orthogonal array shown in Fig. 4.11a. The results of the experiments are (in ascending order) 76, 79, 92, 84, 65, 68, 73, 86 and 74.

  (a) Draw the linear graphs.
  (b) Which factor or factors do you think are most important?
  (c) What settings (1, 2 or 3) for each factor will maximize the parameter?

4.7 A component's time-to-failure PDF is given by

$$f(t) = \frac{1}{2}t^2 e^{-t}, \qquad 0 \le t \le \infty.$$

The lower specification limit is $LSL = 0.25$ and the cost of not meeting the specification is \$100.

(a) Evaluate the expected Taguchi larger-is-better loss function.

(b) What is the probability that the specification will not be met?

4.8 The following L_4 orthogonal array can be used to treat three factors:

Trial	A	B	C
1	1	1	1
2	1	2	2
3	2	1	2
4	2	2	1

Suppose four tests are run to maximize the strength of an adhesive. They are run for two different application pressures (Factor A), two temperatures (Factor B), and two surface roughnesses (Factor C). The results for trials 1 through 4 are 24, 19, 28, and 21 kg/mm².

(a) Draw the linear graphs.

(b) Which is the most important factor?

(c) What are the optimal levels for the three factors?

4.9 A widget manufacturer is trying to improve the process for producing a critical dimension of 10.0 ± 0.0005 cm.

(a) If there is a short-term capability index of $C_p = 1.4$, what fraction of the widgets will fail to meet specifications, assuming the mean is on-target?

(b) If the mean moves off-target by 0.0001 cm, calculate C_{pk} and determine what fraction of the widgets will fail to meet specifications.

4.10 Suppose the specifications on a part dimension are 40 ± 0.01 cm.

(a) If the mean is on target, what must the standard deviation of a normal distribution be if no more than 0.1% of the parts are to be rejected?

(b) What value of C_p is required to meet the criteria of part a?

(c) If the mean moves off target by 0.003 cm, what is the value of C_{pk}?

(d) With the mean off target by 0.003 cm, to what must the value of C_p be increased to in order to produce no more than 0.1% of the parts out-of-specification?

(e) What will be the value of C_{pm} after C_p is increased?

4.11 Suppose that a batch of ball bearings is produced for which the diameters are distributed normally. The acceptance testing procedures remove all those for which the diameter is more than 1.5 standard deviations from the mean value. Therefore, the truncated distribution of the diameters of the delivered ball bearings is

$$f(x) = \begin{cases} \dfrac{A}{\sqrt{2\pi}\,c} \exp\left[-\dfrac{1}{2c^2} (x - \mu)^2 \right], & |x - \mu| < 1.5c, \\ 0, & |x - \mu| > 1.5c. \end{cases}$$

(a) What fraction of the ball bearings is accepted?

(b) What is the value of A?

(c) What fraction of the accepted ball bearings will have diameters between $\mu - c$ and $\mu + c$?

(d) What is the variance of $f(x)$, the PDF of delivered ball bearings?

{Note: numerical integration is required.}

4.12 A large batch of 50 Ohm resistors has a mean resistance of 49.96 Ohms and a standard deviation of 0.70 Ohms. The resistances are normally distributed. The lower and upper specification limits are 48 and 52 Ohms.

(a) Evaluate C_p.

(b) Evaluate C_{pk}.

(c) Evaluate C_{pm}.

(d) What is the expected Taguchi quality loss if the cost of an out-of-specification resistor is $0.80?

(e) What is the signal-to-noise ratio calculated from Eq. 4.21?

4.13 A process is found to have $C_p = 1.5$ and $C_{pk} = 1.0$. What fraction of the parts will not meet the specifications?

4.14 Repeat exercise 4.12 for a batch of 1.0 cm diameter ball bearings with a mean diameter of 0.9996 cm and a standard deviation of 0.0012 cm. The specification limits are 0.9950 and 1.0050 cm and the cost of an out-of-specification bearing is $0.35.

4.15 If a part must meet six independent specifications, estimate the largest failure probability per specification that can be tolerated if the part yield must be at least 90%.

4.16 Suppose the specification on battery output voltage is given by 10.00 ± 0.50 volts. After measuring the voltage of many batteries the distribution is found to be normal, with $\mu = 10.10$ volts and $\sigma = 0.16$ volts.

(a) What is the value of C_p?

(b) What is the value of C_{pk}?

(c) What fraction of the output will have a value greater than the upper tolerance limit?

4.17 Over a short period of time a roller bearing manufacturer finds that 2% of the bearings exceed the USL diameter of 2.01 cm and 2% are less than the LSL of 1.99 cm. If the distribution of diameters is normal:

(a) What is the mean diameter?

(b) What is the standard deviation?

(c) What is C_p for the process?

CHAPTER 5

Data and Distributions

"And the more observations or experiments there are made, the less will the conclusions be liable to error, provided they admit of being repeated under the same circumstances."

Thomas Simpson 1710—1761

5.1 INTRODUCTION

In the preceding chapters some elementary concepts concerning probability and random variables are introduced and utilized in the discussions of a number of issues relating to quality and reliability. Thus far statistics have been discussed only in the context of the simple binomial trials for estimating a failure probability. But statistical analysis of laboratory experiments, prototype tests, and field data is pervasive in reliability engineering. Only through the statistical analysis of such data can reliability models be applied and their validity tested. We now take up the questions of statistics: Given a set of data, how do we infer the properties of the underlying distribution from which the data have been drawn? If, for example, we have recorded the times to failure of a number of devices of the same design and manufacture, what can we surmise about the probability distribution of times-to-failure that would emerge if a very large population of all such devices was to be tested to failure?

Two approaches may be taken to data analysis; nonparametric and parametric. In nonparametric analysis no assumption is made regarding the distribution from which the sample data has been drawn. Rather, distribution-free properties of the data are examined. The construction of histograms from the sample data is probably the most common form of nonparametric analysis. The sample mean, variance, and other sample statistics can also be obtained from the data without reference to a specific distribution. In addition to histograms and sample statistics, we introduce elementary rank statistics in Section 5.2. They provide an approximate graph of the CDF of the random

102

variable even though there is insufficient data to construct a reasonable histogram. Rank statistics also serve as a basis for the probability plotting methods covered in Section 5.3.

Parametric analysis encompasses both the choice of the probability distribution and the evaluation of the distribution parameters. A number of factors guide distribution choice. Frequently, previous experience in fitting distributions to data from very similar tests may strongly favor the choice of a particular distribution. Alternatively, the choice between distributions may be made on the basis of the phenomena. If the sum of many small effects is involved, for instance, the normal distribution may be suitable; if it is a weakest link effect the Weibull distribution may be more appropriate. Corresponding arguments can be made for the exponential, lognormal, extreme-value, and other distribution functions. Finally, the nonparametric analysis tools discussed in Section 5.2 may often provide insight toward the selection of a distribution.

Once a distribution has been selected, the next step is the estimation of the parameters. Probability plotting, described in Section 5.3, has the advantage of providing both parameter estimates and a visual representation of how well the distribution describes the data. Such plotting is particularly valuable when the paucity of data makes more classical methods for parameter estimation problematical. In Section 5.4 we return to the notion of the confidence interval in order to determine the precision with which we can estimate the distribution parameters. Only the most elementary results—those applicable to large sample sizes—are presented, however, for the determination of confidence limits for smaller sample sizes requires statistical techniques that are beyond the scope of an introductory text.

The methods described in Sections 5.2 through 5.4 deal with complete sets of data; that is, data that come from tests that have been run to completion. Important situations exist, however, where results are needed at the earliest possible time. In testing products to failure, for example, decisions must often be reached before the last test specimen has failed. The data is then said to be censored. The methods for handling such data are examined in Chapter 8. A second situation where timely decisions must be made is in statistical process control, where inadvertent changes in manufacturing processes must be detected rapidly to prevent the production of defective items. Section 5.5 contains a brief introduction to the statistical process control techniques by which this is accomplished.

5.2   NONPARAMETRIC METHODS

Nonparametric methods allow us to gain perspective as to the nature of the distribution from which data has been drawn without selecting one particular distribution. When there is a sufficient number of data points, the representation of the distribution by a histogram or with sample statistics can be quite helpful. In many situations, however, the amount of data is insufficient to construct a realistic histogram. It is then useful to approximate the CDF by the technique plotting the median rank—a term that is defined below.

TABLE 5.1 Raw data: 70 Stopping Distance Measurements
in Feet

	A	B	C	D	E	F	G
1	39	54	21	42	66	50	56
2	62	59	40	41	75	63	58
3	32	43	51	60	65	48	61
4	27	46	60	73	36	38	54
5	60	36	35	76	54	55	45
6	71	54	46	47	42	52	47
7	62	55	49	39	40	69	58
8	52	78	56	55	62	32	57
9	45	84	36	58	64	67	62
10	51	36	73	37	42	53	49

Data from E. Pieruschka, *Principles of Reliability*, Prentice-Hall, Englewood Cliffs,
NJ, 1963, p. 5.

Histograms

The histogram may be constructed as follows. We first find the range of the
data (i.e., the maximum minus the minimum value). Knowing the range, we
choose an interval width such that data can be divided into some number N
of groups. Consider, for example, the stopping distance data displayed as
Table 5.1. If the interval for this data is chosen to be 10 ft, a table can be
made up according to how many data points fall in each interval. This is
carried out in Table 5.2, with the data falling into seven intervals. A histogram,
referred to as a frequency diagram, may then be drawn as indicated in Fig. 5.1a.

   In order to glean as much information from the data as possible, the
number of intervals into which the data are divided must be reasonable. If too
few intervals are used, as indicated in Fig. 5.1b, the nature of the distribution is
obscured by the lack of resolution. If the number is too large, as in Fig. 5.1c,
the large fluctuations in frequency hide the nature of the distribution. More
data points allow larger numbers of intervals to be used effectively, and result
in better representation of the distribution. Although there is no precise rule
for determining the optimum number of the intervals, the following rule of

TABLE 5.2 Frequency Table

Class interval, ft	Tally	Frequency
20–29	//	2
30–39	///// ///// /	11
40–49	///// ///// ///// /	16
50–59	///// ///// ///// /////	20
60–69	///// ///// ////	14
70–79	///// /	6
80–89	/	1

Source: Erich Pieruschka, *Principles of Reliability*, © 1963, p. 5, with permission from
Prentice-Hall, Englewood Cliffs, NJ.

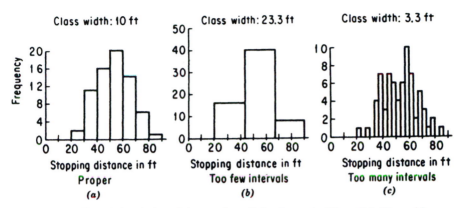

FIGURE 5.1 Effect of the choice of the number of class intervals. (From Eric Pieruschka, *Principles of Reliability*, © 1963, p. 6, with permission from Prentice-Hall, Englewood Cliffs, NJ.)

thumb may be used.* If N is the number of data points and r is the range of the data, a reasonable interval width Δ is

$$\Delta \approx r[1 + 3.3 \log_{10}(N)]^{-1}. \tag{5.1}$$

A crude method for observing how well a known distribution describes a data set consists of plotting the analytical form of the distribution over the histogram. But first, the frequency diagram must be normalized to approximate $f(x)$, the PDF. This is accomplished by requiring that the histogram satisfy the normalization condition Eq. 3.7.

Suppose that $n_1, n_2, \ldots$ are the frequencies with which the data appear in the various intervals, and $n_1 + n_2 + n_3 \ldots = N$. If we want to approximate $f(x)$ by f_i in the ith interval, f_i must be proportional to n_i:

$$f_i = an_i, \tag{5.2}$$

where a is the necessary proportionality constant. For the histogram to satisfy Eq. 3.7, the normalization condition on the PDF, we must have

$$\sum_i f_i \Delta = 1. \tag{5.3}$$

Combining the two equations yields

$$1 = \sum_i an_i \Delta = a\Delta \sum_i n_i = a\Delta N. \tag{5.4}$$

Hence $a = 1/(N\Delta)$, and

$$f_i = \frac{1}{\Delta}\frac{n_i}{N}. \tag{5.5}$$

The histogram that approximates $f(x)$ for the stopping distance data is plotted in Fig. 5.2. For comparison, we have plotted the PDF for a normal distribution;

* H. A. Sturges, "The Choice of a Class Interval," *J. Am. Stat. Assoc.*, **21**, 65–66 (1926); see also E. Pieruschka, *Principles of Reliability*, Prentice-Hall, Englewood Cliffs, NJ, 1963.

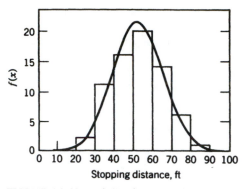

FIGURE 5.2 Normal distribution and histogram for the data in Table 5.1.

the values of μ and σ used in the distribution are estimated from nonparametric sample statistics, which we treat next.

Sample Statistics

The sample statistics treated here are estimates of random variable properties that do not require the form of the underlying probability distribution to be known. We consider estimates for the mean, variance, skewness, and kurtosis defined in Chapter 3. Suppose we have a sample of size N of a random variable x. Then the mean can be estimated with

$$\hat{\mu} = \frac{1}{N}\sum_{i=1}^{N} x_i \tag{5.6}$$

and the variance with

$$\hat{\sigma}^2 = \frac{1}{N}\sum_{i=1}^{N} (x_i - \mu)^2 \tag{5.7}$$

if the mean is known. If the mean is not known, but must be estimated from Eq. 5.6, then the variance is increased to

$$\hat{\sigma}^2 = \frac{1}{N-1}\sum_{i=1}^{N} (x_i - \hat{\mu})^2. \tag{5.8}$$

The same technique which is applied to Eq. 3.20 may be employed to rewrite the variance as

$$\hat{\sigma}^2 = \frac{N}{N-1}\left[\frac{1}{N}\sum_{i=1}^{N} x_i^2 - \left(\frac{1}{N}\sum_{i} x_i\right)^2\right]. \tag{5.9}$$

The estimators for the skewness and kurtosis are, respectively:

$$\widehat{sk} = \frac{\frac{1}{N}\sum_{i=1}^{N} (x_i - \hat{\mu})^3}{\left[\frac{1}{N}\sum_{i=1}^{N} (x_i - \hat{\mu})^2\right]^{3/2}}, \qquad \widehat{ku} = \frac{\frac{1}{N}\sum_{i=1}^{N} (x_i - \hat{\mu})^4}{\left[\frac{1}{N}\sum_{i=1}^{N} (x_i - \hat{\mu})^2\right]^2}. \tag{5.10}$$

These sample statistics are said to be point estimators because they yield a single number, with no specification as to how much in error that number is likely to be. They are unbiased in the following sense. If the same statistic is applied over and over to successive sets of N data points drawn from the same population, the grand average of the resulting values will converge to the true value as the number of data sets goes to infinity. In Section 5.4 the precision of point estimators is characterized by confidence intervals. Unfortunately, with the exception of the mean, given by Eq. 5.6, confidence intervals can only be obtained after the form of the distribution has been specified.

EXAMPLE 5.1

Calculated the mean, variance, skewness, and kurtosis of the stopping power data given in Table 5.1

Solution These four quantities are commonly included as spread-sheet formulae. The data in Table 5.1 is already in spread sheet format. Using Excel-4,* we simply calculate the four sample quantities with the standard formulae as follows:

$$
\begin{array}{llll}
\text{Mean:} & \hat{\mu} = \text{AVERAGE (A1:G10)} & = 52.3 \\
\text{Variance:} & \hat{\sigma}^2 = \text{VAR (A1:G10)} & = 168.47 \\
\text{Skewness:} & \widehat{sk} = \text{SKEW (A1:G10)} & = 0.0814 \\
\text{Kurtosis:} & \widehat{ku} = \text{KURT(A1:G10)} & = -0.268
\end{array}
$$

Note that in applying the formulae to data in Table 5.1, all the data in the rectangle with Column A row 1 on the upper left and Column G row 10 on the lower right is included.

___

Rank Statistics

Often, the number of data points is too small to construct a histogram with enough resolution to be helpful. Such situations occur frequently in reliability engineering, particularly when an expensive piece of equipment must be tested to failure for each data point. Under such circumstances rank statistics provide a powerful graphical technique for viewing the cumulative distribution function (i.e., the CDF). They also serve as a basis for the probability plotting taken up in the following section.

To employ this technique, we first take the samplings of the random variable and rank them; that is, list them in ascending order. We then approximate the CDF at each value of x_i. With a large number N of data points the CDF could reasonably be approximated by

$$\hat{F}(x_i) = \frac{i}{N}, \qquad i = 1, 2, 3, \ldots N, \tag{5.11}$$

where $F(0) = 0$ if the variable is defined only for $x > 0$.

___

* Excel is a registered trademark of the Microsoft Corporation.

If N is not a large number, say less than 15 or 20, there are some shortcomings in using Eq. 5.11. In particular, we find that $F(x) = 1$ for values of x greater than x_N. If a much larger set of data were obtained, say $10N$ values, it is highly likely that several of the samples would have larger values than x_N. Therefore Eq. 5.11 may seriously overestimate $F(x)$. The estimate is improved by arguing that if a very large sample were to be obtained, roughly equal numbers of events would occur in each of the intervals between the x_i, and the number of samples larger than x_N would probably be about equal to the number within one interval. From this argument we may estimate the CDF as

$$\hat{F}(x_i) = \frac{i}{N+1}, \qquad i = 1, 2, 3, \ldots N. \tag{5.12}$$

This quantity can by derived from more rigorously statistical arguments; it is known in the statistical literature as the mean rank. Other statistical arguments may be used to obtain slightly different approximations for $F(x)$. One of the more widely used is the median rank, or

$$\hat{F}(x_i) = \frac{i - 0.3}{N + 0.4}, \qquad i = 1, 2, 3, \ldots N. \tag{5.13}$$

In practice, the randomness and limited amounts of data introduce more uncertainty than the particular form that is used to estimate F. For large values of N, they yield nearly identical results for $F(x)$ after the first few samples. For the most part we shall use Eq. 5.12 as a reasonable compromise between computational ease and accuracy.

EXAMPLE 5.2

The following are the times to failure for 14, six volt flashlight bulbs operated at 12.6 volts to accelerate rate the failure: 72, 82, 97, 103, 113, 117, 126, 127, 127, 139, 154, 159, 199, and 207 minutes. Make a plot of $F(t)$, where t is the time to failure.

Solution　Table 5.3 contains the necessary calculations. The data rank i is in column A, and the failure times in column B. Column C contains $i/(14 + 1)$ (Columns D and E are used for Example 5.5) for each failure time. $F(t_i)$ vs. t_i (i.e., column C vs. column B) is plotted in Fig. 5.3.

5.3　PROBABILITY PLOTTING

Probability plotting is an extremely useful technique. With relatively small sample sizes it yields estimates of the distribution parameters and provides both a graphical picture and a quantitative estimate of how well the distribution fits the data. It often can be used with success in situations where too few data points are available for the parameter estimation techniques discussed in Section 5.4 to yield acceptably narrow confidence intervals. With larger sample sizes probability plotting becomes increasingly accurate for the estimate of parameters.

TABLE 5.3 Spreadsheet for Weibull Probability Plot of Flashlight Bulb Data in Example 5.4

	A	B	C	D	E
	i	t	F(t) = i/(N + 1)	x = LN(t)	y = LN(LN(1/(1 − F)))
1					
2	1	72	0.0667	4.2767	−2.6738
3	2	82	0.1333	4.4067	−1.9442
4	3	97	0.2000	4.5747	−1.4999
5	4	103	0.2667	4.6347	−1.1707
6	5	113	0.3333	4.7274	−0.9027
7	6	117	0.4000	4.7622	−0.6717
8	7	126	0.4667	4.8363	−0.4642
9	8	127	0.5333	4.8442	−0.2716
10	9	127	0.6000	4.8442	−0.0874
11	10	139	0.6667	4.9345	0.0940
12	11	154	0.7333	5.0370	0.2790
13	12	159	0.8000	5.0689	0.4759
14	13	199	0.8667	5.2933	0.7006
15	14	207	0.9333	5.3327	0.9962

Basically, the method consists of transforming the equation for the CDF to a form that can be plotted as

$$y = ax + b. \tag{5.14}$$

Equation 5.12 is used to estimate the CDF at each data point in the resulting nonlinear plot. A straight line is then constructed through the data and the distribution parameters are determined in terms of the slope and intercept.

The procedure is best illustrated with a simple example. Suppose we want to fit the exponential distribution

$$F(x) = 1 - e^{-x/\theta}, \qquad 0 \le x \le \infty \tag{5.15}$$

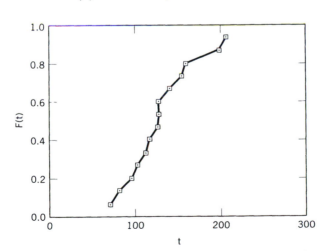

FIGURE 5.3 Graphical estimate of failure time cumulative distribution.

to a series of failure times x_i. We can rearrange this equation by first solving for $1/(1 - F)$ and then taking the natural logarithm to obtain

$$\ln\left[\frac{1}{1 - F(x)}\right] = \frac{1}{\theta}x. \tag{5.16}$$

We next approximate $F(x_i)$ by Eq. 5.12 and plot the resulting values of

$$\frac{1}{1 - F(x_i)} = \frac{1}{1 - \dfrac{i}{N+1}} = \frac{N+1}{N+1-i} \tag{5.17}$$

on semilog paper versus the corresponding x_i. The data should fall roughly along a straight line if they were obtained by sampling an exponential distributions. Comparing Eqs. 5.14 and 5.16, we see that $\theta = 1/a$ can be estimated from the slope of the line. More simply, we note that the left side of Eq. 5.16 is equal to one when $1/(1 - F) = e = 2.72$, and thus at that point $\theta = x$. Since the exponential is a one-parameter distribution, b, the y intercept is not utilized.

EXAMPLE 5.3

The following failure time data is exponentially distributed: 5.2, 6.8, 11.2, 16.8, 17.8, 19.6, 23.4, 25.4, 32.0, and 44.8 minutes. Make a probability plot and estimate θ.

Solution   Since $N = 10$, from Eq. 5.17 we have $1/[1 - F(t_i)] = 11/(11 - i)$ or 1.1, 1.222, 1.373, 1.571, 1.833, 2.2, 2.75, 3.666, 5.5 and 11. In Fig. 5.4 these numbers

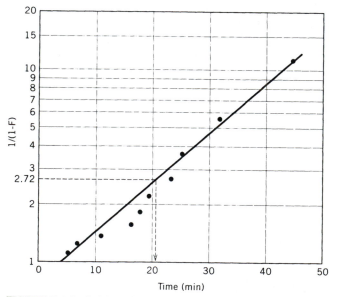

FIGURE 5.4 Probability plot of exponentially distributed data.

have been plotted on semilog paper versus the failure times. After drawing a straight line through the data we note that when $1/(1 - F) = 2.72$, $x = \theta = 21$ min.

Two-parameter distributions require more specialized graph paper if the plots are to be made by hand. The more common of such graph papers and an explanation of their use is included as Appendix D. Approximate curve fitting by eye that is required in the use of these graph papers, however, is becoming increasingly dated, and may soon go the way of the slide rule. With the power of readily available spread sheets, the straight line approximation to the data can be constructed quickly and more accurately, by using least-squares fitting techniques. These techniques, moreover, provide not only the line that "best" fits the data, but also a measure of the goodness of fit. Readily available graphics packages also display the line and data to provide visualization of the ability of the distribution to fit the data. The value of these techniques is illustrated for several distributions in examples that follow. First, however, we briefly explain the least-squares fitting techniques. Whereas the mathematical procedure is automated in spread sheet routines, and thus need not be performed by the user, an understanding of the methods is important for prudent interpretation of the results.

Least Squares Fit

Suppose we have N pairs of data points, (x_i, y_i) that we want to fit to a straight line:

$$y = ax + b, \tag{5.18}$$

where a is the slope and b the y axis intercept as illustrated in Fig. 5.5. In the least squares fitting procedure we minimize the mean value of the square deviation of the vertical distance between the points (x_i, y_i) and the corresponding point (x_i, y) on the straight line:

$$S = \frac{1}{N} \sum_{i=1}^{N} (y_i - y)^2, \tag{5.19}$$

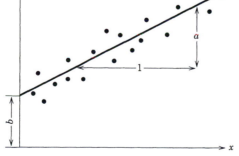

FIGURE 5.5 Least squares fit of data to the function $y = ax + b$.

or using Eq. 5.18 to evaluate y on the line at x_i, we have

$$S = \frac{1}{N}\sum_{i=1}^{N}(y_i - ax_i - b)^2. \tag{5.20}$$

To select the values of a and b that minimize S, we require that the partial derivatives of S with respect to the slope and intercept vanish: $\partial S/\partial a = 0$ and $\partial S/\partial b = 0$. We obtain, respectively

$$\overline{xy} - a\overline{x^2} - b\overline{x} = 0 \tag{5.21}$$

and

$$\overline{y} - a\overline{x} - b = 0, \tag{5.22}$$

where we have defined the following averages:

$$\overline{x} = \frac{1}{N}\sum_{i=1}^{N}x_i, \qquad \overline{y} = \frac{1}{N}\sum_{i=1}^{N}y_i,$$

$$\overline{xy} = \frac{1}{N}\sum_{i=1}^{N}x_iy_i, \qquad \overline{x^2} = \frac{1}{N}\sum_{i=1}^{N}x_i^2, \qquad \overline{y^2} = \frac{1}{N}\sum_{i=1}^{N}y_i^2. \tag{5.23}$$

Equations 5.21 and 5.22 may be solved to yield the unknowns a and b,

$$a = \frac{\overline{xy} - \overline{x}\,\overline{y}}{\overline{x^2} - \overline{x}^2} \tag{5.24}$$

and

$$b = \overline{y} - a\overline{x}. \tag{5.25}$$

If these values of a and b are inserted into Eq. 5.20 the minimum value of S is found to be

$$S = (1 - r^2)(\overline{y^2} - \overline{y}^2), \tag{5.26}$$

where r^2, referred to as the coefficient of determination, is given by

$$r^2 = \frac{(\overline{xy} - \overline{x}\,\overline{y})^2}{(\overline{x^2} - \overline{x}^2)(\overline{y^2} - \overline{y}^2)}. \tag{5.27}$$

The coefficient of determination is a good measure of how well the line is able to represent the data. It is equal to one, if the points all fall perfectly on the line, and zero, if there is no correlation between the data and a straight line. Thus as the representation of the data by a straight line is improved, the value of r^2 becomes closer to one.

The values of a, b, and r^2 may be obtained directly as formulae on spread sheets or other personal computer software. It is nevertheless instructive to use a graphics program to actually see the data. If there are outliers, either from faulty data tabulation or from unrecognized confounding of the experiment from which the data is obtained, they will only be reflected in the tabular results as decreased values of r^2. In contrast, offending points are highlighted on a graph. The value of visualization will become apparent with the examples which follow.

Weibull Distribution Plotting

We are now prepared to employ the least-squares method in probability plotting. We consider first the two-parameter Weibull distribution. The CDF with respect to time is given by

$$F(t) = 1 - \exp[-(t/\theta)^m], \qquad 0 \le t \le \infty. \tag{5.28}$$

The distribution is put in a form for probability plotting by first solving for $1/(1 - F)$,

$$\frac{1}{1 - F(t)} = \exp(t/\theta)^m \tag{5.29}$$

and then taking the logarithm twice to obtain

$$\ln \ln \left[\frac{1}{1 - F(t)} \right] = m \ln t - m \ln \theta. \tag{5.30}$$

This can be cast into the form of Eq. 5.18 if we define

$$y = \ln \ln \left[\frac{1}{1 - F(t)} \right] \tag{5.31}$$

and

$$x = \ln t. \tag{5.32}$$

We find that the shape parameter is just equal to the slope

$$\hat{m} = a, \tag{5.33}$$

whereas the scale parameter is estimated in terms of the slope and the intercept by

$$\hat{\theta} = \exp(-b/a). \tag{5.34}$$

The procedure is best illustrated by providing a detailed solution of an example problem.

EXAMPLE 5.4

Use probability plotting to fit the flashlight bulb failure times given in Example 5.2 to a two parameter Weibull distribution. What are the shape and scale parameters?

Solution   The ranks of the failures, the failure times, and the estimates of $F(t_i)$ are already given in columns A, B and C of Table 5.3. In column D we tabulate $\ln(t_i)$ and in column E, $\ln(\ln(1/(1 - F)))$. Then we plot column E versus column D and

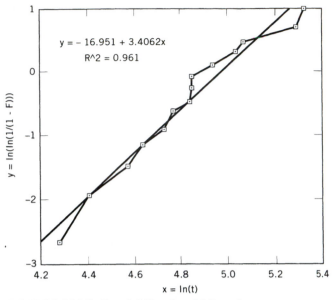

FIGURE 5.6 Weibull probability plot of failure times.

calculate a, b and r^2. The result are shown in Fig. 5.6. Since $a = 3.41$ and $b = -16.95$, we have from Eqs. 5.33 and 5.34: $\hat{m} = 3.41$ and $\hat{\theta} = \exp(+16.95/3.41) = 144$ min.

Extreme Value Distribution Plotting

The procedure for treating extreme-value distributions is quite similar to that employed for Weibull distributions. For example, with the minimum extreme-value distribution, the CDF is given by

$$F(x) = 1 - \exp[-e^{(x-u)/\Theta}], \qquad -\infty < x < \infty \qquad (5.35)$$

in Eq. 3.101. If we solve for $1/(1 - F)$, and take the natural logarithm twice, we obtain

$$\ln \ln \left[\frac{1}{1 - F(x)} \right] = \frac{1}{\Theta} x - \frac{u}{\Theta}. \qquad (5.36)$$

Thus we can make a linear plot with

$$y = \ln \ln \left[\frac{1}{1 - F(x)} \right]. \qquad (5.37)$$

The scale parameter is estimated in terms of the slope as

$$\hat{\Theta} = 1/a \qquad (5.38)$$

and the location parameter as

$$\hat{u} = -b/a, \qquad (5.39)$$

respectively. Likewise, for the maximum extreme value CDF, given by

$$F(x) = \exp[-e^{(x-u)/\Theta}], \qquad -\infty < x < \infty \qquad (5.40)$$

an analogous procedure can be used to determine the rectified equation

$$\ln \ln \left[\frac{1}{F(x)} \right] = -\frac{1}{\Theta} x + \frac{u}{\Theta}, \qquad (5.41)$$

where the distribution parameters may be estimated in terms of the slope and intercept to be

$$\hat{\Theta} = -1/a \qquad (5.42)$$

and

$$\hat{u} = -b/a. \qquad (5.43)$$

EXAMPLE 5.5

Determine whether the failure data in Example 5.2 can be fitted more accurately with a minimum extreme-value distribution than with a Weibull distribution. Estimate the parameters in each case. Employ spread sheet slope, intercept and coefficient formulae.

Solution The necessary values of y_i and x_i, respectively, are already tabulated in Table 5.3, columns E and B, for the minimum extreme value distribution and in columns E and D for the Weibull distribution. Thus for the extreme-value distribution, we obtain

$$r^2 = \text{RSQ}(E2:E15, B2:B15) \qquad = 0.88$$

$$a = \text{SLOPE}(E2:E15, B2:B15) \qquad = 0.025$$

$$b = \text{INTERCEPT}(E2:E15, B2:B15) = -3.76.$$

Thus, from Eqs. 5.38 and 5.39 the extreme value parameters are

$$\hat{\Theta} = 1/a = 1/0.025 = 40 \text{ min., and } \hat{u} = -b/a = 3.76/0.025 = 150.4 \text{ min.}$$

For the Weibull distribution

$$r^2 = \text{RSQ}(E2:E15, D2:BD5) \qquad = 0.96$$

$$a = \text{SLOPE}(E2:E15, D2:D15) \qquad = 3.41$$

$$b = \text{INTERCEPT}(E2:E15, D2:D15) = -16.95$$

Not surprisingly, these are the same values exhibited in Fig. 5.6. From Eq. 5.33 and 5.34, the Weibull parameters are $\hat{m} = a = 3.41$; $\hat{\theta} = exp(-b/a) = exp(16.9\,5/3.41) = 144$ min. The resulting value of $r^2 = 0.88$ for the extreme-value distribution is substantially smaller than that of 0.96 obtained with the Weibull distribution. Therefore the extreme value fit is poorer.

Normal Distribution Plotting

Normal and lognormal distributions find frequent application. However, unlike the Weibull and extreme value distributions they cannot be inverted to obtain y in analytical form. Rather we must rely on inverse operator notation. First consider the normal distribution with the CDF

$$F(x) = \Phi\left(\frac{x - \mu}{\sigma}\right). \tag{5.44}$$

We invert the standard normal distribution to obtain

$$\Phi^{-1}(F) = \frac{1}{\sigma} x - \frac{1}{\sigma} \mu. \tag{5.45}$$

Thus the linear equation $y = ax + b$ is obtained by taking

$$y = \Phi^{-1}(F). \tag{5.46}$$

The standard deviation estimate is then

$$\hat{\sigma} = 1/a \tag{5.47}$$

and the mean

$$\hat{\mu} = -b/a. \tag{5.48}$$

The availability of the standardized normal distribution and its inverse as spreadsheet formulae allows normal data to be analyzed with a minimum of effort. This is illustrated in the following example.

EXAMPLE 5.6

An electronics manufacturer receives 50 ± 2.5 ohm resistors from two suppliers. A sample of 30 resistors is taken from each supplier. The resistance values are measured and tabulated in rank order in columns B and C of Table 5.4. All of the resistance's are noted to fall within the specification limits of $LSL = 47.5$ ohm and $USL = 52.5$ ohms. Assume that the resistors are normally distributed and make probability plots of the two samples. Evaluate the Taguchi loss function, assuming a loss of $1.00 per out-of-specification resistor, and the process capability Cp for each supplier. Which supplier should you choose if there were no difference in price?

Solution   The estimates of $F(x_i) = i/(N + 1)$ are tabulated in columns D and I of Table 5.4. In columns E and J we use the Excel formula NORMSINV for the inverse of the standard normal distribution to tabulate

$$y_i = \Phi^{-1}(F_i) = \text{NORMSINV}(F_i)$$

from Eq. 5.46. The probability plots for suppliers #1 and #2 are shown in Fig. 5.7. The mean and standard deviation of each sample can be calculated from the Eqs. 5.47 and 5.48. They are

$$\hat{\mu} = 59.2/1.19 = 49.7 \text{ and } \hat{\sigma} = 1/1.19 = 0.84 \text{ for } \#1$$

$$\hat{\mu} = 31.4/0.627 = 50.1 \text{ and } \hat{\sigma} = 1/0.627 = 1.59 \text{ for } \#2$$

TABLE 5.4 Spreadsheet for Normal Probability Plot of Resistor Data in Example 5.6

	A	B	C	D	E	F	G	H	I	J
1	i	xi (#1)	xi (#2)	F(xi)	yi	i	xi (#1)	xi (#2)	F(xi)	yi
2	1	48.47	47.67	0.0323	−1.85	16	49.75	50.75	0.5161	0.04
3	2	48.49	47.70 .	0.0645	−1.52	17	49.78	50.60	0.5484	0.12
4	3	48.66	48.00	0.0968	−1.30	18	49.93	50.63	0.5806	0.20
5	4	48.84	48.41	0.1290	−1.13	19	49.96	50.90	0.6129	0.29
6	5	49.14	48.42	0.1613	−0.99	20	50.03	51.02	0.6452	0.37
7	6	49.27	48.44	0.1935	−0.86	21	50.06	51.05	0.6774	0.46
8	7	49.29	48.64	0.2258	−0.75	22	50.07	51.28	0.7097	0.55
9	8	49.30	48.65	0.2581	−0.65	23	50.09	51.33	0.7419	0.65
10	9	49.32	48.68	0.2903	−0.55	24	50.42	51.38	0.7742	0.75
11	10	49.39	48.85	0.3226	−0.46	25	50.44	51.43	0.8065	0.86
12	11	49.43	49.17	0.3548	−0.37	26	50.57	51.60	0.8387	0.99
13	12	49.49	49.72	0.3871	−0.29	27	50.70	51.70	0.8710	1.13
14	13	49.52	49.85	0.4194	−0.20	28	50.77	51.74	0.9032	1.30
15	14	49.54	49.87	0.4516	−0.12	29	50.87	52.06	0.9355	1.52
16	15	49.69	50.07	0.4839	−0.04	30	51.87	52.33	0.9677	1.85

For the Taguchi Loss function is $L_0 = \$1.00$ and $\Delta = (52.5 - 47.5)/2. = 2.5$. Therefore the coefficients given by Eq. 4.6 is $k = \$1.00/2.5^2 = \$0.16$. Hence, from Eq. 4.9, we estimate

$$\overline{L} = \$0.16[0.84^2 + (49.7 - 50)^2] = \$0.13 \text{ for \#1}$$

$$\overline{L} = \$0.16[1.59^2 + (50.1 - 50)^2] = \$0.41 \text{ for \#2.}$$

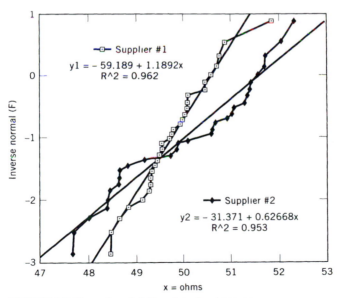

FIGURE 5.7 Normal probability plot of resistances.

From Eq. 4.24 we estimate

$$C_p = 2.5/(3 \times 0.84) = 0.99 \text{ for #1}$$

$$C_p = 2.5/(3 \times 1.59) = 0.52 \text{ for #2.}$$

Since the loss factor is smaller and the process capability higher, #1 is the preferable supplier.

Lognormal Distribution Plotting

Probability plotting with the normal and lognormal distributions is very similar. From Eq. 3.65 we may write the CDF for the lognormal distribution as

$$F(t) = \Phi\left[\frac{1}{\omega}\ln(t/t_o)\right]. \tag{5.49}$$

We invert the standard normal distribution to obtain

$$\Phi^{-1}(F) = \frac{1}{\omega}\ln t - \frac{1}{\omega}\ln t_o. \tag{5.50}$$

The required linear equation is obtained by once again taking

$$y = \Phi^{-1}(F), \tag{5.51}$$

but with $x = \ln t$. The estimates for the lognormal parameters are

$$\hat{\omega} = 1/a \tag{5.52}$$

and

$$\hat{t}_o = \exp(-b/a). \tag{5.53}$$

EXAMPLE 5.7

The fatigue lives of 20 specimens, measured in thousands of stress cycles are found to be 3.1, 6.1, 7.3, 10.4, 15.5, 20.9, 21.7, 21.89, 25.3, 30.5, 31.4, 32.7, 35.4, 35.9, 38.9, 39.6, 40.1, 65.5, 70.9, and 98.7. Use probability plotting to fit a lognormal distribution to the data, and estimate the parameters and the goodness-of-fit.

Solution   The calculations are made in Table 5.5.
The data rank and the failure times are tabulated in columns A and B, the natural logarithms of the failure times are tabulated in column C. In column D the estimates of $F(x_i) = i/(N+1)$ are tabulated. In column E we tabulate $y_i = \Phi^{-1}(F_i)$ from Eq. 5.51. In Fig. 5.8 we have plotted column E versus column C and used least-squares fit to obtain the best straight line through the data. From Eqs. 5.52 and 5.53 we find the parameters to be $\hat{\omega} = 1/a = 1/1.01 = 0.99$ and $\hat{t}_o = \exp(-b/a) = \exp(3.22/1.01) = 24.2$ thousand cycles. The fit is quite good with $r^2 = 0.929$.

TABLE 5.5 Spreadsheet for Lognormal Probability Plot of
Data in Example 5.7

	A	B	C	D	E
1	i	ti	ln(ti)	F(ti)	yi
2	1	3.1	1.1314	0.0476	−1.6684
3	2	6.1	1.8083	0.0952	−1.3092
4	3	7.3	1.9879	0.1429	−1.0676
5	4	10.4	2.3418	0.1905	−0.8761
6	5	15.5	2.7408	0.2381	−0.7124
7	6	20.9	3.0397	0.2857	−0.5659
8	7	21.7	3.0773	0.3333	−0.4307
9	8	21.8	3.0819	0.3810	−0.3030
10	9	25.3	3.2308	0.4286	−0.1800
11	10	30.5	3.4177	0.4762	−0.0597
12	11	31.4	3.4468	0.5238	0.0597
13	12	32.7	3.4874	0.5714	0.1800
14	13	35.4	3.5667	0.6190	0.3030
15	14	35.9	3.5807	0.6667	0.4307
16	15	38.9	3.6610	0.7143	0.5659
17	16	39.6	3.6788	0.7619	0.7124
18	17	40.1	3.6914	0.8095	0.8761
19	18	65.5	4.1821	0.8571	1.0676
20	19	70.9	4.2613	0.9048	1.3092
21	20	98.7	4.5921	0.9524	1.6684

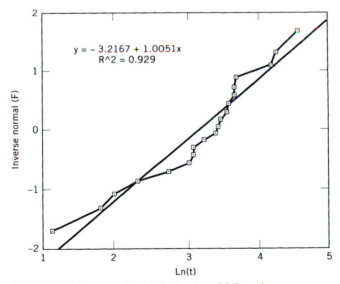

$$y = -3.2167 + 1.0051x$$
$$R^2 = 0.929$$

FIGURE 5.8 Lognormal probability plot of failure times.

Goodness-of-Fit

The forgoing examples illustrate some of the uses of probability plotting in the analysis of quality and reliability data. They also serve as a basis for the extensive use of these methods made in Chapter 8 for the analysis of failure data. With the computations carried out quite simply on a spread sheet or other software, one is not limited to a single analysis. Frequently, it may be advisable to try to fit more than one distribution to the data to determine the best fit. Comparison of the values of r^2 is the most objective criterion for this purpose. Other valuable information is obtained from visual inspection of the graph. Outliers may be eliminated, and if the data tends to fall along a curve instead of a straight line it may provide a clue as to what other distribution should be tried. For example, if normally distributed data is used to make an exponential probability plot, the data will fall along a curve that is concave upward. With some experience, such visual patterns become recognizable, allowing one to estimate which other distribution may be more appropriate.

More formal methods for assessing the goodness-of-fit exist. These establish a quantitative measure of confidence that the data may be fit to a particular distribution. The most accessible of these are the chi-squared test, which is applicable when enough data is available to construct a histogram, and the Kolmogorov–Smirnov (or K–S) test, which is applicable to ungrouped data. These tests are presented in elementary statistics texts but are not directly applicable to the analysis of much reliability data. In their standard form they assume not only that a distribution has been chosen but that the parameters are known; they establish only the level of confidence to which a specific distribution with known parameters fits a given set of data. In contrast, in probability plotting we are attempting both to estimate distribution parameters *and* establish how well the data fit the resulting distribution.

Aside from the simple comparison of r^2 values obtained from probability plotting, establishing goodness-of-fit from estimated parameters requires the use of more advanced maximum likelihood, moment, or other techniques and often involves a significant amount of computation. Such techniques are treated in advanced statistical texts and increasingly incorporated into statistical software packages. The use of these techniques is often justified to maximize the utility of reliability data. They are, however, beyond the scope of what can be included in an introductory reliability text of reasonable length. Instead, we focus next on an elementary treatment of confidence levels of estimated parameters.

5.4   POINT AND INTERVAL ESTIMATES

The mean, variance, and other sample statistics introduced in Section 5.2 are referred to as nonparametric point estimators. They are nonparametric because they may be evaluated without knowing the population distribution from which the sample was drawn, and they are point estimators because they yield a single number. Point estimates can also be made for the parameters of specific distributions, for example, the shape and scale parameters of a Weibull distribution. The corresponding interval estimates, which provide some level of confidence that a parameter's true value lies within a specified range of the point estimate, occupy a pivotal place in statistical analysis.

We begin our examination of interval estimates by expressing the sample static properties in terms of the probability concepts developed in Chapter 3. Suppose we want to estimate a property θ, where θ might be the mean, variance, or skewness, or a parameter associated with a specific distribution. The estimator $\hat{\theta}$ is itself a random variable with the sampling variability characterized by a PDF, referred to as a sampling distribution. Let the sampling distribution be denoted by $f_{\hat{\theta}}(\hat{\theta})$. If we repeatedly form $\hat{\theta}$ from samples of size N, and make a histogram of the values of $\hat{\theta}$, after many trials the sampling distribution $f_{\hat{\theta}}(\hat{\theta})$ will emerge. A sketch of a typical sampling distribution is provided in Fig. 5.9a. If the estimator is unbiased, then $E\{\hat{\theta}\} = \theta$, which is to say that the mean value of the sampling distribution is the true value of θ:

$$\int_{-\infty}^{\infty} \hat{\theta} f_{\hat{\theta}}(\hat{\theta}) \, d\hat{\theta} = \theta. \tag{5.54}$$

Along with the value of the point estimate $\hat{\theta}$, we would like to gain some idea of its precision. For this we calculate a confidence interval as follows. Suppose we pick a value $\theta + A$ on the $\hat{\theta}$ axis in Fig. 5.9b such that the probability that $\hat{\theta} \leq \theta + A$ is $1 - \alpha/2$, where α is typically a small number such as one or five percent. This condition may be written in terms of the sampling distribution as

$$P\{\hat{\theta} \leq \theta + A\} = \int_{-\infty}^{\theta+A} f_{\hat{\theta}}(\hat{\theta}) \, d\hat{\theta} = 1 - \alpha/2. \tag{5.55}$$

As shown in Fig. 5.9b the area under the sampling distribution to the right of $\theta + A$ is $\alpha/2$. Rearranging the inequality on the left, we have

$$P\{\hat{\theta} - A \leq \theta\} = \int_{-\infty}^{\theta+A} f_{\hat{\theta}}(\hat{\theta}) \, d\hat{\theta} = 1 - \alpha/2. \tag{5.56}$$

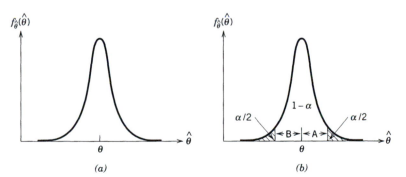

FIGURE 5.9 Sampling distribution.

Likewise, if we choose a value B such that the probability that $\hat{\theta} \geq \theta - B$ is $1 - \alpha/2$ we obtain

$$P\{\hat{\theta} \geq \theta - B\} = \int_{\theta - B}^{\infty} f_{\hat{\theta}}(\hat{\theta}) \, d\hat{\theta} = 1 - \alpha/2, \tag{5.57}$$

and as indicated in Fig. 5.9b, the area under the sampling distribution to the left $\theta - B$ is also $\alpha/2$. Rearranging the inequality on the left, we have

$$P\{\theta \leq \hat{\theta} + B\} = \int_{\theta - B}^{\infty} f_{\hat{\theta}}(\hat{\theta}) \, d\hat{\theta} = 1 - \alpha/2. \tag{5.58}$$

The probability that $\hat{\theta} - B \leq \theta$ *and* $\theta \leq \hat{\theta} + A$ is just the area $1 - \alpha$ under the central section of the sampling distribution, or

$$P\{\hat{\theta} - A < \theta \leq \hat{\theta} + B\} = \int_{\theta - B}^{\theta + A} f_{\hat{\theta}}(\hat{\theta}) \, d\hat{\theta} = 1 - \alpha. \tag{5.59}$$

The lower and upper confidence limits for estimates based on a sample size N are defined as

$$L_{\alpha/2,N} = \hat{\theta} - A \tag{5.60}$$

and

$$U_{\alpha/2,N} = \hat{\theta} + B, \tag{5.61}$$

respectively. Hence the $100(1 - \alpha)$ percent two-sided confidence interval is

$$P\{L_{\alpha/2,N} \leq \theta \leq U_{\alpha/2,N}\} = 1 - \alpha. \tag{5.62}$$

We must be specific about the preceding probability statements, for they define the meaning of confidence intervals. Equation 5.62 may be understood with the aid of Fig. 5.10 as follows. Suppose that a large number of samples each of size N are taken, and $\hat{\theta}$, $L_{\alpha/2,N}$, and $U_{\alpha/2,N}$ are calculated for each sample. These three quantities are random variables and in general will be different for each sample. In Fig. 5.10 we have plotted them for 10 such samples. If $L_{\alpha/2,N}$ and $U_{\alpha/2,N}$ define the 90% confidence interval, then for 90% of the samples of size N the true value of θ will lie within the intervals indicated by the solid vertical lines. Conversely, there is an $\alpha = 0.1$ risk that the true value will lie outside of the confidence interval. For brevity we frequently suppress the subscripts in Eq. 5.60 and 5.61 and denote the lower and upper confidence limits by $\theta^- \equiv L_{\alpha/2,N}$ and $\theta^+ \equiv U_{\alpha/2,N}$.

For the foregoing methodology to be applied to the computation of the confidence interval for a particular parameter, the properties of the corresponding sampling distribution, $f_{\hat{\theta}}(\hat{\theta})$, must be sufficiently well understood. In this respect the situation is quite different for the mean variance, skewness, and kurtosis, which may be defined for any distribution, and the specific

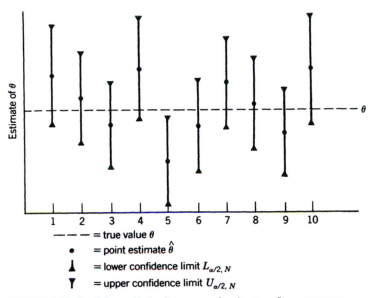

FIGURE 5.10 Confidence limits for repeated estimates of a parameter. See, for example K. C. Kapur and L. R. Lamberson, *Reliability in Engineering Design*, Wiley, NY, 1977.

parameters appearing in the normal, lognormal, Weibull, or other distribution. If the parent distribution is not designated, then a confidence interval can be determined only for the mean, μ, and then only if the sample size is sufficiently large, say $N > 30$. In this situation the sampling distribution becomes normal and, as shown in the following subsection, the confidence interval can be estimated.

If the parent distribution is known, then the point and interval estimates of the distribution parameters become the center of attention. Here, the situation differs markedly depending on whether N, the sample size, is large. For small or intermediate sample sizes taken from a normal distribution, the Student's-t and the Chi-squared sampling distributions can be used to estimate the confidence interval for the mean and variance respectively. The procedures are covered in elementary statistical texts. The more sophisticated procedures required for other parent distributions are found in the more advanced statistical literature, but are increasingly accessible though statistical software packages. Large sample sizes, point estimates, and confidence intervals for distribution parameters may be expressed in more elementary terms; then the sampling distributions approach the normal form, enabling the confidence intervals to be expressed in terms of the standard normal CDF. In subsequent subsections, the results compiled by Nelson* are presented for point estimates and confidence intervals of the normal, lognormal, Weibull, and extreme-value parameters.

* W. Nelson, *Applied Life Data Analysis*, John Wiley & Sons, New York, NY, 1982.

Estimate of the Mean

The sample mean given by Eq. 5.6, in addition to being the most ubiquitous statistic, has a unique property. An interval estimate is associated with the mean that is independent of the distribution from which the sample is drawn. Provided the sample size is sufficiently large, say $N > 30$, the central limit theorem provides a powerful result; the sampling distribution $f_{\hat{\mu}}(\hat{\mu})$ for $\hat{\mu}$ becomes normal with a mean of μ and variance of σ^2/N. Thus,

$$f_{\hat{\mu}}(\hat{\mu}) = \frac{\sqrt{N}}{\sqrt{2\pi}\sigma} \exp\left[-\frac{N}{2\sigma^2}(\hat{\mu} - \mu)^2\right]. \tag{5.63}$$

Replacing θ with $\hat{\mu}$ in Eq. 5.59, we have

$$\int_{\mu-B}^{\mu+A} \frac{\sqrt{N}}{\sqrt{2\pi}\sigma} \exp\left[-\frac{N}{2\sigma^2}(\hat{\mu} - \mu)^2\right] d\hat{\mu} = 1 - \alpha \tag{5.64}$$

or with the substitution $\zeta = \sqrt{N}(\hat{\mu} - \mu)/\sigma$,

$$\int_{-\sqrt{N}B/\sigma}^{\sqrt{N}A/\sigma} \frac{1}{\sqrt{2\pi}} \exp[-\tfrac{1}{2}\zeta^2]\, d\zeta = 1 - \alpha. \tag{5.65}$$

Comparing this integral with the normal CDF given in standard form by Eq. 3.44, we see that

$$\Phi(\sqrt{N}A/\sigma) - \Phi(-\sqrt{N}B/\sigma) = 1 - \alpha. \tag{5.66}$$

The standardized normal distribution is plotted in Fig. 5.11. Recall that A is chosen so that the area under the sampling curve to the right is $\alpha/2$. We designate $z_{\alpha/2}$ to be the value of the reduced variate for which this condition holds. Thus the area to the left of $z_{\alpha/2}$ is given by

$$\Phi(z_{\alpha/2}) = 1 - \alpha/2. \tag{5.67}$$

The symmetry of the normal distribution results in the condition given by Eq. 3.45. Consequently, we also have

$$\Phi(-z_{\alpha/2}) = \alpha/2. \tag{5.68}$$

Thus Eq. 5.66 is satisfied if we take

$$A = B = z_{\alpha/2}\sigma/\sqrt{N}. \tag{5.69}$$

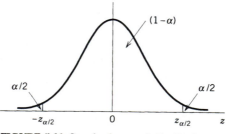

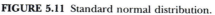

FIGURE 5.11 Standard normal distribution.

If we combine these conditions with Eqs. 5.60 and 5.61, and estimate σ from the sample variance given by Eq. 5.9, the $100(1 - \alpha)$ percent two-sided confidence interval for μ is given by

$$L_{\alpha/2,N} = \hat{\mu} - z_{\alpha/2}\frac{\hat{\sigma}}{\sqrt{N}} \qquad (5.70)$$

and

$$U_{\alpha/2,N} = \hat{\mu} + z_{\alpha/2}\frac{\hat{\sigma}}{\sqrt{N}}. \qquad (5.71)$$

Some of the more commonly used confidence intervals are 80, 90, 95, and 99%. These correspond to risks of $\alpha = 20, 10, 5$ and 1% respectively. The corresponding values of $z_{\alpha/2}$ may be found from the CDF for the normal distribution tabulated in Appendix C. They are, respectively:

$$z_{0.1} = 1.28, \qquad z_{0.05} = 1.648, \qquad z_{0.025} = 1.96 \qquad z_{0.005} = 2.58.$$

EXAMPLE 5.8

Find the 90% and the 95% confidence interval for the mean of the 70 stopping power data given in Table 5.1

Solution   The sample mean and variance obtained in Example 5.2 are $\hat{\mu} = 52.3$ and $\hat{\sigma}^2 = 168.47$. Thus the standard deviation is $\hat{\sigma} = 12.98$. For two-sided 90 percent confidence $z_{\alpha/2} = 1.645$. Thus $z_{\alpha/2}\hat{\sigma}/\sqrt{N} = 1.645 \times 12.98/8.367 = 2.55$ and thus from Eqs. 5.70 and 5.71, $\hat{\mu} = 52.3 \pm 2.55$ with 90 percent confidence. Likewise, for 95 percent confidence, $z_{\alpha/2} = 1.960$ and $z_{\alpha/2}\hat{\sigma}/\sqrt{N} = 1.960 \times 12.98/8.367 = 3.04$. Thus $\hat{\mu} = 52.3 \pm 3.04$ with 95 percent confidence.

To recapitulate, the interval estimate for the mean, μ, is nonparametric in that the distribution from which the sample of N derives need not be normal. The two-sided confidence limits can be used for any distribution so long as the variance exists, and N is sufficiently large, usually greater than $N = 30$. In Eq. 2.86 we applied this result to estimate the confidence interval of the mean of the binomial distribution for a sufficiently large sample size. No distribution-free confidence intervals exist for the variance, skewness or other properties.

Normal and Lognormal Parameters

Since the two parameters appearing in the normal distribution are just the mean and the standard deviation (i.e., the square root of the variance) the unbiased point estimators are given by Eqs. 5.6 and 5.8. For $N > 30$ the central limit theorem is applicable to the mean, and therefore the confidence interval is given by Eqs. 5.70 and 5.71. The $100(1 - \alpha)$ percent two-sided confidence limits are thus

$$\mu^{\pm} = \hat{\mu} \pm z_{\alpha/2} \frac{\hat{\sigma}}{\sqrt{N}}. \tag{5.72}$$

The confidence interval for the standard deviation for $N > 30$, may be estimated as

$$\sigma^{\pm} = \hat{\sigma} \pm z_{\alpha/2} \frac{\hat{\sigma}}{\sqrt{2(N-1)}}. \tag{5.73}$$

EXAMPLE 5.9

Find the point estimate and the 90% confidence interval for the mean and the standard deviation for the population of resistors coming from supplier 1 in Example 5.6.

Solution   We first obtain the mean and the variance, applying the spread sheet formula to Table 5.4

$$\hat{\mu} = \text{AVERAGE}(\text{B3:B17}, \text{G3:G17}) = 49.77$$

$$\hat{\sigma}^2 = \text{VAR}(\text{B3:B17}, \text{G3:G17}) = 0.5732$$

$$\hat{\sigma} = \sqrt{0.5732} = 0.7571$$

Since there are 30 data points, we may use the expressions for large sample size. For the mean we use Eq. 5.72 to obtain

$$\hat{\mu} = 49.77 \pm 1.645 \times 0.7571/\sqrt{30} = 49.77 \pm 0.23$$

For the standard deviation we use Eq. 5.73 to obtain

$$\hat{\sigma} = 0.757 \pm 1.645 \times 0.7571/\sqrt{2 \times 29} = 0.757 \pm 0.164$$

Note that the point estimate of the variance is not identical to that obtained from probability plotting in Example 5.6. The result from plotting, however, does lie within the 90% confidence limit.

The CDF of a random variable y that is lognormally distributed is directly related to the standard normal distribution through the relationship $x = \ln(y)$ yielding the CDF

$$F(y) = \Phi\left[\frac{1}{\omega}\ln(y/y_o)\right]. \tag{5.74}$$

Here, $\ln y_o$, the log mean, is estimated by

$$\ln \hat{y}_o = \frac{1}{N}\sum_i \ln y_i, \tag{5.75}$$

or solving for $\hat{y}_o$ and simplifying

$$\hat{y}_o = \left(\prod_i y_i\right)^{1/N}. \tag{5.76}$$

Likewise we may write

$$\hat{\omega}^2 = \frac{N}{N-1}\left[\frac{1}{N}\sum_i (\ln y_i)^2 - \left(\frac{1}{N}\sum_i \ln y_i\right)^2\right]. \qquad (5.77)$$

The $100(1 - \alpha)$ percent two-sided confidence limits are similarly obtained by transforming Eqs. 5.72 and 5.73

$$y_o^{\pm} = \hat{y}_o \exp(\pm z_{\alpha/2}\hat{\omega}N^{-1/2}), \qquad (5.78)$$

and

$$\omega^{\pm} = \hat{\omega} \pm z_{\alpha/2}\frac{\hat{\omega}}{\sqrt{2(N-1)}}. \qquad (5.79)$$

Extreme Value and Weibull Parameters

Point estimates for the parameters appearing in extreme value and Weibull distributions can also be made. Determining the confidence intervals that can be associated with these parameters is more problematical. In cases where the sample size is not large, say less than 30, tedious and sometimes iterative procedures are employed that are beyond the scope of what space allows us to consider here. For larger sample sizes, rough estimates of the confidence interval are obtainable using the relationships recommended by Nelson.* It is these that appear in what follows.

Extreme value distributions In Eqs 3.92 and 3.93 the mean and the variance of the maximum extreme value distribution are given in terms of the shape and location parameters. If we invert these equations, the Θ and u parameters can be given in terms of the mean and variance:

$$\Theta = \frac{\sqrt{6}}{\pi}\sigma \qquad (5.80)$$

and

$$u = \mu - \gamma\frac{\sqrt{6}}{\pi}\sigma. \qquad (5.81)$$

Accordingly, we may replace μ and σ on the right of these equations by the sample mean and variance; we obtain the following point estimates of the parameters:

$$\hat{\Theta} = \frac{\sqrt{6}}{\pi}\hat{\sigma} \qquad (5.82)$$

and

$$\hat{u} = \hat{\mu} - \gamma\frac{\sqrt{6}}{\pi}\hat{\sigma}. \qquad (5.83)$$

* W. Nelson, *Applied Life Data Analysis*, Wiley, New York, 1982, Ch. 6.

Since Θ in the minimum extreme value distribution is also related to the variance by Eq. 5.82, we may estimate Θ for both minimum and maximum extreme value distributions. As indicated in Chapter 3 the maximum extreme value distribution u, μ, and σ are related by

$$u = \mu + \gamma \frac{\sqrt{6}}{\pi} \sigma. \tag{5.84}$$

Hence replacing μ and σ by their point estimators the parameters yields

$$\hat{u} = \hat{\mu} + \gamma \frac{\sqrt{6}}{\pi} \hat{\sigma}. \tag{5.85}$$

For large values of the sample size, say $N > 30$, Nelson provides the following confidence limit estimates:

$$\Theta^{\pm} = \hat{\Theta} \exp(\pm 1.049\, z_{\alpha/2} N^{-1/2}) \tag{5.86}$$

$$u^{\pm} = \hat{u} + 1.018\, z_{\alpha/2} \hat{\Theta} N^{-1/2}. \tag{5.87}$$

The two-parameter Weibull distribution is obtained from the minimum extreme value distribution by making the transformation $x = \ln y$, whereas in Eqs. 3.106 and 3.107 the Weibull parameters are given in terms of the corresponding minimum extreme-value parameters as $\theta = e^{u}$ and $m = 1/\Theta$. These relationships may be combined with the estimators for u and Θ, given by Eqs. 5.82 and 5.83, to yield

$$\hat{m} = \frac{\pi}{\sqrt{6}\hat{\sigma}} \tag{5.88}$$

and

$$\hat{\theta} = \exp\left(\hat{\mu} + \gamma \frac{\sqrt{6}}{\pi} \hat{\sigma}\right). \tag{5.89}$$

For the Weibull distribution, however, the transformation $x = \ln y$ must also be applied to the definitions of the mean and the variance. Thus we now have the log mean and log variance

$$\hat{\mu} = \frac{1}{N} \sum_{i} \ln y_i \tag{5.90}$$

and

$$\hat{\sigma}^2 = \frac{N}{N-1}\left[\frac{1}{N}\sum_i (\ln y_i)^2 - \left(\frac{1}{N}\sum_i \ln y_i\right)^2\right]. \tag{5.91}$$

With these definitions, $\hat{\mu}$ can by eliminated from Eq. 5.89 to yield

$$\hat{\theta} = \left(\prod_i y_i\right)^{1/N} \exp\left(\gamma \frac{\sqrt{6}}{\pi} \hat{\sigma}\right). \tag{5.92}$$

Approximate confidence intervals for the Weibull parameters can also be obtained by applying the transforms of Eqs. 3.106 and 3.107 to Eqs. 5.86 and 5.87. The result are the following estimates for m and θ confidence intervals, which are applicable for sufficiently large sample size:

$$m^{\pm} = \hat{m} \exp(\pm 1.049 \, z_{\alpha/2} N^{-1/2}) \tag{5.93}$$

and

$$\theta^{\pm} = \hat{\theta} \exp(\pm 1.018 \, z_{\alpha/2} \hat{m}^{-1} N^{-1/2}), \tag{5.94}$$

where the $z_{\alpha/2}$ are determined as before.

EXAMPLE 5.10

The data points in Table 5.6a for voltage discharge are thought to follow a Weibull distribution. Make point estimates of the Weibull shape and scale parameters and determine their 90% confidence limits.

Solution   We tabulate the natural logarithms of the 60 voltage discharges in Table 5.6b. We calculate the log mean and log variance, Eqs. 5.90 and 5.91, from the data in Table 5.6b:

$$\hat{\mu} = \text{AVERAGE}(A1:C20) = 4.101$$

$$\hat{\sigma}^2 = \text{VAR}(A1:C20) \quad\quad = 0.0056$$

TABLE 5.6 Voltage Discharge Data for Example 5.10

	A	B	C
1	63	65	62
2	72	67	70
3	66	68	59
4	75	63	63
5	61	72	69
6	63	70	73
7	70	64	61
8	57	58	66
9	68	68	55
10	74	57	68
11	70	68	64
12	63	64	68
13	64	57	59
14	72	74	69
15	66	72	63
16	62	57	73
17	72	64	66
18	69	64	65
19	64	66	66
20	63	62	65

TABLE 5.7   Natural Logarithms of Voltage
Discharge Data

	A	B	C
1	4.1431	4.1744	4.1271
2	4.2767	4.2047	4.2485
3	4.1897	4.2195	4.0775
4	4.3175	4.1431	4.1431
5	4.1109	4.2767	4.2341
6	4.1431	4.2485	4.2905
7	4.2485	4.1589	4.1109
8	4.0431	4.0604	4.1897
9	4.2195	4.2195	4.0073
10	4.3041	4.0431	4.2195
11	4.2485	4.2195	4.1589
12	4.1431	4.1589	4.2195
13	4.1589	4.0431	4.0775
14	4.2767	4.3041	4.2341
15	4.1897	4.2767	4.1431
16	4.1271	4.0431	4.2905
17	4.2767	4.1589	4.1897
18	4.2341	4.1589	4.1744
19	4.1589	4.1897	4.1897
20	4.1431	4.1271	4.1744

and hence $\hat{\sigma} = 0.075$. Thus from Eqs. 5.88 and 5.89 the shape and scale point estimates are

$$\hat{m} = 3.141/(2.449 \times 0.075) = 17.1$$

$$\hat{\theta} = \exp(4.101 + 0.5772 \times 2.449 \times 0.075/3.141) = 62.5$$

For the 90 percent confidence interval, $z_{\alpha/2} = 1.645$. Thus from Eq. 5.93:

$$m^{\pm} = 17.1 \exp(\pm 1.049 \times 1.645/\sqrt{60})$$

or $m^{+} = 21.4$ and $m^{-} = 13.7$.

From Eq. 5.94:

$$\theta^{\pm} = 62.5 \exp(\pm 1.018 \times 1.645/17.1\sqrt{60})$$

or $\theta^{+} = 63.3$ and $\theta^{-} = 61.7$.

5.5 STATISTICAL PROCESS CONTROL

Thus far we have dealt with the analysis of complete sets of data. In a number of circumstances, however, it is necessary to take data in time sequence and advantageous to analyze that data at the earliest possible time. One example is in life testing where a number of items are tested to failure. Since the time to the last failure may be excessive, it is often desirable to glean information from the times of the first few failures, or even from the fact that there have been none, if that is the situation. We take up the analysis of such tests in Chapter 8.

A second circumstance, which we treat briefly here, arises in statistical process control or SPC. Usually, in initiating the process and bringing it under control, a data base is established to demonstrate that the process follows a normal distribution. Then, as discussed in Chapter 4, it is desirable to ensure that the variability is due only to random, short-term, part-to-part variation. If systematic changes cause the process mean to shift, they must be detected as soon as possible so that corrective actions can be taken and the number of out-of-specification items that are produced is held to a minimum.

One approach to the forgoing problem consists of collecting blocks of data of say 50 to 100 measurements, forming histograms, and calculating the sample mean and variance. This, however, is very inefficient, for if a mean shift takes place many out-of-tolerance items would be produced before the shift could be detected. At the other extreme each individual measurement could be plotted, as has been done for example in Fig. 5.12a and b. In Fig 5.12a all of the data are distributed normally with a constant mean and variance. In Fig. 5.12b, however, a shift in the mean takes place at run number 50. Because of the large random component of part-to-part variability the shift is difficult to detect, particularly after relatively few additional data points have been entered.

More effective detection of shifts in the distribution is obtained by averaging over a small number of measurements, referred to as a rational subgroup. Such averaging is performed over groups of ten measurements in Fig. 5.13. The noise caused by the random variations is damped, making changes in mean more easily detected. At the same time, the delays caused by the grouping are not so large as to cause unacceptable numbers of out-of-tolerance items to escape detection before corrective action can begin. Note that upper- and lower-control limit lines are included to indicated at what point corrective action should be taken. From this simple example it is clear that in setting up a control chart to track a particular statistic, such as the mean or the variance, one must determine (a) the optimal number N of measurements to include in the rational subgroup, and (b) the location of the control limits.

Averaging over rational subgroups has a number of beneficial effects. As discussed in section 5.4, the central limit theorem states that as the number of units, N, included in an average is increased, the sampling distribution will tend toward being normal even though the parent distribution is nonnormal. Furthermore the standard deviation of the sampling distribution will be the $\sigma/\sqrt{N}$, where σ is the standard deviation of the parent distribution. Typically values of N between 4 and 20 are used, depending on the parent distribution. If the parent distribution is close to normal, $N = 4$ may be adequate, for the sampling distribution will already be close to normal. In general, smaller rational subgroups, say $N = 4$, 5, or 6, are frequently used to detect larger changes in the mean while larger subgroups, say 10 or more, are needed to find more subtle deviations. A substantial number of additional considerations come into play in specifying the rational subgroup size. These include the time and expense of making the individual measurements, whether every unit

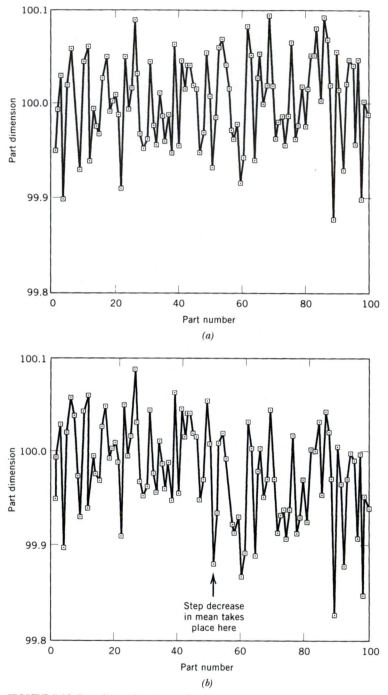

FIGURE 5.12 Part dimension vs. production sequence: (a) no disturbance, (b) change in mean.

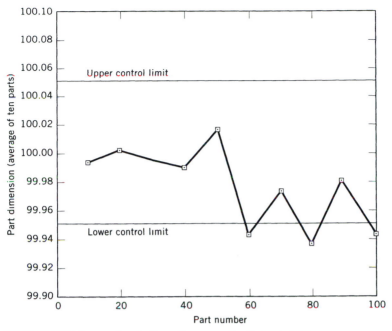

FIGURE 5.13 Averaged part dimension vs production sequence.

is to be measured, or only periodic samplings are to be made, and the cost of producing out of tolerance units, which must be reworked or scrapped.

The specification of the control limits also involves tradeoffs. If they are set too tightly about the process mean, there will be frequent false alarms in which the random part-by-part variability causes a limit to be crossed. In the hypothesis-testing sense these are referred to as Type I errors; they indicate that the distribution is deviating from the in-control distribution, when in fact it is not. Conversely, if the control limits are set too far from the target value, there will be few if any false alarms, but significant changes in the mean may go undetected. These are then Type II errors, for they fail to detect differences from the base distribution.

Control limits are customarily set only when the process is known to be in control and when sufficient data has been taken to determine the process mean and standard deviation with reasonable accuracy. Probability plotting or the chi-squared test may be used to determine how nearly the data fits a normal distribution. The upper- and lower-control limits (UCL and LCL) may then be determined from

$$UCL = \mu + 3\frac{\sigma}{\sqrt{N}} \qquad LCL = \mu - 3\frac{\sigma}{\sqrt{N}} \qquad (5.95)$$

where μ and σ are the mean and standard deviation of the process, and $\sigma/\sqrt{N}$ is the standard deviation of the rational subgroup. The coefficient of three is most often chosen if only part-to-part variation is present. With this value, 0.26% of the sample will fall outside the control limits in the absence of long-term variations. This level of 26 false alarms in 10,000 average computations is considered acceptable.

Note that the *LCL* and *UCL* are not related to the lower- and upper-specification limits (the *LSL* and *USL*) discussed the Chapter 4. Control charts are based only on the process variance and the rational control group size, N, and not on the specifications that must be maintained. Their purpose is to ensure that the process stays in control, and that any problems causing a shift in μ are recognized quickly so that corrective actions may be taken.

EXAMPLE 5.11

A large number of ±5% resistors are produced in a well-controlled process. The process mean is 50.0 ohms and a standard deviation is 0.84 ohms. Set up a control chart for the mean. Assume a rational subgroup of $N = 6$.

Solution From Eq. 5.95 we obtain $UCL = 50 + 3 \times 0.84/\sqrt{6} = 51.0$ ohms $LCL = 50 - 3 \times 0.84/\sqrt{6} = 49.0$ ohms. Note that the ±5% specification limits $USL = 52.5$ and $LSL = 47.5$ are quite different.

The chart discussed thus far is referred to as a Shewhart x chart. Often, it is used in conjunction with a chart to track the dispersion of the process as measured by σ, the process standard deviation. In practice, bootstrap methods may be used to estimate the process standard deviation by taking the ranges of a number of small samples. One then calculates the average range and uses it in turn to estimate σ. Likewise, statistical process control charts may also be employed for attribute data, and a number of more elaborate sampling schemes employing moving averages and other such techniques are covered in texts devoted specifically to quality control.

Bibliography

Crowder, M. J., A. C. Kimber, R. L. Smith, and T. J. Sweeting, *Statistical Analysis of Reliability Data*, Chapman & Hall, London, 1991.

Kapur, K. C., and L. R. Lamberson, *Reliability in Engineering Design*, Wiley, NY, 1977.

Kececioglu, D., *Reliability and Life Testing Handbook*, Vol. I & II, PTR Prentice-Hall, Englewood Cliffs, NJ, 1993.

Lawless, J. F., *Statistical Models and Methods for Lifetime Data*, Wiley, NY, 1982.

Mann, N. R., R. E. Schafer, and N. D. Singpurwalla, *Methods for Statistical Analysis of Reliability and Life Data*, Wiley, NY, 1974.

Mitra, A., *Fundamentals of Quality Control and Improvement*, Macmillan, NY, 1993.

Nelson, W., *Applied Life Data Analysis*, Wiley, NY, 1982.

Exercises

5.1 Consider the following response time data measured in seconds.*

1.48	1.46	1.49	1.42	1.35
1.34	1.42	1.70	1.56	1.58
1.59	1.59	1.61	1.25	1.31
1.66	1.58	1.43	1.80	1.32
1.55	1.60	1.29	1.51	1.48
1.61	1.67	1.36	1.50	1.47
1.52	1.37	1.66	1.44	1.29
1.80	1.55	1.46	1.62	1.48
1.64	1.55	1.65	1.54	1.53
1.46	1.57	1.65	1.59	1.47
1.38	1.66	1.59	1.46	1.61
1.56	1.38	1.57	1.48	1.39
1.62	1.49	1.26	1.53	1.43
1.30	1.58	1.43	1.33	1.39
1.56	1.48	1.53	1.59	1.40
1.27	1.30	1.72	1.48	1.66
1.37	1.68	1.77	1.62	1.33

(a) Compute the mean and the variance.

(b) Use the Sturges formula to make a histogram approximating $f(x)$.

5.2 Fifty measurements of the ultimate tensile strength of wire are given in the accompanying table.

(a) Group the data and make an appropriate histogram to approximate the PDF.

(b) Calculate $\hat{\mu}$ and $\hat{\sigma}^2$ for the distribution from the ungrouped data.

(c) Using $\hat{\mu}$ and $\hat{\sigma}$ from part *b*, draw a normal distribution through the histogram.

Ultimate Tensile Strength

103,779	102,325	102,325	103,799
102,906	104,651	105,377	100,145
104,796	105,087	104,796	103,799
103,197	106,395	106,831	103,488
100,872	100,872	105,087	102,906
97,383	104,360	103,633	101,017
101,162	101,453	107,848	104,651
98,110	103,779	99,563	103,197
104,651	101,162	105,813	105,337

* Data from A. E. Green and A. J. Bourne, *Reliability Technology*, Wiley, NY, 1972.

Ultimate Tensile Strength (continued)

102,906	102,470	108,430	101,744
103,633	105,232	106,540	106,104
102,616	106,831	101,744	100,726
103,924		101,598	

Source: Data from E. B. Haugen, *Probabilistic Mechanical Design,*
Wiley, NY, 1980.

5.3 For the data in Example 5.3:

(a) Calculate the sample mean, variance, skewness, and kurtosis.

(b) Analytically determine the variance, skewness, and kurtosis for an exponential distribution that has a mean equal to the sample mean obtained in part a.

(c) What is the difference between the sample and analytic values of the variance, skewness, and kurtosis obtained in parts a and b?

5.4 The following are sixteen measurements of circuit delay times in microseconds: 2.1, 0.8, 2.8, 2.5, 3.1, 2.7, 4.5, 5.0, 4.2, 2.6, 4.8, 1.6, 3.5, 1.9, 4.6, and 2.1.

(a) Calculate the sample mean, variance, and skewness.

(b) Make a normal probability plot of the data.

(c) Compare the mean and variance from the probability plot with the results from part a.

5.5 Make a Weibull probability plot of the data in Example 5.7 and determine the parameters. Is the fit better or worse than that using a lognormal distribution as in Example 5.7? What criterion did you use to decide which was better?

5.6 The following failure times (in days) have been recorded in a proof test of 20 units of a new product: 2.6, 3.2, 3.4, 3.9, 5.6, 7.1, 8.4, 8.8, 8.9, 9.5, 9.8, 11.3, 11.8, 11.9, 12.3, 12.7, 16.0, 21.9, 22.4, and 24.2.

(a) Make a graph of $F(t)$ vs. t.

(b) Make a Weibull probability plot and determine the scale and shape parameters.

(c) Make a lognormal plot and determine the two parameters.

(d) Determine which of the two distributions provides the best fit to the data, using the coefficient of determination as a criterion.

5.7 Calculate the sample mean, variance, skewness, and kurtosis for the data in Exercise 5.6

5.8 Make a least-squares fit of the following (x, y) data points to a line of the form $y = ax + b$, and estimate the slope and y intercept:

x: 0.54, 0.92, 1.27, 1.35, 1.38, 1.56, 1.70, 1.91, 2.15, 2.16, 2.50, 2.75, 2.90, 3.11, 3.20

y: 28.2, 30.6, 29.1, 24.3, 27.5, 25.0, 23.8, 20.4, 22.1, 17.3, 17.1, 18.5, 16.0, 14.1, 15.6

5.9 Make a normal probability plot for the data in Example 5.6 using Eq. 5.13 instead of 5.12. Compare the means and the standard deviations to the values obtained in Example 5.6.

5.10 (a) Make a normal probability data plot from Exercise 5.1.

(b) Estimate the mean and the variance, assuming that the distribution is normal.

(c) Compare the mean and variance determined from your plot with the values calculated in part *a* of Exercise 5.1.

5.11 Make a lognormal probability plot of the data in Example 5.3 and determine the parameters. How does the value or r^2 compare to that obtained when a Weibull distribution is used to fit the data?

5.12 Make a lognormal probability plot for the voltage discharge data in Example 5.10 and estimate the parameters.

5.13 Make a normal probability plot for the data in Exercise 5.2 and estimate the mean, the variance and r^2.

5.14 Calculate the skewness from the voltage data in Example 5.10. If it is positive (negative) make a maximum (minimum) extreme value plot and estimate the parameters.

5.15 The times to failure in hours on four compressors are 240, 420, 630, and 1080.

(a) Make a lognormal probability plot.

(b) Estimate the most probable time to failure.

5.16 Redo Example 5.3 by making the probability plot with a spread sheet, and compare your estimate of θ with Example 5.3.

5.17 Use Eqs. 5.72 and 5.73 to estimate the 90% and the 95% confidence intervals for the mean and for the variance obtained in Exercise 5.2.

5.18 The following times to failure (in days) result from a fatigue test of 10 flanges:

1.66, 83.36, 25.76, 24.36, 334.68, 29.62, 296.82, 13.92, 107.04, 6.26.

(a) Make a lognormal probability plot.

(b) Estimate the parameters.

(c) Estimate the factor to which the time to failure is known with 90% confidence.

5.19 Suppose you are to set up a control chart for testing the tensile strength of one of each 100 specimens produced. You are to base your calculations on the data given in Exercise 5.2. Calculate the lower and upper control limits for a rational subgroup size of $N = 5$.

5.20 Find the *UCL* and *LCL* for the control chart in Example 5.12 if the rational subgroup is taken as (a) $N = 4$, (b) $N = 8$.

CHAPTER 6

Reliability and Rates of Failure

"Have you heard of the wonderful one-hoss shay,
That was built in such a logical way
It ran a hundred years to a day,
And then, of a sudden, it——"

Oliver Wendell Holmes
The Deacon's Masterpiece

6.1 INTRODUCTION

Generally, reliability is defined as the probability that a system will perform properly for a specified period of time under a given set of operating conditions. Implied in this definition is a clear-cut criterion for failure, from which we may judge at what point the system is no longer functioning properly. Similarly, the treatment of operating conditions requires an understanding both of the loading to which the system is subjected and of the environment within which it must operate. Perhaps the most important variable to which we must relate reliability, however, is time. For it is in terms of the rates of failure that most reliability phenomena are understood.

In this chapter we examine reliability as a function of time, and this leads to the definition of the failure rate. Examining the time dependence of failure rates allows us to gain additional insight into the nature of failures—whether they be infant mortality failures, failures that occur randomly in time, or failures brought on by aging. Similarly, the time–dependence of failures can be viewed in terms of failure modes in order to differentiate between failures caused by different mechanisms and those caused by different components of a system. This leads to an appreciation of the relationship between failure rate and system complexity. Finally, we examine the impact of failure rate on the number of failures that may occur in systems that may be repaired or replaced.

6.2   RELIABILITY CHARACTERIZATION

We begin this section by quantitatively defining reliability in terms of the PDF and the CDF for the time-to-failure. The failure rate and the mean-time-to-failure are then introduced. The failure rate is discussed in detail, for its characteristic shape in the form of the so-called bathtub curve provides substantial insight into the nature of the three classes of failure mechanisms: infant mortality, random failures, and aging.

Basic Definitions

Reliability is defined in Chapter 1 as the probability that a system survives for some specified period of time. It may be expressed in terms of the random variable **t**, the time-to-system-failure. The PDF, $f(t)$, has the physical meaning

$$f(t)\, \Delta t = P\{t < \mathbf{t} \le t + \Delta t\} = \left\{ \begin{array}{c} \text{probability that failure} \\ \text{takes place at a time} \\ \text{between } t \text{ and } t + \Delta t \end{array} \right\}. \tag{6.1}$$

for vanishingly small Δt. From Eq. 3.1 we see that the CDF now has the meaning

$$F(t) = P\{\mathbf{t} \le t\} = \left\{ \begin{array}{c} \text{probability that failure} \\ \text{takes place at a time less} \\ \text{than or equal to } t \end{array} \right\}. \tag{6.2}$$

We define the reliability as

$$R(t) = P\{\mathbf{t} > t\} = \left\{ \begin{array}{c} \text{probability that a system} \\ \text{operates without failure} \\ \text{for a length of time } t \end{array} \right\}. \tag{6.3}$$

Since a system that does not fail for $\mathbf{t} \le t$ must fail at some $\mathbf{t} > t$, we have

$$R(t) = 1 - F(t), \tag{6.4}$$

or equivalently either

$$R(t) = 1 - \int_0^t f(t')\, dt' \tag{6.5}$$

or

$$R(t) = \int_t^\infty f(t')\, dt'. \tag{6.6}$$

From the properties of the PDF, it is clear that

$$R(0) = 1 \tag{6.7}$$

and

$$R(\infty) = 0. \tag{6.8}$$

We see that the reliability is the CCDF of t, that is, $R(t) = \bar{F}(t)$. Similarly, since $F(t)$ is the probability that the system will fail before $\mathbf{t} = t$, it is often referred to as the unreliability or failure probability; at times we may denote the unreliability as

$$\bar{R}(t) = 1 - R(t) = F(t). \tag{6.9}$$

Equation 6.5 may be inverted by differentiation to give the PDF of failure times in terms of the reliability:

$$f(t) = -\frac{d}{dt}R(t). \tag{6.10}$$

Insight is normally gained into failure mechanisms by examining the behavior of the failure rate. The *failure rate*, $\lambda(t)$, may be defined in terms of the reliability or the PDF of the time-to-failure as follows. Let $\lambda(t)\,\Delta t$ be the probability that the system will fail at some time $\mathbf{t} < t + \Delta t$ given that it has not yet failed at $\mathbf{t} = t$. Thus it is the conditional probability

$$\lambda(t)\,\Delta t = P\{\mathbf{t} < t + \Delta t \,|\, \mathbf{t} > t\}. \tag{6.11}$$

Using Eq. 2.5, the definition of a conditional probability, we have

$$P\{\mathbf{t} < t + \Delta t \,|\, \mathbf{t} > t\} = \frac{P\{(\mathbf{t} > t) \cap (\mathbf{t} < t + \Delta t)\}}{P\{\mathbf{t} > t\}}. \tag{6.12}$$

The numerator on the right-hand side is just an alternative way of writing the PDF; that is,

$$P\{(\mathbf{t} > t) \cap (\mathbf{t} < t + \Delta t)\} \equiv P\{t < \mathbf{t} < t + \Delta t\} = f(t)\,\Delta t. \tag{6.13}$$

The denominator of Eq. 6.12 is just $R(t)$, as may be seen by examining Eq. 6.3. Therefore, combining equations, we obtain

$$\lambda(t) = \frac{f(t)}{R(t)}. \tag{6.14}$$

This quantity, the failure rate, is also referred to as the hazard or mortality rate.

The most useful way to express the reliability and the failure PDF is in terms of the failure rate. To do this, we first eliminate $f(t)$ from Eq. 6.14 by inserting Eq. 6.10 to obtain the failure rate in terms of the reliability,

$$\lambda(t) = -\frac{1}{R(t)}\frac{d}{dt}R(t). \tag{6.15}$$

Then multiplying by dt, we obtain

$$\lambda(t)\,dt = -\frac{dR(t)}{R(t)}. \tag{6.16}$$

Integrating between zero and t yields

$$\int_0^t \lambda(t')\, dt' = -\ln[R(t)] \tag{6.17}$$

since $R(0) = 1$. Finally, exponentiating results in the desired expression for the reliability

$$R(t) = \exp\left[-\int_0^t \lambda(t')\, dt'\right]. \tag{6.18}$$

To obtain the probability density function for failures, we simply insert Eq. 6.18 into Eq. 6.14 and solve for $f(t)$:

$$f(t) = \lambda(t) \exp\left[-\int_0^t \lambda(t')\, dt'\right]. \tag{6.19}$$

Probably the single most-used parameter to characterize reliability is the *mean time to failure* (or MTTF). It is just the expected or mean value $E\{t\}$ of the failure time t. Hence

$$\text{MTTF} = \int_0^\infty t f(t)\, dt. \tag{6.20}$$

The MTTF may be written directly in terms of the reliability by substituting Eq. 6.10 into Eq. 6.20 and integrating by parts:

$$\text{MTTF} = -\int_0^\infty t\, \frac{dR}{dt}\, dt = -tR(t)\Big|_0^\infty + \int_0^\infty R(t)\, dt \tag{6.21}$$

Clearly, the $tR(t)$ term vanishes at $t = 0$. Similarly, from Eq. 6.18, we see that $R(t)$ will decay exponentially or faster, since the failure rate $\lambda(t)$ must be greater than zero. Thus $tR(t) \to 0$ as $t \to \infty$. Therefore, we have

$$\text{MTTF} = \int_0^\infty R(t)\, dt. \tag{6.22}$$

EXAMPLE 6.1

An engineer approximates the reliability of a cutting assembly by

$$R(t) = \begin{cases} (1 - t/t_0)^2, & 0 \le t < t_0, \\ 0 & t \ge t_0. \end{cases}$$

(*a*) Determine the failure rate.

(*b*) Does the failure rate increase or decrease with time?

(*c*) Determine the MTTF.

Solution (*a*) From Eq. 6.10,

$$f(t) = -\frac{d}{dt}(1 - t/t_0)^2 = \frac{2}{t_0}(1 - t/t_0), \qquad 0 \le t < t_0.$$

and from Eq. 6.14,

$$\lambda(t) = \frac{f(t)}{R(t)} = \frac{2}{t_0(1 - t/t_0)}, \qquad 0 \le t < t_0.$$

(*b*) The failure rate increases from $2/t_0$ at $t = 0$ to infinity at $t = t_0$.

(*c*) From Eq. 6.22

$$\text{MTTF} = \int_0^{t_0} dt(1 - t/t_0)^2 = t_0/3.$$

The Bathtub Curve

The behavior of failure rates with time is quite revealing. Unless a system has redundant components, such as those discussed in Chapter 9, the failure rate curve usually has the general characteristics of a "bathtub" such as shown in Fig. 6.1. The bathtub curve, in fact, is an ubiquitous characteristic of living creatures as well as of inanimate engineering devices, and much of the failure rate terminology comes from demographers' studies of human mortality distributions. In the biomedical community, for example, reliability is referred to as the survivability and denoted as $S(t)$. Moreover, comparisons of human mortality and engineering failures add insight into the three broad classes of failures that give rise to the bathtub curve.

The short period of time on the left-hand side of Fig. 6.1 is a region of high but decreasing failure rates. This is referred to as the period of *infant mortality*, or early failures. Here, the failure rate is dominated by infant deaths caused primarily by congenital defects or weaknesses. The death rate decreases with time as the weaker infants die and are lost from the population or their defects are detected and repaired. Similarly, defective pieces of equipment, prone to failure because they were not manufactured or constructed properly, cause the high initial failure rates of engineering devices. Missing parts, substandard material batches, components that are out of tolerance, and damage in shipping are a few of the quality weaknesses that may cause excessive failure rates near the beginning of design life.

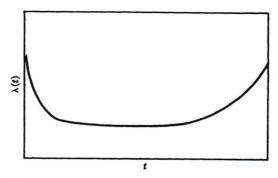

FIGURE 6.1 A "bathtub" curve representing a time-dependent failure rate.

Early failures in engineering devices are nearly synonymous with the "product noise" quality loss stressed in the Taguchi methodology. As discussed in Chapter 4, the preferred method for eliminating such failures is through design and production quality control measures that will reduce variability and hence susceptibility to infant mortality failures. If such measures are inadequate, a period of time may be specified during which the device undergoes wearin.* During this time loading and use are controlled in such a way that weaknesses are likely to be detected and repaired without failure, or so that failures attributable to defective manufacture or construction will not cause inordinate harm or financial loss. Alternately, in environmental stress screening and in proof-testing products are stressed beyond what is expected in normal use so that weak units will fail before they are sold or put in service.

The middle section of the bathtub curve contains the smallest and most nearly constant failure rates and is referred to as the useful life. This flat behavior is characteristic of failures caused by random events and hence referred to as *random failures*. They are likely to stem from unavoidable loads coming from without, rather than from any inherent defect in the device or system under consideration. Consequently, the probability that failure will occur in the next time increment is independent of the system's age. In human populations, deaths during this part of the bathtub curve are likely to be due to accidents or to infectious disease. In engineering devices, the external loading may take a wide variety of forms, depending on the type of system under consideration: earthquakes, power surges, vibration, mechanical impact, temperature fluctuations, and moisture variation are some of the common causes. In the Taguchi quality methodology such loads are referred to as "outer noise."

Random failure can be reduced by improving designs: making them more robust with respect to the environments to which they are subjected. As discussed in detail in Chapter 7 this may be accomplished by increasing the ratio of components capacities relative to the loads placed upon them. The net outcome may be visualized as in Fig. 6.2, where for an assumed operating environment, the failure rate decreases as the component load is reduced. This procedure of deliberately reducing the loading is referred to as derating. The terminology stems from the deliberate reduction of voltages of electrical systems, but it is also applicable to mechanical, thermal, or other classes of loads as well. Conversely, the chance of component failure is decreased if the capacity or strength of the component is increased.

On the right of the bathtub curve is a region of increasing failure rates. During this period of time *aging failures* become dominant. Again, with an obvious analogy to the loss of bone mass, arterial hardening, and other aging effects found in human populations, the failures tend to be dominated by cumulative effects such as corrosion, embrittlement, fatigue cracking, and diffusion of materials. The onset of rapidly increasing failure rates normally forms the basis for determining when parts should be replaced and for speci-

---

* Also referred to as burnin or runin depending on the device under consideration.

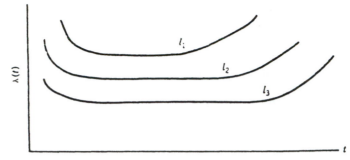

FIGURE 6.2 Time-dependent failure rates at different levels of loading: $l_1 > l_2 > l_3$.

fying the system's design life. Design with more durable components and materials, inspection and preventive maintenance, and control of deleterious environmental stresses are a few of the approaches in the enduring battle to produce longer-lived products. In the Taguchi methodology the causes of deterioration are referred to as "inner noise."

Although Fig. 6.1 displays the general features present in failure rate curves for many types of devices, one of the three mechanisms may be predominant for a particular class of system. Examples of such curves are given in Fig. 6.3. The curve in Fig. 6.3a is representative of much computer and other electronic hardware. In particular, after a rather inconspicuous wearin period, there is a long span of time over which the failure rate is essentially constant. For systems of this type, the primary concerns are with random failures, and with methods for controlling the environment and external loading to minimize their occurrence.

The failure rate curve in Fig. 6.3b is typical of valves, pumps, engines, and other pieces of equipment that are primarily mechanical in nature. Their initial wearin period is followed by a long span of time with a monotonically increasing failure rate. In these systems, for which the primary failure mechanisms are fatigue, corrosion, and other cumulative effects, the central concern is in estimating safe and economical operating lives, and in determining prudent schedules for preventive maintenance and for replacing parts.

Thus far we have not discussed the reliability consequences of logical errors or oversights committed in the design of complex systems. These, for example, may take the form of circuitry errors imbedded in microprocessor

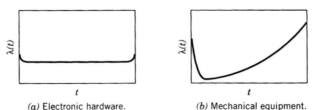

(a) Electronic hardware.   (b) Mechanical equipment.

FIGURE 6.3 Representative failure rates for different classes of systems.

chips, bugs in computer software, or even equation mistakes in engineering reference books. Prototypes normally undergo extensive testing to find and eliminate such errors before a product is put into production. Nevertheless, it may be impossible—or at least impractical—to test a device against all possible combinations of inputs to assure that the correct output is produced in every case. Thus there may exist untested sets of inputs that will cause the system to malfunction. In general, the resulting malfunctions may be expected to occur randomly in time, contributing to the time-independent component of the failure rate curve.

There is sometimes confusion with regard to failure rate definitions for computer software. This results from the common practice of finding and correcting bugs after, as well as before, the software is released for use. Such bugs tend to occur less and less frequently, giving rise to the notion of a decreasing failure rate. But that is not a failure rate in the sense in which it is defined here. In debugging, the software design is modified after each failure, whereas the definition used here is only valid for a product of fixed design. Hardware and software reliability growth attributable to test-fix debugging processes is taken up in Chapter 8.

In the following sections models for representing failure rates with one, or at most a few parameters, are discussed. These are particularly useful when most of the failures are caused by early failures, by random events, or by aging effects. Even when more than one mechanism contributes substantially to the failure rate curve, however, these models can often be used to represent the combined failure modes and their interactions.

6.3   CONSTANT FAILURE RATE MODEL

Random failures that give rise to the constant failure rate model are the most widely used basis for describing reliability phenomena. They are defined by the assumption that the rate at which the system fails is independent of its age. For continuously operating systems this implies a constant failure rate, whereas for demand failures it requires that the failure probability per demand be independent of the number of demands.

The constant failure rate approximation is often quite adequate even though a system or some of its components may exhibit moderate early failures or aging effects. The magnitude of early-failure effects is limited by strict quality control in manufacture and installation and may be further reduced by a wearin period before actual operations are begun. Similarly, in many systems aging effects can be sharply limited by careful preventive maintenance, with timely replacement of the parts or components in which the wear effects are concentrated. Conversely, if components are replaced as they fail, the overall failure rate of a many-component system will appear nearly constant, for the failure of the components will be randomly distributed in time as will the ages of the replacement parts. Finally, even though the system's failure rate may vary in time, we can use a constant failure rate that envelops the curve; this rate will be moderately pessimistic.

In the following sections we first consider the exponential distribution. It is employed when constant failure rates adequately describe the behavior of continuously operating systems. We then examine two demand failure models, one in which the demands take place at equal time intervals and the other in which the demands are randomly distributed in time. Both may be represented as constant failure rates. Finally, we formulate a composite model to describe the behavior of intermittently operating systems that may be subject to both operating and demand modes of failure.

The Exponential Distribution

The constant failure rate model for continuously operating systems leads to an exponential distribution. Replacing the time-dependent failure rate $\lambda(t)$ by a constant λ in Eq. 6.19 yields, for the PDF,

$$f(t) = \lambda e^{-\lambda t}. \tag{6.23}$$

Similarly, the CDF becomes

$$F(t) = 1 - e^{-\lambda t}, \tag{6.24}$$

and from Eq. 6.18 the reliability may be written as

$$R(t) = e^{-\lambda t}. \tag{6.25}$$

Plots of $f(t)$, $R(t)$, and $\lambda(t)$ (the failure rate) are given in Fig. 6.4. With the constant failure rate model, the resulting distributions are described in terms of a single parameter, λ. The MTTF and the variance of the failure times are also given in terms of λ. From Eq. 6.22 we obtain

$$\text{MTTF} = 1/\lambda, \tag{6.26}$$

and the variance is found from Eq. 3.16 to be

$$\sigma^2 = 1/\lambda^2. \tag{6.27}$$

A device described by a constant failure rate, and therefore by an exponential distribution of times to failure, has the following property of "memoryless-ness": The probability that it will fail during some period of time in the future is independent of its age. This is easily demonstrated by the following example.

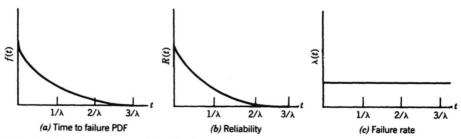

FIGURE 6.4 The exponential distribution.

EXAMPLE 6.2

A device has a constant failure rate of $\lambda = 0.02/\text{hr}$.

(*a*) What is the probability that it will fail during the first 10 hr of operation?

(*b*) Suppose that the device has been successfully operated for 100 hr. What is the probability that it will fail during the next 10 hr of operation?

Solution (*a*) The probability of failure within the first 10 hr is

$$P\{t \leq 10\} = \int_0^{10} f(t)\, dt = F(10) = 1 - e^{-0.02 \times 10} = 0.181.$$

(*b*) From Eq. 2.5, the conditional probability is

$$P\{t \leq 100 \,|\, t > 100\} = \frac{P\{(t \leq 110) \cap (t > 100)\}}{P\{t > 100\}} = \frac{P\{100 \leq t \leq 100\}}{P\{t > 100\}}$$

$$= \int_{100}^{110} \frac{f(t)\, dt}{1 - F(100)}$$

$$= \int_{100}^{110} \frac{0.02 e^{-0.02t}\, dt}{1 - 1 + \exp(-0.02 \times 100)}$$

$$= \frac{\exp(-0.02 \times 100) - \exp(-0.02 \times 110)}{\exp(-0.02 \times 100)}$$

$$= 1 - \exp(-0.02 \times 10) = 0.181.$$

That the probability of failure within a specified time interval is independent of the age of the device should not be surprising. Random failures are normally those caused by external shocks to the device; therefore, they should not depend on past history. For example, the probability that a satellite will fail during the next month owing to meteor impact would not depend on how long the satellite had already been in orbit. It would depend only on the frequency with which meteors pass through the orbit.

Demand Failures

The constant failure rate model has thus far been derived for a continuously operating system. It may also be shown to be applicable to a system exposed to a series of demands or shocks, each one of which has a small probability of causing failure. Suppose that each time a demand is made on a system, the probability of survival is r, giving a corresponding probability of failure of

$$p = 1 - r. \tag{6.28}$$

The term demand here is quite general; it may be the switching of an electric relay, the opening of a valve, the start of an engine, or even the stress on a bridge as a truck passes over it. Whatever the application, there are two salient points. First, we must be able to count or at least infer the number of demands;

and second, the probability of surviving each demand must be independent of the number of previous demands.

We define the reliability R_n as the probability that the system will still be operational after n demands. Let X_n signify the event of success in the nth demand. Then, if the probabilities of surviving each demand are mutually independent, R_n is given by Eq. 2.13 as

$$R_n = P\{X_1\}P\{X_2\}P\{X_3\} \ldots P\{X_n\}, \tag{6.29}$$

or since $P\{X_n\} = r$ for all n,

$$R_n = r^n. \tag{6.30}$$

Then, using Eq. 6.28, we obtain

$$R_n = (1 - p)^n. \tag{6.31}$$

We may put this result in a more useful approximate form. First, note that the exponential of

$$\ln R_n = \ln(1 - p)^n = \ln(1 - p) \tag{6.32}$$

is

$$R_n = \exp[n \ln(1 - p)]. \tag{6.33}$$

If the probability for failure on demand is small, we may make the approximation

$$\ln(1 - p) \approx -p \tag{6.34}$$

for $p \ll 1$, yielding

$$R_n = e^{-np}. \tag{6.35}$$

Since $p \ll 1$ is often a good approximation, we see that the reliability decays exponentially with the number of demands. If the rate at which demands are made on the system is roughly constant, we may express the number of demands occurring before time t as

$$n = \gamma t, \tag{6.36}$$

where γ is the frequency at which demands arrive. Thus if they arrive at time intervals Δt we have $\gamma = 1/\Delta t$. We may then calculate the reliability $R(t)$, defined as the probability that the system will still be operational at time t, as

$$R(t) = e^{-\lambda t}. \tag{6.37}$$

where the failure rate λ is now given by

$$\lambda = \gamma p. \tag{6.38}$$

Equation 6.35 indicates that the exponential distribution arises for systems that are subjected to many independent shocks or demands, each of which creates only a small probability of failure. If we drop the assumption that the demands appear at equal time intervals Δt, and assume that the shocks arrive at random intervals, the same result is obtained without assuming that the probability p of failure per shock is small. Let γ represent the mean number of demands per unit time. Then

$$\mu = \gamma t \tag{6.39}$$

is the mean number of demands over a time interval t. If the demands appear randomly in time obeying a Poisson process, we may represent the probability that there will be n demands per unit time with the Poisson probability mass function given in Eq. 2.59:

$$f(n) = \frac{(\gamma t)^n}{n!} e^{-\gamma t}. \tag{6.40}$$

Since the reliability after n independent demands is just r^n, the reliability at time t will just be the expected value of r^n at t. Using Eq. 2.32 for the expected value we have

$$R(t) = \sum_{n=0}^{\infty} r^n f(n), \tag{6.41}$$

which yields in combination with Eq. 6.40:

$$R(t) = \sum_{n=0}^{\infty} \frac{(r\gamma t)^n}{n!} e^{-\gamma t}. \tag{6.42}$$

We next note that upon moving $e^{-\gamma t}$ outside the sum, we obtain a power series for $e^{r\gamma t}$. Thus the reliability simplifies to

$$R(t) = \exp[(r - 1)\gamma t], \tag{6.43}$$

and upon inserting Eq. 6.28 we again obtain

$$R(t) = e^{-\gamma p t}, \tag{6.44}$$

where the failure rate is given by Eq. 6.38.

EXAMPLE 6.3

A telecommunications leasing firm finds that during the one-year warrantee period, 6% of its telephones are returned at least once because they have been dropped and damaged. An extensive testing program earlier indicated that in only 20% of the drops should telephones be damaged. Assuming that the dropping of telephones in normal use is a Poisson process, what is the MTBD (mean time between drops)? If the telephones are redesigned so that only 4% of drops cause damage, what fraction of the phones will be returned with dropping damage at least once during the first year of service?

Solution (*a*) The fraction of telephones not returned is $R = e^{-\gamma p t}$ or $0.94 = e^{-\gamma \times 0.2 \times 1}$. Therefore

$$\gamma = \frac{1}{0.2 \times 1} \ln \left(\frac{1}{0.94} \right) = 0.3094/\text{year},$$

$$\text{MTBD} = \frac{1}{\gamma} = 3.23 \text{ year.}$$

(*b*) For the improved design $R = e^{-\gamma pt} = e^{-0.3094 \times 0.04 \times 1} = 0.9877$. Therefore the fraction of the phones returned at least once is

$$1 - 0.9877 = 1.23\%.$$

Time Determinations

Careful attention must be given to the determination of appropriate time units. Is it operating time or calendar time? A warrantee of 100,000 miles or ten years, for example, includes both, since the 100,000 miles is converted to an equivalent operating time. Two failure rates are then relevant, one for when the vehicle is operating, and another presumably smaller one for when it is not. A third consideration is the number of start-stop cycles that the vehicle is likely to undergo, for the related stress and thermal cycling may aggravate some failure mechanisms. Whatever the situation, we must clearly state what measure of time is being used. If the reliability is to be expressed in calendar time rather than operating time the duty cycle or capacity factor c, defined as the fraction of time that the engine is running, must also enter the calculations.

Consider as an example a refrigerator motor that runs some fraction c of the time; the failure rate is λ_0 per unit operating time. The contribution to the total failure rate from failures while the refrigerator is operating will then be $c\lambda_0$ per unit calendar time. If the demand failure is also to be taken into account, we must know how many times the motor is turned on. Suppose that the average length of time that the motor runs when it comes on is $\bar{t}_0$. Then the average number of times that the motor is turned on per unit operating time is $1/\bar{t}_0$. The average number of times that it is turned on per unit calendar time is $m = c/\bar{t}_0$. To obtain the total failure rate, we add the demand and operating failure rates. Consequently, the composite failure rate to be used in Eqs. 6.23 through 6.27 is

$$\lambda = \frac{c}{\bar{t}_0} p + c\lambda_0. \tag{6.45}$$

In the foregoing development we have neglected the possibility that the motor may fail while it is not operating, that is, while it is in a standby mode. Often such failure rates are small enough to be neglected. However, for systems that are operated only a small fraction of the time, such as an emergency generator, failure in the standby mode may be quite significant. To take this into account, we define λ_s as the failure rate in the standby mode. Since the system in our example is in the standby mode for a fraction $1 - c$ of the time, we add a contribution of $(1 - c)\lambda_s$ to the composite failure rate in Eq. 6.45:

$$\lambda = \frac{c}{\bar{t}_0} p + c\lambda_0 + (1 - c)\lambda_s. \tag{6.46}$$

EXAMPLE 6.4

A pump on a volume control system at a chemical process plant operates intermittently. The pump has an operating failure rate of 0.0004/hr and a standby failure rate of

0.00001/hr. The probability of failure on demand is 0.0005. The times at which the pump is turned on t_u and turned off t_d over a 24-hr period are listed in the following table.

t_u	0.78	1.69	2.89	3.92	4.71	5.97	6.84	7.76
t_d	1.02	2.11	3.07	4.21	5.08	6.31	7.23	8.12
t_u	8.91	9.81	10.81	11.87	12.98	13.81	14.87	15.97
t_d	9.14	10.08	11.02	12.14	13.18	14.06	15.19	16.09
t_u	16.69	17.71	18.61	19.61	20.56	21.49	22.58	23.61
t_d	16.98	18.04	19.01	19.97	20.91	21.86	22.79	23.89

Assuming that these data are representative, (*a*) Calculate a composite failure rate for the pump under these operating conditions. (*b*) What is the probability of the pump's failing during any 1-month (30-day) period?

Solution (*a*) From the data given we first calculate

$$\sum_{i=1}^{M} t_{di} = 301.50 \quad \text{and} \quad \sum_{i=1}^{M} t_{ui} = 294.36,$$

where $M = 24$ is the number of operations. The average operating time $\bar{t}_0$ of the pump is estimated for the data to be

$$\bar{t}_0 = \frac{1}{M} \sum_{i=1}^{M} (t_{di} - t_{ui}) = \frac{1}{M} \left(\sum_{i=1}^{M} t_{di} - \sum_{i=1}^{M} t_{ui} \right)$$

$$= \frac{1}{24} (301.50 - 294.36) = 0.2975 \text{ hr.}$$

Then the capacity factor is

$$c = \frac{M\bar{t}_0}{24} = \frac{24 \times 0.2975}{24} = 0.2975.$$

Thus the failure rate from Eq. 6.46 is

$$\lambda = \frac{0.2975}{0.2975} \times 0.0005 + 0.2975 \times 0.0004 + (1 - 0.2975) \times 0.00001$$

$$= 6.26 \times 10^{-4} \text{ hr}^{-1}.$$

(*b*) The reliability is

$$R = \exp(-\lambda \times 24 \times 30) = \exp(-0.4507) = 0.637,$$

yielding a 30-day failure probability of

$$1 - R = 0.363.$$

6.4 TIME-DEPENDENT FAILURE RATES

A variety of situations in which the explicit treatment of early failures or aging effects, or both, require the use of time-dependent failure rate models. This may be illustrated by considering the effect of the accumulated operating time T_0 on the probability that a device can survive for an additional time t. Suppose that we define $R(t \mid T_0)$ as the reliability of a device that has previously been operated for a time T_0. We may therefore write

$$R(t \mid T_0) = P\{t' > T_0 + t \mid t' > T_0\}, \tag{6.47}$$

where $t' = T_0 + t$ is the time elapsed at failure since the device was new. From the definition given in Eq. 2.5, we may write the conditional probability as

$$P\{t' > T_0 + t \mid t' > T_0\} = \frac{P\{(t' > T_0 + t) \cap (t' > T_0)\}}{P\{t' > T_0 + t\}}. \tag{6.48}$$

However, since $(t' > T_0 + t) \cap (t' > T_0) = t' > T_0 + t$, we may combine equations to obtain

$$R(t \mid T_0) = \frac{P\{t' > T_0 + t\}}{P\{t' > T_0\}}. \tag{6.49}$$

The reliability of a new device is then just

$$R(t) = R(t \mid T_0 = 0) = P\{t' > t\}, \tag{6.50}$$

and we obtain

$$R(t \mid T_0) = \frac{R(t + T_0)}{R(T_0)}. \tag{6.51}$$

Finally, using Eq. 6.18, we obtain

$$R(t \mid T_0) = \exp\left[-\int_{T_0}^{t+T_0} \lambda(t') \, dt' \right]. \tag{6.52}$$

The significance of this result may be interpreted as follows. Suppose that we view T_0 as a wearin time undergone by a device before being put into service, and t as the service time. Now we ask whether the wearin time decreases or increases the service life reliability of the device. To determine this, we take the derivative of $R(t \mid T_0)$ with respect to the wearin period and obtain

$$\frac{\partial}{\partial T_0} R(t \mid T_0) = -[\lambda(T_0) - \lambda(T_0 + t)] R(t \mid T_0). \tag{6.53}$$

Increasing the wearin period thus improves the reliability of the device only if the failure rate is decreasing [i.e., $\lambda(T_0) > \lambda(T_0 + t)$]. If the failure rate increases with time, wearin only adds to the deterioration of the device, and the service life reliability decreases.

To model early failures or wear effects more explicitly, we must turn to specific distributions of the time to failure. In contrast to the exponential distribution used for random failures, these distributions must have at least two parameters. Although the normal and lognormal distributions are frequently used to model aging effects, the Weibull distribution is probably the most universally employed. With it we may model early failures and random failures as well as aging effects.

The Normal Distribution

To describe the time dependence of reliability problems, we write the PDF for the normal distribution given by Eq. 3.38 with t as the random variable,

$$f(t) = \frac{1}{\sqrt{2\pi}\sigma} \exp\left[-\frac{(t - \mu)^2}{2\sigma^2} \right],$$ (6.54)

where μ is now the MTTF. The corresponding CDF is

$$F(t) = \int_{-\infty}^{t} \frac{1}{\sqrt{2\pi}\sigma} \exp\left[-\frac{(t' - \mu)^2}{2\sigma^2} \right] dt',$$ (6.55)

or in standardized normal form,

$$F(t) = \Phi\left(\frac{t - \mu}{\sigma} \right).$$ (6.56)

From Eq. 6.4 the reliability for the normal distribution is found to be

$$R(t) = 1 - \Phi\left(\frac{t - \mu}{\sigma} \right),$$ (6.57)

and the associated failure rate is obtained by substituting this expression into Eq. 6.14:

$$\lambda(t) = \frac{1}{\sqrt{2\pi}\sigma} \exp\left[-\frac{(t - \mu)^2}{2\sigma^2} \right]\left[1 - \Phi\left(\frac{t - \mu}{\sigma} \right) \right]^{-1}.$$ (6.58)

The failure rate along with the reliability and the PDF for times to failure are plotted in Fig. 6.5. As indicated by the behavior of the failure rate, normal distributions are used to describe the reliability of equipment that is quite different from that to which constant failure rates are applicable. It is useful in describing reliability in situations in which there is a reasonably well-defined wearout time, μ. This may be the case, for example, in describing the life of a tread on a tire or the cutting edge on a machine tool. In these situations the life may be given as a mean value and an uncertainty. When normal distribution is used, the uncertainty in the life is measured in terms of intervals

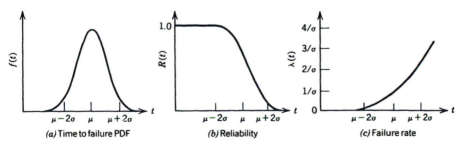

FIGURE 6.5 The normal distribution.

in time. For instance, if we say that there is a 90% probability that the life will fail between, $\mu - \Delta t$ and $\mu + \Delta t$, then

$$P\{\mu - \Delta t \leq t \leq \mu + \Delta t\} = 0.9. \tag{6.59}$$

If the times to failures are normally distributed, it is equally probable that the failure will take place before $\mu - \Delta t$ or after $\mu + \Delta t$. Moreover, we can determine the failure distribution time from the standardized curve. Equation 6.59 implies that

$$\Delta t = 1.645\sigma. \tag{6.60}$$

Therefore, σ can be determined. The corresponding values for several other probabilities are given in Table 6.1. Once μ and σ are known, the reliability can be determined as a function of time from Eq. 6.57.

EXAMPLE 6.5

A tire manufacturer estimates that there is a 90% probability that his tires will wear out between 25,000 and 35,000 miles. Assuming a normal distribution, find μ and σ.

Solution   Assume that 5% of failures are at fewer than 25×10^3 miles and 5% at more than 35×10^3 miles:

$$\Phi(z_1) = 0.05, z_1 = \frac{25 - \mu}{\sigma}, \Phi(z_2) = 0.95, z_2 = \frac{35 - \mu}{\sigma}.$$

From Appendix C, $z_1 = -1.65$, $z_2 = +1.65$. Hence

$$-1.65\sigma = 25 - \mu, + 1.65\sigma = 35 - \mu,$$

and the solutions are $\mu = 30$ thousand miles, $\sigma = 3.03$ thousand miles.

The Lognormal Distribution

As we have indicated, the normal distribution is particularly useful for describing aging when we can specify a time to failure along with an uncertainty, Δt. The lognormal is a related distribution that has been found to be useful in

TABLE 6.1 Confidence Intervals for a Normal Distribution

Standard deviations	Confidence interval, %
$\pm 0.5\sigma$	0.3830
$\pm 1.0\sigma$	0.6826
$\pm 1.5\sigma$	0.8664
$\pm 2.0\sigma$	0.9544
$\pm 2.5\sigma$	0.9876
$\pm 3.0\sigma$	0.9974

describing failure distributions for a variety of situations. It is particularly appropriate under the following set of circumstances. If the time to failure is associated with a large uncertainty, so that, for example, the variance of the distribution is a large fraction of the MTTF, the use of the normal distribution is problematical. However, it still may be possible to state a failure time and to estimate with it the probability that the time to failure lies within some factor, say n, of this value. For example, if it is known that 90% of the failures are within a factor of n of some time t_0,

$$P\left\{\frac{t_0}{n} \le t \le nt_0\right\} = 0.9. \tag{6.61}$$

As indicated in Chapter 3, the lognormal distribution describes such situations. The PDF for the time to failure is then

$$f(t) = \frac{1}{\sqrt{2\pi}\omega t} \exp\left\{-\frac{1}{2\omega^2}\left[\ln\left(\frac{t}{t_0}\right)\right]^2\right\}, \tag{6.62}$$

and the corresponding CDF

$$F(t) = \Phi\left[\frac{1}{\omega}\ln(t/t_0)\right]. \tag{6.63}$$

Now, however, t_0 is not the MTTF; rather, they are related as indicated in Chapter 3, by

$$\text{MTTF} = \mu = t_0 \exp(\omega^2/2). \tag{6.64}$$

Similarly, the variance of $f(t)$ is not equal to ω^2, but rather to

$$\sigma^2 = t_0^2 \exp(\omega^2)[\exp(\omega^2) - 1]. \tag{6.65}$$

When the time to failure is known to within a factor of n, t_0 and ω may be determined as follows. If it is assumed that 90% of the failures occur between $t_- = t_0/n$ and $t_- = t_0/n$, then t_0 is the geometric mean,

$$t_0 = [t_- \times t_+]^{1/2} \tag{6.66}$$

and

$$\omega = \frac{1}{1.645}\ln n. \tag{6.67}$$

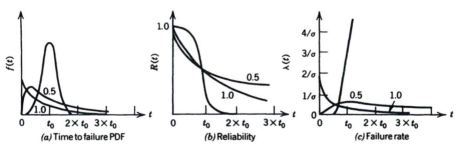

FIGURE 6.6 The lognormal distribution.

The PDF for the time to failure, reliability, and failure rate $\lambda(t)$ for the lognormal distribution are plotted in Fig. 6.6. Note that the failure rate can be increasing or decreasing depending on the value of ω. The lognormal distribution is frequently used to describe fatigue and other phenomena caused by aging or wear and results in failure rates that increase with time.

EXAMPLE 6.6

It is known that 90% of the truck axles of a particular type will suffer fatigue failure between 120,000 and 180,000 miles. Assuming that the failures may be fit to a lognormal distribution,

(a) To what factor n is the fatigue life known with 90 percent confidence?
(b) What are the parameters t_0 and ω of the lognormal distribution?
(c) What is the MTTF?

Solution   (a) For 90% certainty, $t_0 n = 180$ and $t_0/n = 120$. Taking the quotients of these equations yields

$$n^2 = \frac{180}{120}$$

$$n = 1.2247.$$

(b) Taking the products of $t_0 n$ and t_0/n, we have

$$t_0^2 = 180 \times 120$$

$$t_0 = 146.97 \times 10^3 \text{ miles.}$$

For 90% confidence Eq. 6.67 gives

$$\omega = \frac{1}{1.645} \ln n = \frac{\ln(1.2247)}{1.645} = 0.1232.$$

(c) From Eq. 6.64,

$$\text{MTTF} = 146.97 \times \exp(\tfrac{1}{2} \times 0.1232^2) = 148.09 \times 10^3 \text{ miles.}$$

The Weibull Distribution

The Weibull distribution is one of the most widely used in reliability calculations, for with an appropriate choice of parameters a variety of failure rate behaviors can be modeled. These include, as a special case, the constant failure rate, in addition to failure rates modeling both wearin and wearout phenomena. The Weibull distribution may be formulated in either a two- or a three-parameter form. We treat the two-parameter form first.

The two-parameter Weibull distribution, introduced in Chapter 3, assumes that the failure rate is in the form of a power law:

$$\lambda(t) = \frac{m}{\theta}\left(\frac{t}{\theta}\right)^{m-1}. \tag{6.68}$$

From this failure rate we may use Eq. 6.19 to obtain the PDF:

$$f(t) = \frac{m}{\theta}\left(\frac{t}{\theta}\right)^{m-1}\exp\left[-\left(\frac{t}{\theta}\right)^{m}\right]. \tag{6.69}$$

Then, integrating over the time variable from zero to t, we obtain the CDF to be

$$F(t) = 1 - \exp[-(t/\theta)^{m}] \tag{6.70}$$

and since $R = 1 - F$, the reliability is

$$R(t) = \exp[-(t/\theta)^{m}]. \tag{6.71}$$

The mean and the variance of the Weibull distribution may be shown to be

$$\mu = \theta\Gamma(1 + 1/m) \tag{6.72}$$

and

$$\sigma^2 = \theta^2[\Gamma(1 + 2/m) - \Gamma(1 + 1/m)^2]. \tag{6.73}$$

In these expressions the complete gamma function $\Gamma(v)$ is given by the integral of Eq. 3.78 where a graph is also provided.

Figure 6.7 shows the properties of $\lambda(t)$, $f(t)$ and $R(t)$ for a number of values of m. From these figures and the foregoing equations it is clear that the Weibull distribution provides a good deal of flexibility in fitting failure rate data. When $m = 1$, the exponential distribution corresponding to a constant failure rate is obtained. For values of $m < 1$ failure rates are typical of wearin phenomena decrease, and for $m > 1$ failure rates are typical of aging effects and increase. Finally, as m becomes large, say $m > 4$, a normal PDF is approximated.

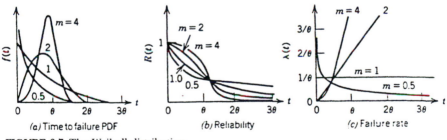

(a) Time to failure PDF   (b) Reliability   (c) Failure rate

FIGURE 6.7 The Weibull distribution.

EXAMPLE 6.7

A device has a decreasing failure rate characterized by a two-parameter Weibull distribution with $\theta = 180$ years and $m = \frac{1}{2}$. The device is required to have a design-life reliability of 0.90.

(*a*) What is the design life if there is no wearin period?

(*b*) What is the design life if the device is first subject to a wearin period of one month?

Solution   (*a*) $R(T) = \exp[-(T/\theta)^m]$. Therefore, $T = \theta\{\ln[1/R(T)]\}^{1/m}$. Then

$$T = 180[\ln(1/0.9)]^2 = 2.00 \text{ years.}$$

(*b*) The reliability with wearin time T_0 is given by Eq. 6.51. With the Weibull distribution it becomes

$$R(t\,|\,T_0) = \frac{\exp\left[-\left(\dfrac{t+T_0}{\theta}\right)^m\right]}{\exp\left[-\left(\dfrac{T_0}{\theta}\right)^m\right]}.$$

Setting $t = T$, the design life, we solve for T,

$$T = \theta\left\{\ln\left[\frac{1}{R(T)}\right] + \left(\frac{T_0}{\theta}\right)^m\right\}^{1/m} - T_0$$

$$= 180\left[\ln\left(\frac{1}{0.9}\right) + \left(\frac{1}{12 \times 180}\right)^{1/2}\right]^2 - \frac{1}{12}$$

$$= 2.81 \text{ years.}$$

Thus a wearin period of 1 month adds nearly 10 months to the design life.

The three-parameter Weibull distribution is useful in describing phenomena for which some threshold time must elapse before there can be failures. To obtain this distribution, we simply translate the origin to the right by an amount t_0 on the time axis. Thus we have

$$\lambda(t) = \begin{cases} 0, & t < t_0 \\ \dfrac{m}{\theta}\left(\dfrac{t - t_0}{\theta}\right)^{m-1} & t \geq t_0 \end{cases},$$

$$f(t) = \begin{cases} 0, & t < t_0 \\ \dfrac{m}{\theta}\left(\dfrac{t - t_0}{\theta}\right)^{m-1} \exp\left[-\left(\dfrac{t - t_0}{\theta}\right)^m\right] & t \geq t_0 \end{cases}, \qquad (6.74)$$

$$F(t) = \begin{cases} 0, & t < t_0 \\ 1 - \exp\left[-\left(\dfrac{t - t_0}{\theta}\right)^m\right] & t \geq t_0 \end{cases}.$$

The variance is the same as for the two-parameter distribution given in Eq. 6.73, and the mean is obtained simply by adding t_0 the right-hand side of Eq. 6.72.

6.5 COMPONENT FAILURES AND FAILURE MODES

In Sections 6.3 and 6.4 the quantitative behavior of reliability is modeled for situations with constant and time-dependent failure rates, respectively. In real systems, however, failures occur through a number of different mechanisms, causing the failure rate curve to take a bathtub shape too complex to be described by any single one of the distributions discussed thus far. The mechanisms may be physical phenomena within a single monolithic structure, such as the tread wear, puncture, and defective sidewalls in an automobile tire. Or physically distinct components of a system, such as the processor unit, disk drives, and memory of a computer may fail. In either case it is usually possible to separate the failures according to the mechanism or the components that caused them. It is then possible, provided that the failures are independent, to generalize and treat the system reliability in terms of mechanisms or component failures. We refer to these collectively as independent failure modes.

Failure Mode Rates

Whether we refer to component failure or failure modes—and the distinction is sometimes blurred—we may analyze the reliability of a system in terms of the component or mode failures provided they are independent of one another. Independence requires that the probability of failure of any mode is not influence by that of any other mode. The reliability of a system with M different failure modes is

$$R(t) = P\{X_1 \cap X_2 \cap \ldots \cap X_M\}, \qquad (6.75)$$

where X_i, is the event in which the ith failure mode does *not* occur before time t. If the modes are independent we may write the system reliability as the product of the mode survival probabilities:

$$R(t) = P\{X_1\}P\{X_2\} \cdots P\{X_M\}. \qquad (6.76)$$

where the mode i reliability is

$$R_i(t) = P\{X_i\}, \qquad (6.77)$$

yielding

$$R(t) = \prod_i R_i(t). \qquad (6.78)$$

Naturally, if mode i is the failure of component i, then $R_i(t)$ is just the component reliability.

For each mode we may define a PDF for time to failure, $f_i(t)$, and an associated failure rate, $\lambda_i(t)$. The derivation is exactly the same as in Section 6.2 yielding

$$R_i(t) = 1 - \int_0^t f_i(t') \, dt', \qquad (6.79)$$

$$\lambda_i(t) = \frac{f_i(t)}{R_i(t)}, \tag{6.80}$$

$$R_i(t) = \exp\left[-\int_0^t \lambda_i(t')\,dt'\right] \tag{6.81}$$

and

$$f_i(t) = \lambda_i(t)\exp\left[-\int_0^t \lambda_i(t')\,dt'\right]. \tag{6.82}$$

Combining Eq. 6.76 and 6.77 with Eq. 6.81 then yields:

$$R(t) = \exp\left[-\int_0^t \lambda(t')\,dt'\right], \tag{6.83}$$

where

$$\lambda(t) = \sum_i \lambda_i(t). \tag{6.84}$$

Thus, to obtain the system reliability, we simply add the mode failure rates.

Consider a system with a failure rate that results from the contributions of independent modes. Suppose some modes are associated with failure rates that decrease with time, while the failure rates of others are either constant or increase with time. Weibull distributions are particularly useful for modeling such modes. If we write

$$\int_0^t \lambda(t')\,dt' = \left(\frac{t}{\theta_a}\right)^{m_a} + \left(\frac{t}{\theta_b}\right)^{m_b} + \left(\frac{t}{\theta_c}\right)^{m_c} \tag{6.85}$$

and take $0 < m_a < 1$, $m_b = 1$, and $m_c > 1$, the three terms correspond, respectively, to contributions to the failure-rate contributions that decrease, remain flat, and increase with time. These are associated with early failures, random failures, and wear failures, respectively. Thus the shape of the bathtub curve can be expressed as a superposition of Weibull failure rates. It is not valid to think of these individual terms as arising from Eqs. 6.78 through 6.84 unless each of them results from independent failure modes or the failures of different components. When they arise as the result of a single cause, the contributions from infant mortality, random and aging effects are strongly interactive. In these cases Eq. 6.85 may be a useful empirical representation of the failure rate curve so long as the individual terms are not identified uniquely with infant mortality, random, or aging failures. We shall consider the interactions which give rise to the bathtub curve in more detail in Chapter 7, where they are related to loading and capacity.

For situations in which independent failure modes may be approximated by constant failure rates, $\lambda_i(t) \to \lambda_i$, the reliability is given by Eq. 6.25 with

$$\lambda = \sum_i \lambda_i, \tag{6.86}$$

and Eq. 6.26 may be used to determine the system's mean time to failure. If we define the mode mean time to failure as

$$\text{MTTF}_i = 1/\lambda_i, \tag{6.87}$$

the system mean time to failure is related by

$$\frac{1}{\text{MTTF}} = \sum_i \frac{1}{\text{MTTF}_i}. \tag{6.88}$$

Component Counts

The ability to add failure rates is most widely applied in situations in which each failure mode corresponds to a component or part failure. Often, failure rate data may be available at a component level but not for an entire system. This is true, in part, because several professional organizations collect and publish failure rate estimates for frequently used items, whether they be diodes, switches, and other electrical components; pumps, valves, and similar mechanical devices; or a number of other types of components. At the same time the design of a new system may involve new configurations and numbers of such standard items. The foregoing equations then allow reliability estimates to be made before the new design is built and tested. In this chapter we consider only systems without redundancy. Consequently, failure of any component implies system failure. In systems with redundant components, the idea of a failure mode is still applicable in a more general sense. We reserve the treatment of such systems to Chapter 9.

When component failure rates are available, the most straightforward, but crudest, estimate of reliability comes from the parts count method. We simply count the number n_j of parts of type j in the system. The system's failure rate is then

$$\lambda = \sum_j n_j \lambda_j \tag{6.89}$$

where the sum is over the part types in the system.

EXAMPLE 6.8

A computer-interface circuit card assembly for airborne application is made up of interconnected components in the quantities listed in the first column of Table 6.2. If the assembly must operate in a 50°C environment, the component failure rates are given in column 2 of Table 6.2. Calculate

(a) the assembly failure rate,

(b) the reliability for a 12-hr mission, and

(c) the MTTF.

Solution (a) We have calculated the total failure rate $n_j \lambda_j$ for each component type with Eq. 6.89 and listed them in the third column of Table 6.2. For a nonredundant system the assembly failure rate is just the sum of these numbers, or, as indicated, $\lambda = 21.6720 \times 10^{-6}/\text{hr}$.

(b) The 12-hr reliability is calculated from $R = e^{-\lambda t}$ to be

$$R(12) = \exp(-21.672 \times 12 \times 10^{-6}) = 0.9997.$$

(c) For constant failure rates the MTTF is

$$\text{MTTF} = \frac{1}{\lambda} = \frac{10^6}{21.672} = 46{,}142 \text{ hr.}$$

TABLE 6.2 Components and Failure Rates for Computer Circuit Card*

Component type	Quantity	Failure rate/10^6 hr	Total failure rate/10^6 hr
Capacitor tantalum	1	0.0027	0.0027
Capacitor ceramic	19	0.0025	0.0475
Resistor	5	0.0002	0.0010
J—K, M—S flip flop	9	0.4667	4.2003
Triple Nand gate	5	0.2456	1.2286
Diff line receiver	3	0.2738	0.8214
Diff line driver	1	0.3196	0.3196
Dual Nand gate	2	0.2107	0.4214
Quad Nand gate	7	0.2738	1.9166
Hex invertor	5	0.3196	1.5980
8-bit shift register	4	0.8847	3.5388
Quad Nand buffer	1	0.2738	0.2738
4-bit shirt register	1	0.8035	0.8035
And-or-inverter	1	0.3196	0.3196
PCB connector	1	4.3490	4.3490
Printed wiring board	1	1.5870	1.5870
Soldering connections	1	0.2328	0.2328
Total	67		21.6720 ◀

* Reprinted from 'Mathematical Modelling' by A. H. K. Ling, *Reliability and Maintainability of Electronic Systems*, edited by Arsenault and Roberts with the permission of the publisher Computer Science Press, Inc., 1803 Research Boulevard, Rockville, Maryland 20850, USA.

The parts count method, of course, is no better than the available failure rate data. Moreover, the failure rates must be appropriate to the particular conditions under which the components are to be employed. For electronic equipment, extensive computerized data bases have been developed that allow the designer to take into account the various factors of stress and environment, as well as the quality of manufacture. For military procurement such procedures have been formalized as the parts stress analysis method.

In parts stress analysis each component failure rate, λ_i, is expressed as a base failure rate, λ_b, and as a series of multiplicative correction factors:

$$\lambda_i = \lambda_b \Pi_E \Pi_Q \ldots \Pi_N \tag{6.90}$$

The base failure rate, λ_b, takes into account the temperature at which the component operates as well as the primary electrical stresses (i.e., voltage, current, or both) to which it is subjected. Figure 6.8 shows qualitatively the effects these variables might have on a particular component type.

The correction factors, indicated by the Πs in Eq. 6.90, take into account environmental, quality, and other variables that are designated as having a significant impact on the failure rate. For example, the environmental factor Π_E accounts for environmental stresses other than temperature; it is related to the vibration, humidity, and other conditions encountered in operation. For purposes of military procurement, there are 11 environmental categories, as listed in Table 6.3. For each component type there is a wide range of values of Π_E for example, for microelectronic devices Π_E ranges from 0.2 for "Ground, benign" to 10.0 for "Missile launch."

Similarly, the quality multiplier Π_Q takes into account the level of specification, and therefore the level of quality control under which the component has been produced and tested. Typically, $\Pi_Q = 1$ for the highest levels of specification and may increase to 100 or more for commercial parts procured under minimal specifications. Other multiplicative corrections also are used. These include Π_A the application factor to take into account stresses found in particular applications, and factors to take into account cyclic loading, system complexity, and a variety of other relevant variables.

6.6   REPLACEMENTS

Thus far we have considered the distribution of the failure times given that the system is new at $t = 0$. In many situations, however, failure does not constitute the end of life. Rather, the system is immediately replaced or repaired and operation continues. In such situations a number of new pieces of information became important. We may want to know the expected number

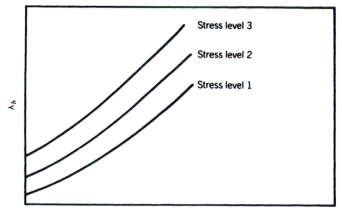

FIGURE 6.8 Failure rate versus temperature for different levels of applied stress (power, voltage, etc.).

TABLE 6.3 Environmental Symbol Identification and Description

Environment	π_ε symbol	Nominal environmental conditions	π_ε value[a]
Ground, benign	G_B	Nearly zero environmental stress with optimum engineering operation and maintenance.	0.2
Space, flight	S_F	Earth orbital. Approaches G_B conditions without access for maintenance. Vehicle neither under powered flight nor in atmospheric reentry.	0.2
Ground, fixed	G_F	Conditions less than ideal: installation in permanent racks with adequate cooling air, maintenance by military personnel, and possible installation in unheated buildings.	1.0
Ground, mobile (and portable)	G_M	Conditions less favorable than those for G_F, mostly through vibration and shock. The cooling air supply may be more limited and maintenance less uniform.	4.0
Naval, sheltered	N_S	Surface ship conditions similar to G_F but subject to occasional high levels of shock and vibration.	4.0
Naval, unsheltered	N_U	Nominal surface shipborne conditions but with repetitive high levels of shock and vibration.	5.0
Airborne, inhabited	A_I	Typical cockpit conditions without environmental extremes of pressure, temperature, shock and vibration.	4.0
Airborne, uninhabited	A_U	Bomb-bay, tail, or wing installations, where extreme pressure, temperature, and vibration cycling may be aggravated by contamination from oil, hydraulic fluid, and engine exhaust.	6.0
Missile, launch	M_L	Severe noise, vibration, and other stresses related to missile launch, boosting space vehicles into orbit, vehicle reentry, and landing by parachute. Conditions may also apply to installation near main rocket engines during launch operations.	10.0

[a] Values for monolithic microelectronic devices.

Source: From R. T. Anderson, *Reliability Design Handbook* RDH-376, Rome Air Development Center, Griffiss Air Force Base NY, 1976.

of failures over some specified period of time in order to estimate the costs of replacement parts. More important, it may be necessary to estimate the probability that more than a specific number of failures N will occur over a period of time. Such information allows us to maintain an adequate inventory of repair parts.

In modeling these situations, we restrict our attention to the constant failure rate approximation. In this the failure rate is often given in terms of the *mean time between failures* (MTBF), as opposed to the mean time to failure, or MTTF. In fact, they are both the same number if, when a system fails it is assumed to be repaired immediately to an as-good-as-new condition. In what follows we use the constant failure rate model to derive $p_n(t)$, the probability of there being n failures during a time interval of length t. The derivation leads again to the Poisson distribution introduced in Chapter 2. From it we can calculate numbers of failures and replacement requirements.

We first consider the times at which the failures take place, and therefore the number that occur within any given span of time. Suppose that we let **n** be a discrete random variable representing the number of failures that take place between $t = 0$ and a time t. Let

$$p_n(t) = P\{\mathbf{n} = n \,|\, t\} \tag{6.91}$$

be the probability that exactly n failures have taken place before time t. Clearly, if we start counting failures at time zero, we must have

$$p_0(0) = 1, \tag{6.92}$$

$$p_n(0) = 0, \qquad n = 1, 2, 3, \dots, \infty. \tag{6.93}$$

In addition, at any time

$$\sum_{n=0}^{\infty} p_n(t) = 1. \tag{6.94}$$

For small Δt, let failure $\lambda \, \Delta t$ be the probability that the $(n + 1)$th failure will take place during the time increment between t and $t + \Delta t$, given that exactly n failures have taken place before time t. Then the probability that no failure will occur during Δt is $1 - \lambda \, \Delta t$. From this we see that the probability that no failures have occurred before $t + \Delta t$ may be written as

$$p_0(t + \Delta t) = (1 - \lambda \, \Delta t) p_0(t). \tag{6.95}$$

Then noting that

$$\frac{d}{dt} p_n(t) = \lim_{\Delta t \to 0} \frac{p_n(t + \Delta t) - p_n(t)}{\Delta t}, \tag{6.96}$$

we obtain the simple differential equation

$$\frac{d}{dt} p_0(t) = -\lambda p_0(t). \tag{6.97}$$

Using the initial condition, Eq. 6.92, we find

$$p_0(t) = e^{-\lambda t}. \tag{6.98}$$

With $p_0(t)$ determined, we may now solve successively for $p_n(t)$, $n = 1$, 2, 3, ... in the following manner. We first observe that if n failures have taken place before time t, the probability that the $(n + 1)$th failure will take place between t and $t + \Delta t$ is $\lambda \, \Delta t$. Therefore, since this transition probability is independent of the number of previous failures, we may write

$$p_n(t + \Delta t) = \lambda \, \Delta t \, p_{n-1}(t) + (1 - \lambda \, \Delta t) p_n(t). \tag{6.99}$$

The last term accounts for the probability that no failure takes place during Δt. For sufficiently small Δt we can ignore the possibility of two or more failures taking place.

Using the definition of the derivative once again, we may reduce Eq. 6.99 to the differential equation

$$\frac{d}{dt} p_n(t) = -\lambda p_n(t) + \lambda p_{n-1}(t). \tag{6.100}$$

This equation allows us to solve for $p_n(t)$ in terms of $p_{n-1}(t)$. To do this we multiply both sides by the integrating factor $\exp(\lambda t)$. Then noting that

$$\frac{d}{dt} [e^{\lambda t} p_n(t)] = e^{\lambda t} \left[\frac{d}{dt} p_n(t) + \lambda p_n(t) \right], \tag{6.101}$$

we have

$$\frac{d}{dt} [e^{\lambda t} p_n(t)] = \lambda p_{n-1}(t) e^{\lambda t}. \tag{6.102}$$

Multiplying both sides by dt and integrating between 0 and t, we obtain

$$e^{\lambda t} p_n(t) - p_n(0) = \lambda \int_0^t p_{n-1}(t') e^{\lambda t'} dt'. \tag{6.103}$$

But, since from Eq. 6.93 $p_n(0) = 0$, we have

$$p_n(t) = \lambda e^{-\lambda t} \int_0^t p_{n-1}(t') e^{\lambda t'} dt'. \tag{6.104}$$

This recursive relationship allows us to calculate the p_n successively. For p_1, insert Eq. 6.98 on the right-hand side and carry out the integral to obtain

$$p_1(t) = \lambda t e^{-\lambda t}. \tag{6.105}$$

Repeating this procedure for $n = 2$ yields

$$p_2(t) = \frac{(\lambda t)^2}{2} e^{-\lambda t}, \tag{6.106}$$

and so on. It is easily shown that Eq. 6.104 is satisfied for all $n \geq 0$ by

$$p_n(t) = \frac{(\lambda t)^n}{n!} e^{-\lambda t}, \tag{6.107}$$

and these quantities in turn satisfy the initial conditions given by Eqs. 6.92 and 6.93.

The probabilities $p_n(t)$ are the same as the Poisson distribution $f(n)$, provided that we set $\mu = \lambda t$. We may therefore use Eqs. 2.27 through 2.29 to determine the mean and the variance of the number n of events occurring over a time span t. Thus the expected number of failures during time t is

$$\mu_n \equiv E\{n\} = \lambda t, \tag{6.108}$$

and the variance of n is

$$\sigma_n^2 = \lambda t. \tag{6.109}$$

Of course, since $p_n(t)$ are the probability mass functions of a discrete variable **n**, we must have, according to Eq. 2.22,

$$\sum_{n=0}^{\infty} p_n(t) = 1. \tag{6.110}$$

The number of failures can be related to the mean time between failures by

$$\mu_n = \frac{t}{\text{MTBF}}. \tag{6.111}$$

We have derived the expression relating μ_n and the MTBF assuming a constant failure rate. It has, however, much more general validity.* Although the proof is beyond the scope of this book, it may be shown that Eq. 6.111 is also valid for time-dependent failure rates in the limiting case that $t \gg \text{MTBF}$. Thus, in general, the MTBF may be determined from

$$\text{MTBF} = \frac{t}{n}, \tag{6.112}$$

where n, the number of failures, is large.

We may also require the probability that more than N failures have occurred. It is

$$P\{\mathbf{n} > N\} = \sum_{n=N+1}^{\infty} \frac{(\lambda t)^n}{n!} e^{-\lambda t}. \tag{6.113}$$

Instead of writing this infinite series, however, we may use Eq. 6.110 to write

$$P\{\mathbf{n} > N\} = 1 - \sum_{n=0}^{N} \frac{(\lambda t)^n}{n!} e^{-\lambda t}. \tag{6.114}$$

EXAMPLE 6.9

In an industrial plant there is a dc power supply in continuous use. It is known to have a failure rate of $\lambda = 0.40/\text{year}$. If replacement supplies are delivered at 6-month intervals, and if the probability of running out of replacement power supplies is to be limited to 0.01, how many replacement power supplies should the operations engineer have on hand at the beginning of the 6-month interval.

Solution   First calculate the probability that the supply will have more than n failures with $t = 0.5$ year,

$$\lambda t = 0.4 \times 0.5 = 0.2; \qquad e^{-0.2} = 0.819.$$

Now use Eq. 6.114

$$P\{\mathbf{n} > 0\} = 1 - e^{-\lambda t} = 0.181,$$

$$P\{\mathbf{n} > 1\} = 1 - e^{-\lambda t}(1 + \lambda t) = 0.018,$$

$$P\{\mathbf{n} > 2\} = 1 - e^{-\lambda t}[1 + \lambda t + \tfrac{1}{2}(\lambda t)^2] = 0.001.$$

---

* See, for example, R. E. Barlow and F. Proschan, *Mathematical Theory of Reliability*, Wiley, New York, 1965.

There is less than a 1% probability of more than two power supplies failing. Therefore, two spares should be kept on hand.

Bibliography

Anderson, R. T., *Reliability Design Handbook,* U. S. Department of Defense Reliability Analysis Center, 1976.

Billinton, R., and R. N. Allan, *Reliability Evaluation of Engineering Systems,* Plenum Press, NY, 1983.

Bazovsky, I., *Reliability Theory and Practice,* Prentice-Hall, Englewood Cliffs, NJ, 1961.

Dillon, B. S., and C. Singh, *Engineering Reliability,* Wiley, NY, 1981.

Reliability Prediction of Electronic Equipment MIL-HDBK-217D, U. S. Department of Defense, 1982.

Shooman, M. L., *Probabilistic Reliability: An Engineering Approach,* Krieger, Malabar, FL, 1990.

Exercises

6.1 The PDF for the time-to-failure of an appliance is

$$f(t) = \frac{32}{(t+4)^3}, \qquad t > 0,$$

where t is in years

(a) Find the reliability of $R(t)$.

(b) Find the failure rate $\lambda(t)$.

(c) Find the MTTF.

6.2 The reliability of a machine is given by

$$R(t) = \exp[-0.04t - 0.008\ t^2]\ (t \text{ in years}).$$

(a) What is the failure rate?

(b) What should the design life be to maintain a reliability of at least 0.90?

6.3 The failure rate for a high-speed fan is given by

$$\lambda(t) = (2 \times 10^{-4} + 3 \times 10^{-6}t)/\text{hr},$$

where t is in hours of operation. The required design-life reliability is 0.95.

(a) How many hours of operation should the design life be?

(b) If, by preventive maintenance, the wear contribution to the failure rate can be eliminated, to how many hours can the design life be extended?

(c) By placing the fan in a controlled environment, we can reduce the constant contribution to $\lambda(t)$ by a factor of two. Then, without preventive maintenance, to how many hours may the design life be extended?

(d) What is the extended design life when both reductions from (b) and (c) are made?

6.4 If the CDF for times to failure is

$$F(t) = 1 - \frac{100}{(t + 10)^2}$$

(a) Find the failure rate as a function of time.

(b) Does the failure rate increase or decrease with time?

6.5 Repeat Exercise 6.3, but fix the design life at 100 hr and calculate the design-life reliability for conditions (a), (b), (c), and (d).

6.6 An electronic device is tested for two months and found to have a reliability of 0.990; the device is also known to have a constant failure rate.

(a) What is the failure rate?

(b) What is the mean-time-to-failure?

(c) What is the design life reliability for a design life of 4 years?

(d) What should the design life be to achieve a reliability of 0.950?

6.7 A logic circuit is known to have a decreasing failure rate of the form

$$\lambda(t) = \tfrac{1}{20}t^{-1/2}/\text{year},$$

where t is in years.

(a) If the design life is one year, what is the reliability?

(b) If the component undergoes wearin for one month before being put into operation, what will the reliability be for a one-year design life?

6.8 A device has a constant failure rate of 0.7/year.

(a) What is the probability that the device will fail during the *second* year of operation?

(b) If upon failure the device is immediately replaced, what is the probability that there will be more than one failure in 3 years of operation?

6.9 The failure rate on a new brake drum design is estimated to be

$$\lambda(t) = 1.2 \times 10^{-6} \exp(10^{-4}t)$$

per set, where t is in kilometers of normal driving. Forty vehicles are each test-driven for 15,000 km.

(a) How many failures are expected, assuming that the vehicles with failed drives are removed from the test?

(b) What is the probability that more than two vehicles will fail?

6.10 The failure rate for a hydraulic component is given empirically by

$$\lambda(t) = 0.001(1 + 2e^{-2t} + e^{t/40})/\text{year}$$

where t is in years. If the system is installed at $t = 0$, calculate the probability that it will have failed by time t. Plot your results for 40 years.

6.11 A home computer manufacturer determines that his machine has a constant failure rate of $\lambda = 0.4$ year in normal use. For how long should the warranty be set if no more than 5% of the computers are to be returned to the manufacturer for repair?

6.12 What fraction of items tested are expected to last more than 1 MTTF if the distribution of times-to-failure is

(a) exponential,
(b) normal,
(c) lognormal with $\omega = 2$,
(d) Weibull with $m = 2$?

6.13 A one-year guarantee is given based on the assumption that no more than 10% of the items will be returned. Assuming an exponential distribution, what is the maximum failure rate that can be tolerated?

6.14 There is a contractual requirement to demonstrate with 90% confidence that a vehicle can achieve a 100-km mission with a reliability of 99%. The acceptance test is performed by running 10 vehicles over a 50,000-km test track.

(a) What is the contractual MTTF?
(b) What is the maximum number of failures that can be experienced on the demonstration test without violating the contractual requirement? (*Note:* Assume an exponential distribution, and review Section 2.5.)

6.15 The reliability for the Rayleigh distribution is

$$R(t) = e^{-(t/\theta)^2}.$$

Find the MTTF in terms of θ.

6.16 Suppose the CDF for time to failure is given by

$$R(t) = \begin{cases} 1 - at^2. & t < 1/\sqrt{a} \\ 0, & t > 1/\sqrt{a} \end{cases}$$

Determine the following:

(a) the PDF $f(t)$,
(b) the failure rate,
(c) the MTTF.

6.17 Suppose that amplifiers have a constant failure rate of $\lambda = 0.08/$month. Suppose that four such amplifiers are tested for 6 months. What is the probability that more than one of them will fail? Assume that when they fail, they are not replaced.

6.18 A device has a constant failure rate with a MTTF of 2 months. One hundred of the devices are tested to failure.

(a) How many of the devices do you expect to fail during the second month?

(b) Of the devices which survive two months, what fraction do you expect to fail during the third month?

(c) If you are allowed to stop the test after 80 failures, how long do you expect the test to last?

6.19 A manufacturer determines that the average television set is used 1.8 hr/day. A one-year warranty is offered on the picture tube having a MTTF of 2000 hr. If the distribution is exponential, what fraction of the tubes will fail during the warranty period?

6.20 Ten control circuits are to undergo simultaneous accelerated testing to study the failure modes. The accelerated failure rate has previously been estimated to be constant with a value of 0.04 days^{-1}.

(a) What is the probability that there will be at least one failure during the first day of the test?

(b) What is the probability that there will be more than one failure during the first week of the test?

6.21 The reliability of a cutting tool is given by

$$R(t) \equiv \begin{cases} (1 - 0.2t)^2, & 0 \le t \le 5, \\ 0, & t > 5, \end{cases}$$

where t is in hours.

(a) What is the MTTF?

(b) How frequently should the tool be changed if failures are to be held to no more than 5%?

(c) Is the failure rate decreasing or increasing? Justify your result.

6.22 A motor-operated valve has a failure rate λ_0 while it is open and λ_c while it is closed. It also has a failure probability p_0 to open on demand and a failure probability p_c to close on demand. Develop an expression for the composite failure rate similar to Eq. 6.46 for the valve.

6.23 A failure PDF for an appliance is assumed to be a normal distribution with $\mu = 5$ years and $\sigma = 0.8$ years. Set the design life for

(a) a reliability of 90%,

(b) a reliability of 99%.

6.24 A designer assumes a 90% probability that a new piece of machinery will fail at some time between 2 years and 10 years.

  (a) Fit a lognormal distribution to this belief.

  (b) What is the MTTF?

6.25 The life of a rocker arm is assumed to be 4 million cycles. This is known to a factor of two with 90% probability. If the reliability is to be 0.95, how many cycles should the design life be?

6.26 Two components have the same MTTF; the first has a constant failure rate λ_0 and the second follows a Rayleigh distribution, for which

$$\int_0^t \lambda(t')\, dt' = \left(\frac{t}{\theta}\right)^2.$$

  (a) Find θ in terms of λ_0.

  (b) If for each component the design-life reliability must be 0.9, how much longer (in percentage) is the design life of the second (Rayleigh) component?

6.27 Night watchmen carry an industrial flashlight 8 hr per night, 7 nights per week. It is estimated that on the average the flashlight is turned on about 20 min per 8-hr shift. The flashlight is assumed to have a constant failure rate of 0.08/hr while it is turned on and of 0.005/hr when it is turned off but being carried.

  (a) In working hours, estimate the MTTF of the light.

  (b) What is the probability of the light's failing during one 8-hr shift?

  (c) What is the probability of its failing during one month (30 days) of 8-hr shifts?

6.28 Consider the two components in Exercise 6.26.

  (a) For what design-life reliability are the design lives of the two components equal?

  (b) On the same graph plot reliability versus time for the two components.

6.29 The two-parameter Weibull distribution with $m = 2$ is known as the Rayleigh distribution. For a nonredundant system made of N components, each described by the same Rayleigh distribution, find the system MTTF in terms of N and the component θ.

6.30 If waves hit a platform at the rate of 0.4/min and the "memoryless" failure probability is 10^{-6}/wave, estimate the failure rate in days^{-1}.

6.31 The one-month reliability on an indicator lamp is 0.95 with the failure rate specified as constant. What is the probability that more than two spare bulbs will be needed during the first year of operation? (Ignore replacement time.)

6.32 A part for a marine engine with a constant failure rate has an MTTF of two months. If two spare parts are carried,

(a) What is the probability of surviving a six-month cruise without losing the use of the engine as a result of part exhaustion?

(b) What is the result for part *a* if three spare parts are carried?

6.33 In Exercise 6.27, suppose that there are three watchmen on duty every night for 8 hr.

(a) How many flashlight failures would you expect in one year?

(b) Assuming that the failures are not caused by battery or bulb wearout (these are replaced frequently), how many spare flashlights would be required to be on hand at the beginning of the year, if the probability of running out of spares is to be less than 10%?

6.34 An electronics manufacture mixes 1,000 capacitors with an MTTF of 3 months and 2,000 capacitors with an MTTF of 6 months. Assuming that the capacitors have constant failures rates:

(a) What is the PDF for the combined population?

(b) Use Eq. 6.15 to derive an expression for the failure rate of the combined population.

(c) What is the failure rate at $t = 0$?

(d) Does the failure rate increase or decrease with time?

(e) What is the failure rate at very long times?

6.35 A servomechanism has an MTBF of 2000 hr. with a constant failure rate.

(a) What is the reliability for a 125-hr mission?

(b) Neglecting repair time, what is the probability that more than one failure will occur during a 125-hr mission?

(c) That more than two failures will occur during a 125-hr mission?

6.36 Assume that the occurrence of earthquakes strong enough to be damaging to a particular structure is governed by the Poisson distribution. If the mean time between such earth quakes is twice the design life of the structure:

(a) What is the probability that the structure will be damaged during its design life?

(b) What is the probability that it will suffer more than one damaging earthquake during its design life?

(c) Calculate the failure rate (i.e., damage rate due to earthquakes).

6.37 A relay circuit has an MTBF of 0.8 yr. Assuming random failures,

(a) Calculate the probability that the circuit will survive one year without failure.

(b) What is the probability that there will be more than two failures in the first year?

(c) What is the expected number of failures per year?

6.38 Demonstrate that Eq. 6.106 satisfies Eq. 6.104.

6.39 The MTBF for punctures of truck tires is 150,000 miles. A truck with 10 tires carries 1 spare.

(a) What is the probability that the spare will be used on a 10,000-mile trip?

(b) What is the probability that more than the single spare will be required on a 10,000-mile trip?

6.40 Widgets have a constant failure rate with MTTF = 5 days. Ten widgets are tested for one day.

(a) What is the expected number of failures during the test?

(b) What is the probability that *more than one* will fail during the test?

(c) For how long would you run the test if you wanted the expected number of failures to be five?

CHAPTER 7

Loads, Capacity, and Reliability

"Now in the building of chaises, I tell you what,
There is always, somewhere, a weakest spot,—
In hub, tire, felloe, in spring or thill,
In panel, or crossbar, or floor, or sill,
In screw, bolt, thoroughbrace,—lurking, still,
Find it somewhere you must and will,—
Above or below; or within or without,—
And that's the reason, beyond a doubt,
That a chaise breaks down, but doesn't wear out."

Oliver Wendell Holmes
The Deacons's Masterpiece

7.1 INTRODUCTION

In the preceding chapters failure rates were used to emphasize the strong dependence of reliability on time. Empirically, these failure rates are found to increase with system complexity and also with loading. In this chapter we explore the concepts of loads and capacity and examine their relationship to reliability. This examination allows us both to relate reliability to traditional design approaches using safety factors, and to gain additional insight into the relations between failure rates, infant mortality, random failures and aging.

Safety factors and margins are defined in the following way: Suppose we define l as the load on a system, structure, or piece of equipment and c as the corresponding capacity. The safety factor is then defined as

$$v = \frac{c}{l}.$$ (7.1)

Alternately, the safety margin may be used. It is defined by

$$m = c - l.$$ (7.2)

Failure then occurs if the safety factor falls to a value less than one, or if the safety margin becomes negative.

The concepts of load and capacity are employed most widely in structural engineering and related fields, where the load is usually referred to as stress and the capacity as strength. However, they have much wider applicability. For example, if a piece of electric equipment is under consideration, we may speak of electric load and capacity. A telecommunications system load and capacity may be measured in terms of telephone calls per unit time, and for an energy conversion system thermal units for load and capacity may be used. The point is that a wide variety of applications can be formulated in terms of load and capacity. For a given application, however, l and c must have the same units.

In the traditional approach to design, the safety factor or margin is made large enough to more than compensate for uncertainties in the values of both the load and the capacity of the system under consideration. Thus, although these uncertainties cause the load and the capacity to be viewed as random variables, the calculations are deterministic, using for the most part the best estimates of load and capacity. The probabilistic analysis of loads and capacities necessary for estimating reliability clarifies and rationalizes the determination and use of safety factors and margins. This analysis is particularly useful for situations in which no fixed bound can be put on the loading, for example, with earthquakes, floods and other natural phenomena, or for situations in which flaws or other shortcomings may result in systems with unusually small capacities. Similarly, when economics rather than safety is the primary criteria for setting design margins, the trade-off of performance versus reliability can best be studied by examining the increase in the probability of failure as load and capacity approach one another.

The expression for reliability in terms of the random variables $\mathbf{l}$ and $\mathbf{c}$ comes from the notion that there is always some small probability of failure that decreases as the safety factor is increased. We may define the failure probability as

$$p = P\{\mathbf{l} \geq \mathbf{c}\}.$$ (7.3)

In this context the reliability is defined as the nonfailure probability or

$$r = 1 - p,$$ (7.4)

which may also be expressed as

$$r = P\{\mathbf{l} < \mathbf{c}\}.$$ (7.5)

In treating loads and capacities probabilistically, we must exercise a great deal of care in expressing the types of loads and the behavior of the capacity. If this is done, we may use the resulting formalism not only to provide a probabilistic relation between safety factors and reliability, but also to gain a better understanding of the relations between loading, capacities, and the time dependence of failure rates as exhibited, for example, in the bathtub curve.

In Section 7.2 we develop reliability expressions for a single loading and then, in section 7.3, relate the results to the probabilistic interpretation of safety factors. In Section 7.4 we take up repetitive loading to demonstrate how the time-dependence of failure rate curves stems from the interactions of variable loading with capacity variability and deterioration. In Section 7.5 a failure rate model for the bathtub curve in synthesized in which variable capacity, variable loading, and capacity deterioration, respectively, are related to infant mortality, random failures and aging.

7.2   RELIABILITY WITH A SINGLE LOADING

In this section we derive the relations between load, capacity, and reliability for systems that are loaded only once. The resulting reliability does not depend on time, for the reliability is just the probability that the system survives the application of the load. Nevertheless, before the expressions for the reliability can be derived, the restrictions on the nature of the loads and capacity must be clearly understood.

Load Application

In referring to the load on a system, we are in fact referring to the maximum load from the beginning of application until the load is removed. Figure 7.1 indicates the time dependence of several loading patterns that may be treated as single on loading l, provided that appropriate restrictions are met.

Figure 7.1*a* represents a single loading of finite duration. Missiles during launch, flashbulbs, and any number of other devices that are used only once have such loadings. Such one-time-only loads are also a ubiquitous feature of manufacturing processes, occurring for instance when torque is applied to a bolt or pressure is applied to a rivet. Loading often is not applied in a smooth manner, but rather as a series of shocks, as shown in Fig. 7.1*b*. This behavior would be typical of the vibrational loading on a structure during an earthquake and of the impact loading on an aircraft during landing. In many situations, the extreme value of many short-time loadings may be treated as a single loading provided that there is a definite beginning and end to the disturbance giving rise to it.

The duration of the load in Figs. 7.1*a* and *b* is short enough that no weakening of the system capacity takes place. If no decrease in system capacity is possible, the situations shown in Figs. 7.1*c* and *d* may also be viewed as single loadings, even though they are not of finite duration. The loading shown in Fig. 7.1*c* is typical of the dead loads from the weight of structures;

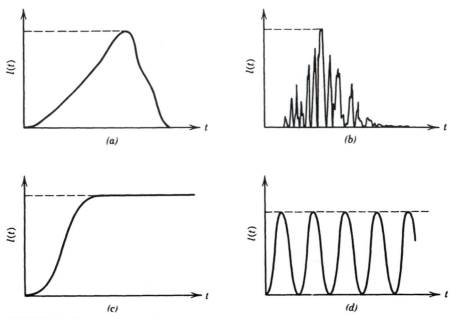

FIGURE 7.1 Time-dependent loading patterns.

these increase during construction and then remain at a constant value. This formulation of the loading is widely used in structural analysis when the load-bearing capacity not only may remain constant, but may in some instances increase somewhat with time because of the curing of concrete or the work-hardening of metals.

Subject to the same restrictions, the patterns shown in Fig. 7.1d may be viewed as a single loading. Provided the peaks are of the same magnitude, the system will either fail the first time the load is applied or will not fail at all. Under such cyclic loading, however, the assumption that the system capacity will not decrease with time should be suspect. Metal fatigue and other wear effects are likely to weaken the capacity of the system gradually. Similarly, if the values of peak magnitudes vary from cycle to cycle, we must consider the time dependence of reliability explicitly, as in Section 7.4.

Thus far we have assumed that a system is subjected to only one load and that reliability is determined by the capacity of the system as a whole to resist this load. In reality, a system is invariably subjected to a variety of different loads; if it does not have the capacity to sustain any one of these, it will fail. An obvious example is a piece of machinery or other equipment, each of whose components are subjected to different loads; failure of any one component will make the system fail. A more monolithic structure, such as a dam, is subject to static loads from its own weight, dynamic loads from earthquakes, flood loadings, and so on. Nevertheless, the considerations that follow remain applicable, provided that the loads are considered in terms of the probability of a particular failure mode or of the loading of a particular component. If the failure modes can be assumed to be approximately independent of one another, the reliability of the overall system can be calculated as the product of the failure mode reliabilities, as discussed in Chapter 6.

Definitions

To derive an expression for the reliability, we must first define independent PDFs for the load, **l**, and for the capacity, **c**. Let

$$f_l(l) \; dl = P\{l \le \mathbf{l} \le l + dl\} \tag{7.6}$$

be the probability that the load is between l and $l + dl$. Similarly, let

$$f_c(c) \; dc = P\{c \le \mathbf{c} < c + dc\} \tag{7.7}$$

be the probability that the capacity has a value between c and $c + dc$. Thus $f_l(l)$ and $f_c(c)$ are the necessary PDFs; we include the subscripts to avoid any possible confusion between the two. The corresponding CDFs may also be defined. They are

$$F_c(c) = \int_0^c f_c(c') \; dc', \tag{7.8}$$

$$F_l(l) = \int_0^l f_l(l') \; dl'. \tag{7.9}$$

We first consider a system with a known capacity c and a distribution of possible loads, as shown in Fig. 7.2a. For fixed c, the reliability of the system is just the probability that $\mathbf{l} < c$, which is the shaded area in the figure. Thus

$$r(c) = \int_0^c f_l(l) \; dl. \tag{7.10}$$

The reliability, therefore, is just $F_l(c)$, the CDF of the load evaluated at c. Clearly, for a system of known capacity, the reliability is equal to one as $c \to \infty$, and to zero as $c \to 0$.

Now suppose that the capacity also involves uncertainty; it is described by the PDF $f_c(c)$. The expected value of the reliability is then obtained from

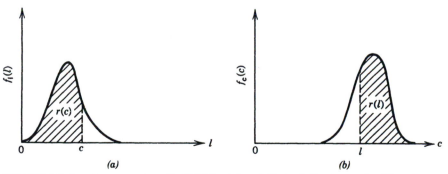

FIGURE 7.2 Area interpretation of reliability: (a) variable load, fixed capacity; (b) variable capacity, fixed load.

averaging over the distribution of capacities:

$$r = \int_0^\infty r(c) f_c(c) \, dc. \tag{7.11}$$

Substituting in Eq. 7.10, we have

$$r = \int_0^\infty \left[\int_0^c f_l(l) \, dl \right] f_c(c) \, dc. \tag{7.12}$$

The failure probability may then be determined from Eq. 7.4 to be

$$p = 1 - \int_0^\infty \left[\int_0^c f_l(l) \, dl \right] f_c(c) \, dc. \tag{7.13}$$

Alternately, we may substitute the condition on the load PDF,

$$\int_0^c f_l(l) \, dl = 1 - \int_c^\infty f_l(l) \, dl, \tag{7.14}$$

into Eq. 7.12. Then, using the condition

$$\int_0^\infty f_c(c) \, dc = 1, \tag{7.15}$$

we obtain for the failure probability

$$p = \int_0^\infty \left[\int_c^\infty f_l(l) \, dl \right] f_c(c) \, dc. \tag{7.16}$$

As shown in Fig. 7.3, the probability of failure is loosely associated with the overlap of the PDFs for load and capacity in the sense that if there is no overlap, the failure probability is zero and $r = 1$.

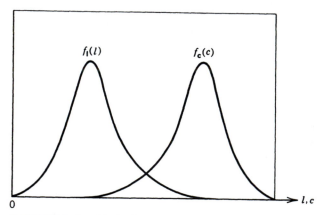

FIGURE 7.3 Graphical reliability interpretation with variable load and capacity.

EXAMPLE 7.1

The bending moment on a match stick during striking is estimated to be distributed exponentially. It is found that match sticks of a given strength break 20% of the time. Therefore, the manufacturer increases the strength of the matches by 50%. What fraction of the strengthened matches are expected to break as they are struck?

Solution   Assume that the strength (capacity) is known; then for the standard matches we have

$$0.8 = r = \int_0^c f_l(l) \, dl = \int_0^c \lambda e^{-\lambda l} \, dl = 1 - e^{-\lambda c}.$$

Therefore, $e^{-\lambda c} = 0.2$ or $\lambda c = -ln(0.2)$, where λ is the unknown parameter of the exponential loading distribution. For the strengthened matches

$$r' = \int_0^{1.5c} f_l(l) \, dl = \int_0^{1.5c} \lambda e^{-\lambda l} \, dl = 1 - e^{-1.5\lambda c},$$

$$p' \equiv 1 - r' = \exp[+ 1.5 \times ln(0.2)] = 0.2^{1.5} = 0.089.$$

Thus about 9% of the strengthened matches are expected to break.

Another derivation of r and p is possible. Although the derivation may be shown to yield results that are identical to Eqs. 7.12 and 7.13, the intermediate results are useful for different sets of circumstances. To illustrate, let us consider a system with known load but uncertain capacity represented by the distribution $f_c(c)$. The reliability for this system with known load is then given by the shaded area in Fig. 7.2*b*.

$$r(l) = \int_l^\infty f_c(c) \, dc, \tag{7.17}$$

or equivalently,

$$r(l) = 1 - \int_0^l f_c(c) \, dc. \tag{7.18}$$

For a system in which the load is also represented by a distribution, the expected value of the reliability is obtained by averaging over the load distribution,

$$r = \int_0^\infty f_l(l) r(l) \, dl, \tag{7.19}$$

or more explicitly

$$r = \int_0^\infty f_l(l) \left[\int_l^\infty f_c(c) \, dc \right] dl. \tag{7.20}$$

Similarly, we may consider the variation of the capacity first in deriving an expression for the failure probability. For a system with a fixed load the failure probability will be the unshaded area under the curve in Fig. 7.2*b*:

$$p(l) = \int_0^l f_c(c)\, dc. \tag{7.21}$$

Then, averaging over the distribution of loads, we have

$$p = \int_0^\infty f_1(l) \left[\int_0^l f_c(c)\, dc \right] dl. \tag{7.22}$$

It is easily shown that Eqs. 7.12 and 7.20 are the same. First write Eq. 7.12 as the double integral

$$r = \int_0^\infty \left[\int_0^c f_c(c) f_1(l)\, dl \right] dc, \tag{7.23}$$

where the shaded domain of integration appears in Fig. 7.4. If we reverse the order of integration, taking the c integration first, we have

$$r = \int_0^\infty \left[\int_l^\infty f_c(c) f_1(l)\, dc \right] dl. \tag{7.24}$$

Putting $f_1(l)$ outside the integral over c, we obtain Eq. 7.20.

To recapitulate, Eqs. 7.12 and 7.20 may be shown to be identical, as may Eqs. 7.16 and 7.22. However, the intermediate results for $r(c)$, $p(c)$, $r(l)$, and $p(l)$ are useful when considering systems whose capacity varies little compared to their load, or vice versa.

7.3   RELIABILITY AND SAFETY FACTORS

In the preceding section reliability for a single loading is defined in terms of the independent PDFs for load and capacity. Similarly, it is possible to define

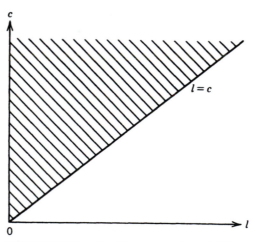

FIGURE 7.4 Domain of integration for reliability calculation.

safety factors in terms of these distributions. Two of the most widely accepted definitions are as follows. In the central safety factor the values of load and capacity in Eq. 7.1 are taken to be the mean values

$$\bar{l} = \int_{-\infty}^{\infty} lf_l(l) \, dl, \tag{7.25}$$

$$\bar{c} = \int_{-\infty}^{\infty} cf_c(c) \, dc. \tag{7.26}$$

Thus the safety factor is

$$v = \bar{c}/\bar{l}. \tag{7.27}$$

There is a second alternative if we express the safety factor in terms of the most probable values l_0 and c_0 at the load and capacity distributions. The safety factor in Eq. 7.1 is then

$$v = c_0/l_0. \tag{7.28}$$

These definitions are naturally associated with loads and capacities represented in terms of normal or of lognormal distributions, respectively. Then the reliability can be expressed in terms of the safety factor along with measures of the uncertainty in load and capacity. Other distributions may also be used in relating reliability to safety factors. Such is the case with the extreme-value distribution. With such analysis the effects of design changes and quality control can be evaluated. Design determines the mean, $\bar{c}$, or most probable value, c_0, of the capacity, whereas the degree of quality control in manufacture or construction influences primarily the variance of $f_c(c)$ about the mean. Similarly, the conditions under which operations take place determine the load distribution $f_l(l)$ as well as the mean value $\bar{l}$.

Normal Distributions

The normal distribution is widely used for relating safety factors to reliability, particularly when small variations in materials and dimensional tolerances and the inability to determine loading precisely make capacity and load uncertain. The normal distribution is appropriate when variability in loads, capacity, or both is caused by the sum of many effects, no one of which is dominant. An appropriate example is the load and capacity of an elevator large enough to carry several people. Since the load is the sum of the weights of the people, the variability of the weight is likely to be very close to a normal distribution for the reasons discussed in Chapter 3. The variability in the weight of any one person is unlikely to have an overriding effect on the total load. Similarly, if the elevator cable is made up of many independent strands of wire, its capacity will be the sum of the strengths of the individual strands. Since the variability in strength of any one strand will not have much effect on the cable capacity, the normal distribution may be used to model the cable capacity.

Suppose that the load and capacity are represented by normal distributions,

$$f_l(l) = \frac{1}{\sqrt{2\pi}\,\sigma_l} \exp\left[-\tfrac{1}{2} \frac{(l - \bar{l})^2}{\sigma_l^2} \right] \tag{7.29}$$

and

$$f_c(c) = \frac{1}{\sqrt{2\pi}\,\sigma_c} \exp\left[-\tfrac{1}{2} \frac{(c - \bar{c})^2}{\sigma_c^2} \right], \tag{7.30}$$

where the mean values of the load and capacity are denoted by $\bar{l}$ and $\bar{c}$, and the corresponding standard deviations are σ_l and σ_c. Substituting these expressions into Eq. 7.12, we obtain for the reliability

$$r = \int_{-\infty}^{\infty} \frac{1}{\sqrt{2\pi}\,\sigma_c} \exp\left[-\tfrac{1}{2} \frac{(c - \bar{c})^2}{\sigma_c^2} \right]$$
$$\times \left\{ \int_{-\infty}^{c} \frac{1}{\sqrt{2\pi}\,\sigma_l} \exp\left[-\tfrac{1}{2} \frac{(l - \bar{l})^2}{\sigma_l^2} \right] dl \right\} dc. \tag{7.31}$$

This expression* for the reliability may be reduced to a much simpler form involving only a single normal integral. To accomplish this, however, involves a significant amount of algebraic manipulation. We begin by transforming variables to the dimensionless quantities

$$x = (c - \bar{c})/\sigma_c, \tag{7.32}$$

$$y = (l - \bar{l})/\sigma_l. \tag{7.33}$$

Equation 7.31 may then be rewritten as

$$r = \frac{1}{2\pi} \int_{-\infty}^{\infty} \left\{ \int_{-\infty}^{(\sigma_c x + \bar{c} - \bar{l})/\sigma_l} \exp[-\tfrac{1}{2}(x^2 + y^2)] \, dy \right\} dx. \tag{7.34}$$

This double integral may be viewed geometrically as an integral over the shaded part of the $x - y$ plane shown in Figure 7.5. The line demarking the edge of the region of integration is determined by the upper limit of the y integration in Eq. 7.34:

$$y = \frac{1}{\sigma_l} (\sigma_c x + \bar{c} - \bar{l}). \tag{7.35}$$

By rotating the coordinates through the angle θ, we may rewrite the reliability as a single standardized normal function. To this end we take

$$x' = x \cos \theta + y \sin \theta \tag{7.36}$$

---

* Note that we have extended the lower limits on the integrals to $-\infty$ in order to accommodate the use of normal distributions. The effect on the result is negligible for $\bar{c} \gg \sigma_c$ and $\bar{l} \gg \sigma_l$.

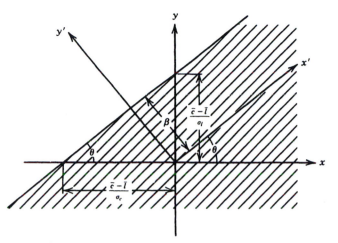

FIGURE 7.5 Domain of integration for normal load and capacity.

and

$$y' = -x \sin \theta + y \cos \theta. \tag{7.37}$$

It may then be shown that

$$x^2 + y^2 = x'^2 + y'^2 \tag{7.38}$$

and

$$dx\, dy = dx'\, dy', \tag{7.39}$$

allowing us to write the reliability as

$$r = \frac{1}{2\pi} \int_{-\infty}^{\infty} \left\{ \int_{-\infty}^{\beta} \exp[-\tfrac{1}{2}(x'^2 + y'^2)]\, dy' \right\} dx'. \tag{7.40}$$

The upper limit on the y' integration is just the distance β shown in Fig. 7.5. With elementary trigonometry, β may be shown to be a constant given by

$$\beta = \frac{\bar{c} - \bar{l}}{(\sigma_c^2 + \sigma_l^2)^{1/2}}. \tag{7.41}$$

The quantity β is referred to as the safety or reliability index. Since β is a constant, the order of integration may be reversed. Then, since

$$\frac{1}{\sqrt{2\pi}} \int_{-\infty}^{\infty} e^{-\tfrac{1}{2}x'^2}\, dx' = \Phi(\infty) = 1, \tag{7.42}$$

the remaining integral, in y', may be written as a standardized normal CDF to yield the reliability in terms of the safety index β:

$$r = \Phi(\beta). \tag{7.43}$$

The results of this equation may be put in a more graphic form by expressing them in terms of the safety factor, Eq. 7.27. A standard measure

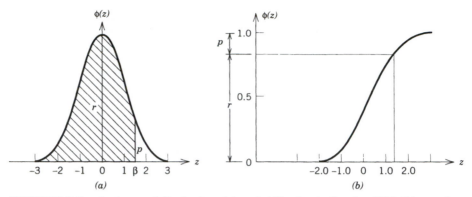

FIGURE 7.6 Standard normal distribution: (*a*) probability density function PDF, (*b*) cumulative distribution function (CDF).

of the dispersion about the mean is the coefficient of variation, defined as the standard deviation divided by the mean:

$$\rho = \sigma/\mu. \tag{7.44}$$

Thus we may write

$$\rho_c = \sigma_c/\bar{c} \tag{7.45}$$

and

$$\rho_l = \sigma_l/\bar{l}. \tag{7.46}$$

With these definitions we may express the safety index in terms of the central safety factor and the coefficients of variation:

$$\beta = \frac{v - 1}{(\rho_c^2 v^2 + \rho_l^2)^{1/2}}. \tag{7.47}$$

In Figure 7.6 the standardized normal distribution is plotted. The area under the curve to the left of β is the reliability r; the area to the right is the failure probability p. In Fig. 7.6b the CDF for the normal distribution is plotted. Thus, given a value of β, we can calculate r and p. Conversely, if the reliability is specified and the coefficients of variation are known, we may determine the value of the safety factor. In Figure 7.7 the relation between safety factor and probability of failure is indicated for some representative values of the coefficients of variation.

EXAMPLE 7.2

Suppose that the coefficients of variation are $\rho_c = 0.1$ and $\rho_l = 0.15$. If we assume normal distributions, what safety factor is required to obtain a failure probability of no more than 0.005?

Solution $p = 0.005$; $r = 0.995$; $r = \Phi(\beta) = 0.995$. Therefore, from Appendix C, $\beta = 2.575$. We must solve Eq. 7.47 for v. We have

$$\beta^2(\rho_c^2 v^2 + \rho_l^2) = (v - 1)^2 \quad \text{or} \quad (1 - \beta^2\rho_c^2)v^2 - 2v + (1 - \beta^2\rho_l^2) = 0.$$

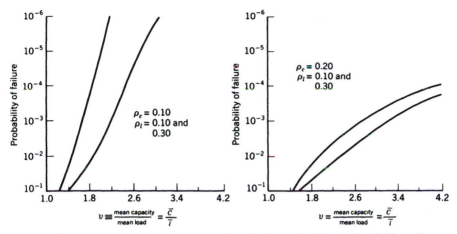

FIGURE 7.7 Probability of failure for normal load and capacity (From Gary C. Hart, *Uncertainty Analysis, Loads, and Safety in Structural Engineering,* © 1982, p. 107, with permission from Prentice-Hall, Englewood Cliffs, NJ.)

Solving this quadratic equation in v, we have

$$v = \frac{2 \pm [4 - 4(1 - \beta^2 \rho_l^2)(1 - \beta^2 \rho_c^2)]^{1/2}}{2(1 - \beta^2 \rho_c^2)}$$

or

$$v = \frac{2 \pm 2(1 - 0.8508 \times 0.9337)^{1/2}}{2 \times 0.9336} = \frac{1 \pm 0.4534}{0.9337}$$

$$= 1.56,$$

since the second solution, 0.5853, will not satisfy Eq. 7.47.

In using Eqs. 7.43 and 7.47 to estimate reliability, we assume that the load and capacity are normally distributed and that the means and variances can be estimated. In practice, the paucity of data often does not allow us to say with any certainty what the distributions of load and capacity are. In these situations, however, the sample mean and variance can often be obtained. They can then be used to calculate the reliability index defined by Eq. 7.47; often the reliability can be estimated from Eq. 7.43. Such approaches are referred to as second-moment methods, since only the zero and second moments of the load and capacity distributions need to be estimated.

Second-moment methods* have been widely employed, for they represent the logical next step beyond the simple use of safety factors in that they also account for the variance of the distributions. Such methods must be employed with care, however, for when the distributions deviate greatly from normal

---

* C. A. Cornell, "Structural Safety Specifications Based on Second-Moment Reliability," *Symposium of the International Association of Bridge and Structural Engineers,* London, 1969; see also A. H.-S. Ang, and W. H. Tang, *Probability Concepts in Engineering Planning and Design,* Vol. 2, Wiley, New York, 1984.

distributions, the resulting formulas may be in serious error. This may be seen from the different expressions for reliability when lognormal or extreme-value distributions are employed.

Lognormal Distributions

The lognormal distribution is useful when the uncertainty about the load, or capacity, or both, is relatively large. Often it is expressed as having 90% confidence that the load or the capacity lies within some factor, say two, of the best estimates l_0 or c_0. In Chapter 3 the properties of the lognormal distribution were presented. As indicated there, the lognormal distribution is most appropriate when the value of the variable is determined by the product of several different factors. For load and capacity, we rewrite Eq. 3.63 for the PDFs as

$$f_l(l) = \frac{1}{\sqrt{2\pi}\omega_l l} \exp\left\{-\frac{1}{2\omega_l^2}\left[\ln\left(\frac{l}{l_0}\right)\right]^2\right\}, \qquad 0 < l \leq \infty, \qquad (7.48)$$

and

$$f_c(c) = \frac{1}{\sqrt{2\pi}\omega_c c} \exp\left\{-\frac{1}{2\omega_c^2}\left[\ln\left(\frac{c}{c_0}\right)\right]^2\right\}, \qquad 0 < c \leq \infty. \qquad (7.49)$$

If Eqs. 7.48 and 7.49 are substituted into Eq. 7.12, the resulting expression for the reliability is

$$
\begin{aligned}
r = \int_0^\infty & \frac{1}{\sqrt{2\pi}\omega_c c} \exp\left\{-\frac{1}{2\omega_c^2}\left[\ln\left(\frac{c}{c_0}\right)\right]^2\right\} \\
& \times \left(\int_0^c \frac{1}{\sqrt{2\pi}\omega_l l} \exp\left\{-\frac{1}{2\omega_l^2}\left[\ln\left(\frac{l}{l_0}\right)\right]^2\right\} dl\right) dc.
\end{aligned}
\qquad (7.50)
$$

Note, however, that with the substitutions

$$y = \frac{1}{\omega_l}\ln\left(\frac{l}{l_0}\right) \qquad (7.51)$$

and

$$x = \frac{1}{\omega_c}\ln\left(\frac{c}{c_0}\right), \qquad (7.52)$$

we obtain

$$r = \frac{1}{2\pi}\int_{-\infty}^\infty \left\{\int_{-\infty}^{(1/\omega_l)[\omega_c x + \ln(c_0/l_0)]} \exp[-\tfrac{1}{2}(x^2 + y^2)]\, dy\right\} dx. \qquad (7.53)$$

The forms of the reliability in Eq. 7.34 and in this equation are identical if in the upper limit of the y integration we substitute ω_l and ω_c for σ_l and σ_s, respectively, and replace $\bar{c} - \bar{l}$ with $\ln(c_0/l_0)$. Thus the reliability still has the form of a standardized normal distribution given by Eq. 7.43. Now, however, the argument β is given by

$$\beta = \frac{\ln(c_0/l_0)}{(\omega_c^2 + \omega_l^2)^{1/2}}. \tag{7.54}$$

EXAMPLE 7.3

Suppose that both the load and the capacity on a device are known within a factor of two with 90% confidence. What value of the safety factor, c_0/l_0, must be used if the failure probability is to be no more than 1.0%?

Solution   For $\Phi(\beta) = r = 1 - p = 0.99$ we find from Appendix C that $\beta = 2.33$. From Eq. 3.73 for 90% confidence with a factor of $n = 2$ uncertainty, we have for both load and capacity $\omega_c = \omega_l = \omega = (1/1.645)\ln(n) = (1/1.645)\ln(2) = 0.4214$. Solve Eq. 7.54 for c_0/l_0:

$$\frac{c_0}{l_0} = \exp[\beta(\omega_c^2 + \omega_l^2)^{1/2}] = \exp(\beta\sqrt{2}\omega)$$

$$= \exp(2.33 \times 1.414 \times 0.4214) = 4.01. \quad \blacktriangleleft$$

Combined Distributions

In general, it is difficult to evaluate analytically the expressions given for reliability when the load and capacity are given by different distributions. However, when the load or capacity is given by an extreme value distribution and the other by a normal distribution, both analytical results and some insight can be obtained.

Consider first a system whose capacity is approximated by the minimum extreme-value distribution introduced in Chapter 3, but about whose loading there is only a small amount of uncertainty. This situation is depicted in Fig. 7.8a. We assume that $\bar{l}$, the mean value of the load, is much smaller than the

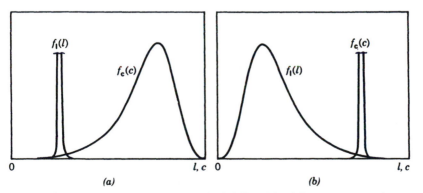

FIGURE 7.8 Graphical representations of reliability: (*a*) minimum extreme-value distribution for capacity, (*b*) maximum extreme-value distribution for loading.

mean, $\bar{c} \equiv u - \Theta\gamma$, of the minimum extreme-value distribution that represents the capacity: $\bar{l} \ll \bar{c}$. For known loading the reliability is given by Eq. 7.18. Thus using CDF from Eq. 3.101, we have

$$r(l) = \exp[-e^{(l-u)/\Theta}], \tag{7.55}$$

which for small enough values of l (i.e., $l \ll u$) becomes

$$r(l) \approx 1 - \exp\left(\frac{l-u}{\Theta}\right). \tag{7.56}$$

Now suppose that we want to take into account some natural variation in the loading on the system. If this is represented by a distribution with small variance of the load about the mean, Eq. 7.19 may be employed to express the reliability as

$$r = 1 - \int_0^\infty f_l(l) \exp\left(\frac{l-u}{\Theta}\right) dl. \tag{7.57}$$

Again, it must be assumed that the variance of the load is not large, $\sigma_l \ll \bar{c} - \bar{l}$, so that the expansion, Eq. 7.56, is valid over the entire range of l where $f_l(l)$ is significantly greater than zero. We obtain for the reliability

$$r = 1 - \exp\left[\frac{1}{2}\left(\frac{\sigma_l}{\theta}\right)^2\right] \exp\left(\frac{\bar{l}-u}{\Theta}\right), \tag{7.57}$$

where $u \equiv \bar{c} + \Theta\gamma \gg \bar{l}$ and γ is Euler's constant.

In the converse situation the capacity has only a small degree of uncertainty, whereas the loading is represented by a maximum extreme-value distribution, again with the stipulation that $\bar{c} \gg \bar{l}$. This situation is depicted in Fig. 7.8b. The reliability at known capacity is first obtained by substituting the maximum extreme-value distribution from Eq. 3.99 into Eq. 7.10,

$$r(c) = F_l(c) = \exp[-e^{-(c-u)/\Theta}], \tag{7.58}$$

or for large c,

$$r(c) \approx 1 - e^{-(c-u)/\Theta}. \tag{7.59}$$

Thus, from Eq. 7.11, we have

$$r = \int_0^\infty f_c(c)\left[1 - \exp\left(\frac{u-c}{\Theta}\right)\right] dc, \tag{7.60}$$

provided that the variance in $f_c(c)$ is small enough that Eq. 7.59 is valid. The resulting reliability is

$$r = 1 - \exp\left[-\frac{1}{2}\left(\frac{\sigma_c}{\Theta}\right)^2\right] \exp\left(\frac{u-\bar{c}}{\Theta}\right). \tag{7.61}$$

where $u \equiv \bar{l} - \Theta\gamma \ll \bar{c}$ and γ is Euler's constant.

7.4 REPETITIVE LOADING

We have considered time only implicitly, or not at all, in conjunction with load-capacity interference theory. Load has been represented as the maximum load over the life of the device or system. Therefore with longer lives the load distribution in Fig. 7.3, would shift to the right, causing the reliability to decrease. Likewise, aging effects have been taken into account only in the conservatism in which the capacity distribution is chosen; it should take weakening with age into account.

Time, however, is arguably the most important variable in many reliability considerations. The bathtub curve representation of failure rate curve pictured in Fig. 6.1 is ubiquitous in characterizing the reliability losses that cause infant mortality, random failures and aging. In this and the following section we demonstrate how load and capacity interact under repetitive loading and result in these three failure mechanisms. Specifically, infant mortality is closely associated with capacity variability, random failures with loading variability, and aging with capacity deterioration. These associations provide a rational for the bathtub shapes of failure rate curves and clarify the relationship between the three failure classes and the corresponding causes of quality loss enumerated by Taguchi: product noise, outer noise, and inner noise.

Loading Variability

Consider a system subject to repetitive loading, and assume that the magnitude of each load is determined by a random variable l, described by a probability density $f_l(l)$. Suppose, for now, that we specify a system with a known capacity $c(t)$ at time t. The probability that a load occurring at time t will cause system failure is then just the probability that $l > c(t)$, or

$$p = \int_{c(t)}^{\infty} f_l(l)\, dl. \tag{7.62}$$

Repetitive loading may occur at either equal or random time intervals, as pictured in Figs. 7.9a or 7.9b respectively. The model that follows is based on random intervals, although when the mean time between loads becomes small the two models yield nearly identical results. We model the random times at which the loads occur by specifying that during a vanishingly small time increment, Δt, the probability of load occurrence is $\gamma\, \Delta t$, where Δt is so small that $\gamma\, \Delta t \ll 1$. The probability of a load occurring at any time is then independent of the time at which the last loading occurred; the loading is then said to be Poisson distributed in time with a frequency γ. The probability of a load that is large enough to cause failure occurring between t and $t + \Delta t$ is thus $p\gamma\, \Delta t$ or, using Eq. 7.62,

$$\gamma \int_{c(t)}^{\infty} f_l(l)\, dl\, \Delta t. \tag{7.63}$$

The system, however, can fail only once. Thus it will fail between t and $t + \Delta t$ only if it has survived to time t *and* the failing load occurs during Δt.

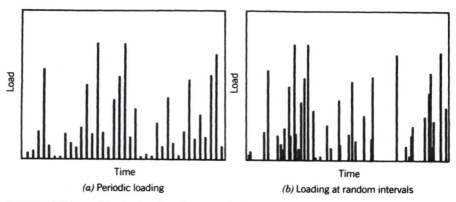

FIGURE 7.9 Repetitive loads of random magnitudes. (a) Periodic loading, (b) Loading at random intervals.

But $R(t)$, the reliability, is just the probability that the system has survived to t. Thus the failure probability during Δt is $R(t)p\gamma\,\Delta t$. Likewise the reliability at $t + \Delta t$ is just the probability that the system survived to t *and* that no failure load occurred during Δt. Since we take the *and* to represents independent events, we may write

$$R(t + \Delta t) = \left[1 - \gamma \int_{c(t)}^{\infty} f_l(l)\,dl\,\Delta t \right] R(t). \qquad (7.64)$$

Rearranging terms yields

$$\frac{R(t + \Delta t) - R(t)}{\Delta t} = -\gamma \int_{c(t)}^{\infty} f_l(l)\,dl\,R(t). \qquad (7.65)$$

Taking the limit as $\Delta t \to 0$ then yields the same form as Eq. 6.15,

$$\lambda(t) = -\frac{1}{R(t)}\frac{d}{dt}R(t), \qquad (7.66)$$

where the failure rate is given in terms of the load distribution as

$$\lambda(t) = \gamma \int_{c(t)}^{\infty} f_l(l)\,dl. \qquad (7.67)$$

This equation clearly indicates that if the capacity of the system is time-independent, so that $c(t) \to c_0$, then time also disappears from the failure rate, yielding the constant failure rate model

$$\lambda = \gamma \int_{c_0}^{\infty} f_l(l)\,dl, \qquad (7.68)$$

and the common exponential distribution $R(t) = exp(-\lambda t)$ results.

EXAMPLE 7.4

A microwave transmission tower is to be constructed at a location where an average of 15 lightning strikes per year are expected. The mean value of the peak current is estimated to be 20,000 amperes, and the peak currents are modeled by an exponential distribution. The MTTF is to be no less than 10 years.

(a) What value of the failure rate is acceptable?

(b) For what peak amperage must the protection system be designed?

> *Solution*   (a) For a constant failure rate phenomena we have
>
> $$\lambda = 1/\text{MTTF} = 1/10 = 0.1 \text{ yr}^{-1}$$

(b) From Eq. 3.88 we may write the exponential load distribution as $F_l(l) = 1 - e^{-l/\bar{l}}$ where the mean load $\bar{l} = 20,000$ and $\gamma = 15/\text{yr}$. Using the relationship between $f_l(l)$ and $F_l(l)$ we may write Eq. 7.68 as

$$\lambda = \gamma \int_{c_0}^{\infty} f_l(l)\, dl = \gamma\,[1 - F_l(c_0)] = \gamma \exp(-c_0/\bar{l}).$$

Since MTTF $= 1/\lambda$ we have

$$\text{MTTF} = \frac{1}{\gamma} \exp(c_0/\bar{l})$$

or inverting,

$$(c_0/\bar{l}) = \ln\,(\gamma \text{MTTF}) = \ln\,(15 \cdot 10) = 5.0$$

or

$$c_0 = 20,000 \cdot 5.0 = 100,000 \text{ Amperes}$$

Aging is present if the capacity decreases with time. We represent this deterioration as

$$c(t) = c_0 - g(t), \tag{7.69}$$

where c_0 is the initial capacity, at $t = 0$, and $g(t)$ is a monotonically increasing function of time, with $g(0) = 0$. Clearly, if the capacity decreases as time elapses, the failure rate will grow, since the lower limit on the integral in Eq. 7.67 then moves toward zero. The rate at which the failure rate increases, however, will be sensitive to the loading distribution as well as to $c(t)$.

Once the failure rate is known, the reliability can be obtained from Eq. 6.18. Thus

$$R(t\,|\,c_0) = \exp\left[-\int_0^t dt'\,\gamma \int_{c(t')}^{\infty} f_l(l)\, dl\right], \tag{7.70}$$

where $c(t)$ is given by Eq. 7.69.

EXAMPLE 7.5

Assume that the capacity of the microwave tower in Example 7.4 deteriorates at a constant rate of 1% per year.

(a) What is the 10 year % decrease in capacity?

(b) What is the 10 year % increase in failure rate?

(c) What is the probability that a damaging lightning strike will take place in the first 10 years without deterioration, and

(d) with deterioration?

> *Solution*   (a) Let $c(t) = c_0(1 - \alpha t)$, where $\alpha = 0.01/\text{yr}$. After 10 years the capacity decrease is $0.01 \times 10 = 10\%$.

(b) Replacing c_0 by $c(t)$ in Example 7.4 we have

$$\lambda(t) = \gamma \exp[-c_0(1 - \alpha t)/\bar{l}] = \lambda(0)\exp(\alpha t c_0/\bar{l}).$$

Since $\alpha t = 0.1$ and $(c_0/\bar{l}) = 5.0$, we have

$$\lambda(10) = \lambda(0) e^{0.1 \times 5.0} = 1.65\, \lambda(0).$$

Thus the increase is 65%.

(c) $1 - R(10) = 1 - e^{-\lambda(0)t} = 1 - e^{-0.1 \times 10} = 0.632$

(d) $\int_0^t \lambda(t')\, dt' = \lambda(0) \int_0^t e^{\alpha t' c_0/\bar{l}}\, dt' = \lambda(0)\,(\alpha c_0/\bar{l})^{-1}(e^{\alpha t c_0/\bar{l}} - 1)$

$$\int_0^{10} \lambda(t')\, dt' = 0.1\,(0.01 \times 5.0)^{-1}(e^{0.1 \times 5.0} - 1) = 1.3$$

$$1 - R(10) = 1 - \exp\left(-\int_0^{10} \lambda(t')\, dt'\right) = 1 - e^{-1.3} = 0.727$$

Variable Capacity

We next consider situations where not every unit of a system or device has exactly the same initial capacity. In reality they would not, since variability in manufacturing processes inevitably leads to some variability in capacity. We model this variability by letting c_0 become a random variable which is described by the probability density function $f_c(c_0)$. We next consider the ensemble of such units, each with its own capacity. The system reliability is then an ensemble average over c_0:

$$R(t) = \int_0^\infty dc_0 f_c(c_0)\, R(t\,|\,c_0). \tag{7.71}$$

Inserting Eq. 7.70 then yields

$$R(t) = \int_0^\infty dc_0 f_c(c_0)\, \exp\left[-\int_0^t dt'\, \gamma \int_{c(t')}^\infty f_l(l)\, dl\right]. \tag{7.72}$$

To focus on the effect of variable capacity on failure rates, we ignore deterioration for the moment by setting $c(t) = c_0$ and assume some fraction, say p_d, of the systems under consideration are flawed in a serious way. This situation may be modeled by writing the PDF of capacities in terms of the Dirac delta functions as

$$f_c(c_0) = (1 - p_d)\delta(c_0 - c_r) + p_d\delta(c_0 - c_d). \tag{7.73}$$

The first term on the right-hand side corresponds to the probability that the system will be a properly built system with target design capacity of c_r. By using the Dirac delta function, we are assuming that the capacity variability of the properly built systems can be ignored. The second term corresponds to the probability that the system will be defective and have a reduced capacity $c_d < c_r$. Such a situation might arise, for example, if a critical component were to be left out of a small fraction of the systems in assembly, or if, in construction, members were not properly assembled with some probability p_d.

The reliability is obtained by first substituting Eq. 7.73 into 7.72 and using the Dirac delta function property given in Eq. 3.56 to evaluate the integrals,

$$R(t) = (1 - p_d) \exp(-\lambda_r t) + p_d \exp(-\lambda_d t), \tag{7.74}$$

where for brevity, we have defined the failure rates

$$\lambda_r = \gamma \int_{c_r}^{\infty} f_l(l) \, dl \tag{7.75}$$

and

$$\lambda_d = \gamma \int_{c_d}^{\infty} f_l(l) \, dl. \tag{7.76}$$

Since the failure rate must increase with decreased capacity, $\lambda_r < \lambda_d$. We now use the definition of the time-dependent failure rate given in Eq. 7.66 to obtain, after evaluating the derivative,

$$\lambda(t) = \lambda_r \left\{ \frac{1 + \dfrac{p_d}{1 - p_d} \dfrac{\lambda_d}{\lambda_r} \exp[-(\lambda_d - \lambda_r)t]}{1 + \dfrac{p_d}{1 - p_d} \exp[-(\lambda_d - \lambda_r)t]} \right\}. \tag{7.77}$$

The decreasing failure rate associated with infant mortality may be seen to appear as a result of the presence of the units with substandard capacities. For clarity we consider the extreme example of a system for which the probability of defective construction is small, $p_d \ll 1$, but for which the defect greatly increases the failure rate, $\lambda_d \gg \lambda_r$. In this case Eq. 7.77 reduces to

$$\lambda(t) = \lambda_r \left(1 + p_d \frac{\lambda_d}{\lambda_r} e^{-\lambda_d t} \right). \tag{7.78}$$

Thus the failure rate decreases from a value of $\approx \lambda_r + p_d \lambda_d$ at zero time to the value of λ_r for the unflawed systems that remain after all defective units have failed.

EXAMPLE 7.6

A servomechanism is designed to have a constant failure rate and a design-life reliability of 0.99, in the absence of defects. A common manufacturing defect, however, is known to cause the failure rate to increase by a factor of 100. The purchaser requires the design-life reliability to be at least 0.975.

(*a*) What fraction of the delivered servomechanisms may contain the defect if the reliability criterion is to be met?

(*b*) If 10% of the servomechanisms contain the defect, how long must they be worn in before delivery to the purchaser?

Solution   (*a*) Without the defect, the failure rate $\lambda_r \equiv \lambda(c_r)$ may be found in terms of the design life T by $R_0(T) = e^{-\lambda_r T}$; then

$$\lambda_r T = \ln\left[\frac{1}{R(T)}\right] = \ln\left(\frac{1}{0.99}\right) = 0.01005.$$

To determine p, the acceptable fraction of units with defects, solve Eq. 7.74; with $t = T$ for p_d:

$$p_d = \frac{1 - R(T)\exp[+\lambda_r T]}{1 - \exp[-(\lambda_d - \lambda_r)T]}.$$

With $\lambda_d \equiv \lambda(c_d) = 100\,\lambda_r$, $R(T) = 0.975$, and $\lambda_r T = 0.01005$,

$$p_d = \frac{1 - 0.975e^{+0.01005}}{1 - e^{-94 \times 0.01005}} = 0.024.$$

(*b*) Recall the definition for reliability with wearin from Eq. 6.51 Combining Eq. 7.74 with this expression, we have, for a wearin period T_w;

$$R(T \mid T_w) = \frac{(1 - p_d)\exp[-\lambda_r(T + T_w)] + p_d\exp[-\lambda_d(T + T_w)]}{(1 - p_d)\exp(-\lambda_r T_w) + p_d\exp(-\lambda_d T_w)}.$$

Solve for T_w:

$$T_w = \frac{1}{\lambda_d - \lambda_r}\ln\left[\frac{p_d}{1 - p_d}\frac{R(T \mid T_w)\exp(-\lambda_d T)}{\exp(-\lambda_r t) - R(T \mid T_w)}\right].$$

With $R(T \mid T_w) = 0.975$, $p_d = 0.1$, $\lambda_r T = 0.01005$, and $\lambda_d T = 1.005$,

$$T_w = \frac{T}{99}\ln\left(\frac{0.1}{1 - 0.1}\frac{0.975 - e^{-100 \times 0.01005}}{e^{-0.01005} - 0.975}\right)$$

$$= 0.015\,T \text{ or } 1\tfrac{1}{2}\% \text{ of the design life.}$$

7.5   THE BATHTUB CURVE—RECONSIDERED

The preceding examples illustrate the constant failure rate that results from loading variability, the increasing failure rates resulting from the combined effects of loading variability and product deterioration, and the decreasing failure rates from loading and initial capacity variability. We next look at the three classes of failure individually and in combination to show how the bathtub curve arises. Table 7.1 lists the eight combinations that may be considered. We next write a general expression for the failure rate that includes all three modes. Since the failure rate is defined in terms of the reliability by Eq. 7.66, we may insert Eq. 7.72 for the reliability and perform the derivative

TABLE 7.1 Failure Modes and Their Interactions

Case	1	2	3	4	5	6	7	8
I. Infant Mortality	no	no	no	yes	no	yes	yes	yes
II. Random Failures	no	no	yes	no	yes	no	yes	yes
III. Aging	no	yes	no	no	yes	yes	no	yes

to yield

$$\lambda(t) = \frac{\gamma \int_0^\infty dc_0 f_c(c_0) \int_{c(t)}^\infty f_l(l)\,dl \exp\left[-\gamma \int_0^t dt' \int_{c(t')}^\infty f_l(l)\,dl\right]}{\int_0^\infty dc_0 f_c(c_0) \exp\left[-\gamma \int_0^t dt' \int_{c(t')}^\infty f_l(l)\,dl\right]}. \quad (7.79)$$

Equations 7.69, 7.72 and 7.79 constitute a reliability model in which infant mortality, random failures, and aging are represented explicitly in terms of capacity variability, loading variability, and capacity degradation.

The relationships are summarized in the first two columns of Table 7.2. Any phenomenon may be eliminated from consideration as indicated in the third column. The fourth column exhibits the particular load and capacity distributions used in the numerical examples that follow. These are normal distributions of load and capacity; in these, we use $\nu = 1.5$ for the safety factor, with $\rho_l = 0.15$ and $\rho_c = 0.10$ for the load and capacity coefficients of variation. We examine the failure modes and their interactions by considering individually each of the eight combinations enumerated in Table 7.1. For each case, load and capacity are plotted versus time in Fig. 7.10 for schematic realizations of the stochastic loading process. The normal distribution plotted on the vertical axis is used to denote cases with variable capacity; the vertical lines denote loading magnitudes at random time intervals.

Single Failure Modes

Of the eight cases, the first is trivial since, as indicated in Fig. 7.10, the absence of both variability and aging leads to a vanishing failure rate and a reliability

TABLE 7.2 Failure Mode Characterization

	Failure mode	Governing property	Mode absent	Mode* present
I.	Infant Mortality (variable capacity)	$f_c(c_0)$	$f_c(c_0) = \delta(c_0 - \bar{c}_0)$	$f_c(c_0) = \phi[(c_0 - \bar{c}_0)/\sigma_c]$
II.	Random Failures (variable load)	$f_l(l)$	$f_l(l) = \delta(l - \bar{l})$	$f_l(l) = \phi[(l - \bar{l})/\sigma_l]$
III.	Aging (deteriorating capacity)	$g(t)$	$g(t) = 0$	$g(t) = \alpha c_0 (t/t_0)^m$

* $\phi(u) \equiv (2\pi)^{-1/2} \exp(-\tfrac{1}{2}u^2)$

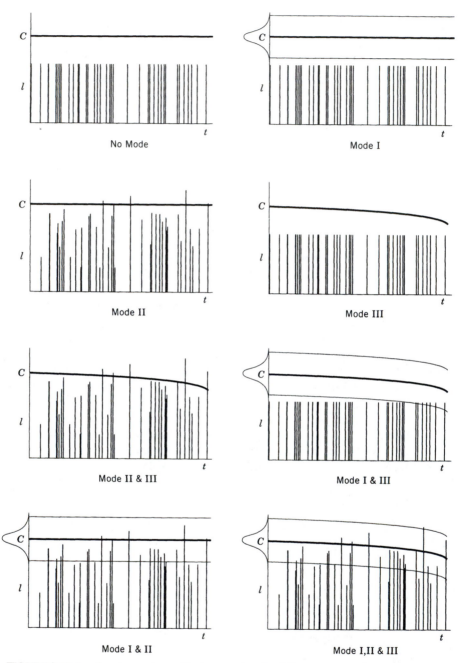

FIGURE 7.10 Load and capacity realizations vs. time for failure mode combinations. (I-infant mortality, II-random, III-aging)

equal to one. In cases two and three there is no capacity variability, and therefore Eqs. 7.72 and 7.79 reduce to Eqs. 7.70 and 7.67. In case two only mode III, aging, is present. Thus the loading is represented by the Dirac delta function, and we may further reduce the Eqs. 7.67 and 7.70 to

$$\lambda(t) = \begin{cases} 0, & t < t_f \\ \gamma, & t > t_f \end{cases}. \tag{7.80}$$

where $t_f = g^{-1}(c_0 - \bar{l})$. Thus,

$$R(t) = \begin{cases} 1, & t < t_f \\ e^{-\gamma(t-t_f)}, & t > t_f \end{cases}. \tag{7.81}$$

This system does not fail before time t_f, but at the first loading thereafter, causing the rapid exponential decay in the reliability. In case three, where only mode II, random failure, due to load variability is present, we replace $c(t)$ by c_0 in Eq. 7.70 to obtain a constant failure rate and the characteristic exponential decay of the reliability.

In case four where only mode I, infant mortality, caused by variable capacity, is present the situation is somewhat more complex. Setting $c(t)$ equal to c_0 and using the Dirac delta function for loading in Eqs. 7.72 and 7.79, we obtain

$$R(t) = 1 - (1 - e^{-\gamma t}) \int_0^{\bar{l}} f_c(c_0)\, dc_0 \tag{7.82}$$

and a corresponding failure rate of

$$\lambda(t) = \frac{\gamma e^{-\gamma t} \int_0^{\bar{l}} f_c(c_0)\, dc_0}{1 - (1 - e^{-\gamma t}) \int_0^{\bar{l}} f_c(c_0)\, dc_0}. \tag{7.83}$$

In this situation the fraction of the system population for which $c_0 < \bar{l}$ fails at the first loading, causing the reliability to drop sharply and then stabilize; the failure rate decreases exponentially at a very rapid rate.

In each of the preceding three cases only one failure mode is present. The modes are compared through the schematic diagrams of reliability and failure rate given in Fig. 7.11*a* and 7.11*b*. The failure rate curves, in particular,

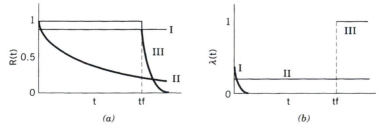

FIGURE 7.11 Effects of single failure modes: (a) reliability, (b) failure rate.

are instructive since they show that the cases of pure infant mortality, random failures and aging failures to some extent resemble the bathtub curve. The differences, however, are striking. The infant mortality contribution drops quickly to zero, since if the system does not fail at the first loading it does not fail at all. Unlike bathtub curves, the failure rate from aging is zero until t_f, at which time it jumps to a value of γ, causing the reliability to drop sharply to zero. Thus it is clear that simple superposition of the failure rates depicted in Fig. 7.11 do not accurately represent the bathtub curve. To obtain realistic results we must also examine the interactions between failure modes.

Combined Failure Modes

Next, we consider combinations of two failure modes. Equations 7.70 and 7.67 describe case five, which combines random failures and aging, modes II and III. Aging is modeled by a power law

$$g(t) = 0.1 c_0 (t/t_0)^m, \tag{7.84}$$

where we take $\gamma t_0 = 100$. In Fig. 7.12 the failure rate is shown to be increasing with time with a behavior which is closely correlated to exponent m in the aging model.

In case six, infant mortality and aging modes I and III, occur together in the absence of random failures. The reliability and failure rate are obtained by replacing the load PDF in Eqs. 7.72 and 7.79 by a Dirac delta function. The reduced expressions are

$$R(t) = 1 - (1 - e^{-\gamma t}) \int_0^{\bar{l}} f_c(c_0)\, dc_0 - \int_{\bar{l}}^{\bar{l}+g(t)} \{1 - e^{-\gamma[t - g^{-1}(c_0 - \bar{l})]}\} f_c(c_0)\, dc_0 \tag{7.85}$$

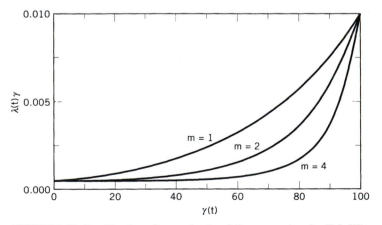

FIGURE 7.12 Combined random and aging failure rates (modes II & III) vs. time for several values of m.

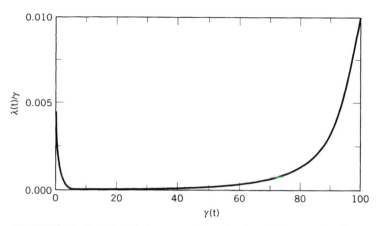

FIGURE 7.13 Combined infant mortality and aging failure rates (modes I & III) vs. time.

for the reliability and

$$\lambda(t) = \frac{\gamma e^{-\gamma t} \left[\int_0^{\bar{l}} f_c(c_0) \, dc_0 + \int_{\bar{l}}^{\bar{l}+g(t)} e^{\gamma g^{-1}(c_0 - \bar{l})} f_c(c_0) \, dc_0 \right]}{1 - (1 - e^{-\gamma t}) \int_0^{\bar{l}} f_c(c_0) \, dc_0 - \int_{\bar{l}}^{\bar{l}+g(t)} \{1 - e^{-\gamma[t - g^{-1}(c_0 - \bar{l})]}\} f_c(c_0) \, dc_0} \tag{7.86}$$

for the failure rate. The failure rate is plotted in Fig. 7.13. This situation resembles that encountered frequently in fatigue testing, where the loading magnitude is carefully controlled. After that fraction of the population for which the initial capacity is less than the load is removed at the first loading, the failure rate is vanishingly small until the effects of aging become significant.

In case seven infant mortality and random failures, modes I and II, are present in the absence of aging. Results obtained by setting $c(t) = c_0$ in Eqs.

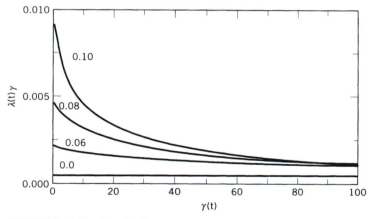

FIGURE 7.14 Combined infant mortality and random failure rates (modes I & II) vs. time for several values of ρ_r.

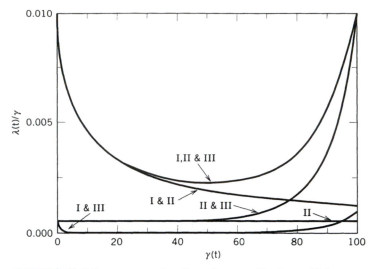

FIGURE 7.15 Failure rates vs. time for various combinations of failure modes.

7.72 and 7.79 are shown in Fig. 7.14. The interaction of infant mortality and random failure modes causes the characteristic decreasing failure rate frequently observed in electronic equipment.

Finally, we consider the eighth case where all three failure modes are present, using Eqs. 7.72 and 7.79 for reliability and failure rate. The bathtub curve characteristics are shown in Fig. 7.15 where we have also included curves for various combinations of two failure modes. These are obtained by removing one failure mode, but keeping the remaining parameters fixed. These results illuminate the origins of the three failure modes: infant mortality with capacity variability, random failures with loading variability, and aging with capacity deterioration. Moreover, while changes in load or capacity distribution often have large effects on the quantitative behavior of the failure rate cures, the qualitative behavior remains essentially the same. The model indicates, however, that the interactions between the three modes are very important in determining the failure rate cure. Thus only if the three failure modes arise from independent failure mechanisms or in different components is it legitimate simply to sum the failure rate contributions.

Bibliography

Ang, A. H-S., and W. H. Tang, *Probability Concepts in Engineering Planning and Design*, Vol. 1, Wiley, NY, 1975.

Brockley, D., (ed.) *Engineering Safety*, McGraw-Hill, London, 1992.

Freudenthal, A. M., J. M. Garrelts, and M. Shinozuka, "The Analysis of Structural Safety," *Journal of the Structural Division ASCE ST 1*, 267–325 (1966).

Gumbel, E. J., *Statistics of Extremes*, Columbia University Press, NY, 1958.

Haugen, E. B., *Probabilistic Mechanical Design*, Wiley, NY, 1980.

Haviland, R. D., *Engineering Reliability and Long Life Design*, Van Nostrand, NY, 1964.

Kapur, K. C., and L. R. Lamberson, *Reliability in Engineering Design*, Wiley, NY, 1977.

Lewis, E. E., and H-C Chen, "Load–Capacity Interference and the Bathtub Curve," *IEEE Trans. Reliability* **43**, 470–475 (1994).

Rao, S. S., *Reliability-Based Design*, McGraw-Hill Inc. New York, 1992.

Thoft–Chirstensen, P., and M. J. Baker, *Structural Reliability Theory and Its Application*, Springer–Verlag, Berlin, 1982.

Exercises

7.1 A design engineer knows that one-half of the lightning loads on a surge protection system are greater than 500 V. Based on previous experience, such loads are known to follow the PDF:

$$f(v) = \gamma e^{-\gamma v}, \qquad 0 \le v < \infty.$$

(a) Estimate γ per volt.

(b) What is the mean load?

(c) For what voltage should the system be designed if the failure probability is not to exceed 5%?

7.2 Given the following distributions of capacity and load, determine the failure probability:

$$f_c(c) = 5c^4 \qquad 0 < c < 1$$
$$= 0 \qquad \text{otherwise}$$
$$f_l(l) = 2 \qquad 0 < 1 < 1/2$$
$$= 0 \qquad \text{otherwise}$$

7.3 Suppose that the PDFs for load and capacities are

$$f_l(l) = \gamma e^{-\lambda l}, \qquad 0 \le l \le \infty,$$

$$f_c(c) = \begin{cases} 0, & 0 \le c < a. \\ 1/a, & a \le c \le 2a, \\ 0, & 2a < c \le \infty. \end{cases}$$

Determine the reliability; evaluate all integrals.

7.4 The impact loading on a railroad coupling is expressed as an exponential distribution:

$$f_l(l) = \beta e^{-\beta l}.$$

The coupling is designed to have a capacity $\mathbf{c} = c_m$. However, because of material flaws, the PDF for the capacity is more accurately expressed as

$$f_c(c) = \begin{cases} \dfrac{\alpha e^{\alpha c}}{\exp(\alpha c_m) - 1}, & 0 \leqslant c \leqslant c_m, \\ 0, & c > c_m. \end{cases}$$

(a) Determine the reliability for a single loading, assuming that the flaws can be neglected.

(b) Recalculate a using the capacity distribution with the flaws included.

(c) Show that the result of b reduces to that of a as $\alpha \rightarrow \infty$.

(d) Show that for $\alpha = 0$, the reliability is

$$r = 1 - \frac{1}{\beta c_m} [1 - e^{-\beta c_m}].$$

7.5 It is estimated that the capacity of a newly designed structure is $\bar{c} = 10,000$ kips, $\sigma_c = 6000$ kips, normally distributed. The anticipated load on the structure will be $\bar{l} = 5000$ kips, with an uncertainty of $\sigma_l = 1500$ kips, also normally distributed. Find the *un*reliability of the structure.

7.6 A structural code requires that the reliability index of a cable must have a value of at least $\beta = 5.0$. If the load and capacity may be considered to be normally distributed with coefficients of variation of $\rho_l = 0.2$ and $\rho_c = 0.1$ respectively, what safety factor must be used?

7.7 Steel cable strands have a normally distributed strength with a mean of 5000 lb and a standard deviation of 150 lb. The strands are incorporated into a crane cable that is proof-tested at 50,000 lb. It is specified that no more than 2% of the cables may fail the proof test. How many strands should be incorporated into the cable, assuming that the cable strength is the sum of the strand strengths?

7.8 Substitute the normal distributions for load and capacity, Eqs. 7.29 and 7.30, into the reliability expression, Eq. 7.20. Show that the resulting integral reduces to Eqs. 7.41 and 7.43.

7.9 The twist strength of a standard bolt is 23 N $\cdot$ m with a standard deviation of 1.3 N $\cdot$ m. The wrenches used to tighten such bolts have an uncertainty of $\sigma = 2.0$ N $\cdot$ m in their torsion settings. If no more than 1 bolt in 1000 may fail from excessive tightening, what should the setting be on the wrenches? (Assume normal distributions.)

7.10 Suppose that a car hits potholes spaced at random distances at a rate of 20/hour. The loading on the wheel bolts caused by these potholes is exponentially distributed.

$$f_l(l) = 0.6 \exp(-0.6l), \qquad 0 \leq l \leq \infty$$

What will the failure rate be if the bolt capacity is designed to be exactly eight times the mean value of the pothole loading?

7.11 Suppose that both load and capacity are known to a factor of two with 90% confidence. Assuming lognormal distributions, determine the safety factor c_0/l_0 necessary to obtain a reliability of 0.995.

7.12 Show in detail that Eq. 7.61 follows from Eqs. 7.30 and 7.60.

7.13 The loading on industrial fasteners of fixed capacity is known to follow an exponential distribution. Thirty percent of the fasteners fail. If the fasteners are redesigned to double their capacity, what fraction will be expected to fail?

7.14 Consider a pressure vessel for which the capacity is defined as p, the maximum internal pressure that the vessel can withstand without bursting. This pressure is given by $p = \tau_0 \sigma_m/2R$, where τ_0 is the unflawed thickness, σ_m is the stress at which failure occurs, and R is the radius. Suppose that the vessel thickness is $\tau(\geq \tau_0)$, but the distribution crack depths are the same as those given in Exercise 3.9.

(a) Show that the PDF for capacity is

$$
f_p(p) \equiv
\begin{cases}
\dfrac{2R}{\gamma \sigma_m} \dfrac{1}{e^{\tau/\gamma} - 1} \exp\left(\dfrac{2R}{\gamma \sigma_m} p\right), & 0 \leq p \leq \dfrac{\tau \sigma_m}{2R}, \\[2ex]
0, & p > \dfrac{\tau \sigma_m}{2R}.
\end{cases}
$$

(b) Normalize to $\tau \sigma_m/2R = 1$, then plot $f_p(p)$ for $\gamma = \tau$, 0.5τ, and 0.1τ.

(c) Physically interpret the results of your plots.

7.15 In Exercise 7.14, suppose that the vessel is proof-tested at a pressure of $p = \tau \sigma_m/4R$. What is the probability of failure if

(a) $\gamma = 0.5\tau$?

(b) $\gamma = 0.1\tau$?

7.16 A system under a constant load, l, has a known capacity that varies with time as $c(t) = c_0(1 - 0.02\,t)$. The safety factor at $t = 0$ is 2.

(a) Sketch $R(t)$

(b) What is the MTTF?

(c) What is the variance of the time to failure?

7.17 Suppose that steel wire has a mean tensile strength of 1200 lb. A cable is to be constructed with a capacity of 10,000 lb. How many wires are required for a reliability of 0.999

(a) if the wires have a 2% coefficient of variation?

(b) If the wires have a 5% coefficient of variation?

(*Note:* Assume that the strengths are normally distributed and that the cable strength is the sum of the wire strengths.)

7.18 Consider a chain consisting of N links that is subjected to M loads. The capacity of a single link is described by the PDF $f_c(c)$. The PDF for any one of the loads is described by $f_l(l)$. Derive an expression in terms of $f_c(c)$ and $f_l(l)$ for the probability that the chain will fail from the M loadings.

7.19 Suppose that the CDF for loading on a cable is

$$F_l(l) = 1 - \exp\left[-\left(\frac{l}{500}\right)^3\right],$$

where l is in pounds. To what capacity should the cable be designed if the probability of failure is to be no more than 0.5%?

7.20 Suppose, that the design criteria for a structure is that the probability of an earthquake severe enough to do structural damage must be no more than 1.0% over the 40-year design life of the building.

(a) What is the probability of one or more earthquakes of this magnitude or greater occurring during any one year?

(b) What is the probability of the structure being subjected to more than one damaging earthquake over its design life?

7.21 Assume that the column in Exercise 3.21 is to be built with a safety factor of 1.6. If the strength of the column is normally distributed with a 20% coefficient of variation, what is the probability of failure?

7.22 Prove that Eqs. 7.72 and 7.79 reduce to Eqs. 7.82 and 7.83 under the assumptions of constant loading and no capacity deterioration.

7.23 The impact load on a landing gear is known to follow an extreme-value distribution with a mean value of 2500 and a variance of 25×10^4. The capacity is approximated by a normal distribution with a mean value of 15,000 and a coefficient of variation of 0.05. Find the probability of failure per landing.

7.24 Prove that Eqs. 7.72 and 7.79 reduce to Eqs. 7.85 and 7.86 under the assumption of constant loading.

7.25 A dam is built with a capacity to withstand a flood with a return period (i.e. mean time between floods) of 100 years. What is the probability that the capacity of the dam will be exceeded during its 40-year design life?

7.26 Suppose that the capacity of a system is given by

$$f_c(c) = \frac{1}{\sqrt{2\pi}\sigma_c}\exp\left\{-\frac{1}{2\sigma_c^2}[c - \bar{c}(t)]^2\right\},$$

where

$$\bar{c}(t) = c_0(1 - \alpha t).$$

If the system is placed under a constant load l,

(a) Find $f(t)$, the PDF for time to failure.

(b) Put $f(t)$ into a standard normal form and find σ_t and the MTTF.

7.27 A manufacturer of telephone switchboards was using switching circuits from a single supplier. The circuits were known to have a failure rate of 0.06/year. In its new board, however, 40% of the switching circuits came from a new supplier. Reliability testing indicates that the switchboards have a composite failure rate that is initially 80% higher than it was with circuits from the single supplier. The failure rate, however, appears to be decreasing with time.

(a) Estimate the failure rate of the circuits from the new supplier.

(b) What will the failure rate per circuit be for long periods of time?

(c) How long should the switchboards be worn in if the average failure rate of circuits should be no more than 0.1/year?

Note: See Example 7.6

7.28 Suppose that a system has a time-independent failure rate that is a linear function of the system capacity c,

$$\lambda(c) = \lambda_0[1 + b(c_m - c)], \qquad b > 0,$$

where c_m is the design capacity of the system. Suppose that the presence of flaws causes the PDF or capacity of the system to be given by $f_c(c)$ in Exercise 7.4.

(a) Find the system failure rate.

(b) Show that it decreases with time.

7.29 The most probable strength of a steel beam is given by $24N^{-0.05}$ kips, where N is the number of cycles. This value is known to within 25% with 90% confidence.

(a) How many cycles will elapse before the beam loses 20% of its strength?

(b) Suppose that the cyclic load on the beam is 10 kips. How many cycles can be applied before the probability of failure reaches 10%?

Note: Assume a lognormal distribution.

CHAPTER 8

Reliability Testing

"One must learn by doing a thing; for though you think you know it, you have not certainty until you try."

Sophocles

8.1 INTRODUCTION

Reliability tests employ a number of the statistical tools introduced in Chapter 5. In contrast to Chapter 5, where emphasis was placed on the more fundamental nature of the statistical estimators, here we examine more closely how the gathering of data and its analysis is used for reliability prediction and verification through the various stages of design, manufacturing, and operation. In reality, the statistical methods that may be employed are often severely restricted by the costs of performing tests with significant sample sizes and by restrictions on the time available to complete the tests.

Reliability testing is constrained by cost, since often the achievement of a statistical sample which is large enough to obtain reasonable confidence intervals may be prohibitively expensive, particularly if each one of the products tested to failure is expensive. Accordingly, as much information as possible must be gleaned from small statistical samples, or in some cases from even a single failure. The use of failure mode analysis to isolate and eliminate the mechanism leading to failure may result in design enhancement long before sufficient data is gathered to perform formal statistical studies.

Testing is also constrained by the time available before a decision must be made in order to proceed to the next phase of the product development cycle. Frequently, one cannot wait the life of the product for it to fail. On specified dates, designs must be frozen, manufacturing commenced and the product delivered. Even where larger sample sizes are available for testing, the severe constraints on testing time lead to the prevalence of censoring and acceleration. In censoring, a reliability test is terminated before all of the units have failed. In acceleration, the stress cycle frequency or stress intensity is increased to obtain the needed failure data over a shorter time period.

These cost and time restrictions force careful consideration of the purpose for which the data is being obtained, the timing as to when the results must be available, and the required precision. These considerations frequently lead to the employment of different methods of data analysis at different points in the product cycle. One must carefully consider what reliability characteristics are important for determining the adequacy of the product. For example, the time-to-failure may be measured in at least three ways:

1. operating time
2. number of on-off cycles
3. calendar time.

If the first two are of primary interest, the test time can be shortened by applying compressed time accelerations, whereas if the last is of concern then intensified stress testing must be used. These techniques are discussed in detail in Section 8.5.

During the conceptual and detailed design stages, before the first prototype is built, reliability data plays a crucial role. Reliability objectives and the determination of associated component reliability requirements enter the earliest conceptual design and system definition. The parts count method, treated in Chapter 6, and similar techniques may be used to estimate reliability from the known failure rate characteristics of standard components. Comparisons to similar existing systems and a good deal of judgment also must be used during the course of the detailed design phase.

Tests may be performed by suppliers early in the design phase on critical components even before system prototypes are built. Thus aircraft, automotive, and other engines undergo extensive reliability testing before incorporation into a vehicle. On a smaller scale, one might decide which of a number of electric motor suppliers to utilize in the design of a small appliance by running reliability tests on the motors. Depending on the design requirement and the impact of failure, such tests may range from quite simple binomial tests, in which one or more of the motors is run continuously for the anticipated life of the machine, to more exhaustive statistical analysis of life testing procedures.

Completion of the first product prototypes allows operating data to be gained, which in turn may be used to enhance reliability. At this stage the test-fix-test-fix cycle is commonly applied to improve design reliability before more formal measures of reliability are applied. As more prototypes become available, environmental stress testing may also be employed in conjunction with failure mode analysis to refine the design for enhanced reliability. These reliability enhancement procedures are discussed in Section 8.2.

As the design is finalized and larger product sample sizes become available, more extensive use of the life testing procedures discussed in Sections 8.3 through 8.6 may be required for design verification. During the manufacturing phase, qualification and acceptance testing become important to ensure that the delivered product meets the reliability standards to which it was designed. Through aggressive quality improvement, defects in the manufacturing process must be eliminated to insure that manufacturing variability does not give rise to unacceptable numbers of infant-mortality failures. Finally, the collection of reliability data throughout the operational life of a system is an important task, not only for the correction of defects that may become apparent only with extensive field service, but also for the setting and optimization of maintenance schedules, parts replacement, and warranty policies.

Data is likely to be collected under widely differing circumstances ranging from carefully controlled laboratory experiments to data resulting from field failures. Both have their uses. Laboratory data are likely to provide more information per sample unit, both in the precise time to failure and in the mechanism by which the failures occur. Conversely, the sample size for field data is likely to be much larger, allowing more precise statistical estimates to be made. Equally important, laboratory testing may not adequately represent the environmental condition of the field, even though attempts are made to do so. The exposures to dirt, temperature, humidity, and other environmental loading encountered in practice may be difficult to predict and simulate in the laboratory. Similarly, the care in operation and quality of maintenance provided by consumers and field crews is unlikely to match that performed by laboratory personnel.

8.2 RELIABILITY ENHANCEMENT PROCEDURES

Reliability studies during design and development are extremely valuable, for they are available at a time when design modifications or other corrections can be made at much less expense than later in the product life cycle. With the building of the first prototypes hands-on operational experience is gained. And as the limitations and shortcomings of the analytical models used for design optimization are revealed, reliability is enhanced through experimentally-based efforts to eliminate failure modes. The number of prototype models is not likely to be large enough to apply standard statistical techniques to evaluate the reliability, failure rate, or related quantities as a function of time. Even if a sample of sufficient size could be obtained, life testing would not in general be appropriate before the design is finalized. If one ran life tests on the initial design, the results would likely underestimate the reliability of the improved model that finally emerged from the prototype testing phase.

The two techniques discussed in this section are often employed as an integral part of the design process, with the failures being analyzed and the design improved during the course of the testing procedure. In contrast, the life testing methods discussed in Sections 8.3 and 8.4 may be used to improve the next model of the product, change the recommended operation procedures, revise the warrantee life, or for any number of other purposes. They are not appropriate, however, while changes are being made to the design.

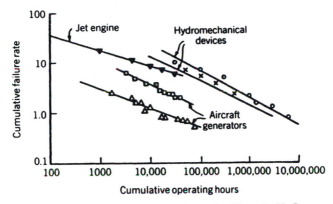

FIGURE 8.1 Duane's data on a log-log scale. [From L. H. Crow, "On Tracking Reliability Growth," *Proceedings 1975 Reliability and Maintainability Symposium*, 438–443 (1975).]

Reliability Growth Testing

Newly constructed prototypes tend to fail frequently. Then, as the causes of the failures are diagnosed and actions taken to correct the design deficiencies, the failures become less frequent. This behavior is pervasive over a variety of products, and has given rise to the concept of reliability growth. Suppose we define the following

T = total operation time accumulated on the prototype

$n(T)$ = number of failures from the beginning of operation through time T.

Duane[*] observed that if $n(T)/T$ is plotted versus T on log-log paper, the result tends to be a straight line, as indicated in Fig. 8.1, no matter what type of equipment is under consideration. From such empirical relationships, referred to as a Duane plots, we may make rough estimates of the growth of the time between failures and therefore also extrapolate a measure of how much reliability is likely to be gained from further cycles of test and fix.

Since Duane plots are straight lines, we may write

$$\ln[n(T)/T] = -\alpha \ln(T) + b, \tag{8.1}$$

or solving for $n(T)$,

$$n(T) = KT^{1-\alpha}, \tag{8.2}$$

where $K = e^b$. Note that if $\alpha = 0$ there is no improvement in reliability, for the number of failures expected is proportional to the testing time. For α greater than zero the expected failures become further and further apart as

[*] J. J. Duane, "Learning Curve Approach to Reliability Modeling," *IEEE. Trans. Aerospace 2* 563 (1964).

the cumulative test time T increases. An upper theoretical limit is $\alpha = 1$, since with this value, Eq. 8.2 indicates that the number of failures is independent of the length of the test.

Suppose we define the rate at which failures occur as just the time derivative of the number of failures, $n(T)$ with respect to the total testing time:

$$\Lambda(T) = \frac{d}{dT} n(T). \tag{8.3}$$

Note that Λ is *not* the same as the failure rate λ discussed at length earlier, since now each time a failure occurs, a design modification is made. Understating this difference, we may combine Eqs. 8.2 and 8.3 to obtain

$$\Lambda(T) = (1 - \alpha) KT^{-\alpha}, \tag{8.4}$$

indicating the decreasing behavior of $\Lambda(T)$ with time.

EXAMPLE 8.1

A first prototype for a novel laser powered sausage slicer is built. Failures occur at the following numbers of minutes: 1.1, 3.9, 8.2, 17.8, 79.7, 113.1, 208.4 and 239.1. After each failure the design is refined to avert further failures from the same mechanism. Determine the reliability grown coefficient α for the slicer.

Solution The necessary calculations are shown on the spread sheet, Table 8.1. A least-squares fit made of column D versus column C. We obtain $a =$ SLOPE(D2:D9,C2:C9) = -0.654. Thus, from Eq. 8.1: $\alpha = 0.654$. The straight-line fit is quite good since we obtain a coefficient of determination that is close to one: $r^2 =$ RSQ(D2:D9,C2:C9) = 0.988.

For the test-fix cycle to be effective in reliability enhancement, each failure must be analyzed and the mechanism identified so that corrective design modifications may be implemented. In product development, these may take the form of improved parts selection, component parameter modifications for increased robustness, or altered system configurations. The procedure is limited by the small sample size—often one—and by the fact that the prototype

TABLE 8.1 Spreadsheet for Reliability Growth Estimate in Example 8.1

	A	B	C	D
	n	T	ln(T)	ln(n/T)
1				
2	1.0	1.1	0.0953	−0.0953
3	2.0	3.9	1.3610	−0.6678
4	3.0	8.2	2.1041	−1.0055
5	4.0	17.8	2.8792	−1.4929
6	5.0	79.7	4.3783	−2.7688
7	6.0	113.1	4.7283	−2.9365
8	7.0	208.4	5.3395	−3.3935
9	8.0	239.1	5.4769	−3.3974

may be operated under laboratory conditions. As failures become increasingly far apart, a point of diminishing returns is reached in which those few that do occur are no longer associated with identifiable design defects. Two strategies may be employed for further reliability enhancement. The first consists of operating the prototypes outside the laboratory under realistic field conditions where the stresses on the system will be more varied. The second consists of artificially increasing the stresses on laboratory prototypes to levels beyond those expected in the field. This second procedure falls under the more general heading of environmental stress testing.

In addition to the development of hardware, Duane plots are readily applied to computer software. As software is run and bugs are discovered and removed, their occurrence should become less frequent, indicating reliability growth. This contrasts sharply to the life-testing methods discussed in the following sections; they must be applied to a population of items of fixed design and therefore are not directly applicable to debugging processes for either hardware prototypes or software.

Reliability growth estimates are applicable to the development and debugging of industrial processes as well as to products. Suppose a new production line is being brought into operation. At first, it is likely that shutdowns will be relatively frequent due to production of out-of-specification products, machinery breakdowns and other causes. As experience is gained and the processes are brought under control, unscheduled shutdowns should become less and less frequent. The progressive improvement can be monitored quantitatively with a Duane plot in terms of hours of operation.

Environmental Stress Testing

Environmental stress testing is based on the premise that increasing the stress levels of temperature, vibration, humidity, or other variables beyond those encountered under normal operational conditions will cause the same failure modes to appear, but at a more rapid rate. The combination of increased stress levels with failure modes analysis often provides a powerful tool for design enhancement. Typically, the procedure is initiated by identifying the key environmental factors that stress the product. Several of the prototype units are then tested for a specified period of time at the stress limits for normal operation. As a next step, voltage, vibration, temperature, or other identified factors are increased in steps beyond the specification limits until failures occur. Each failure is analyzed, and action is taken to correct it. At some level, small increases in stress will cause a dramatic increase in the number of failures. This indicates that fundamental design limits of the system have been exceeded, and further increases in stress are not indicative of the robustness of the design.

Stress tests also may be applied to products taken off the production line during early parts of a run. At this point, however, the changes are typically made to the fabrication or assembly process and with the component suppliers rather than with product design. In contrast to the stress *testing* discussed thus far, whose purpose it is to improve the product design or manufacturing process, environmental stress *screening* is a form of proof or acceptance test. To perform such screening all units are operated at elevated stress levels for some specified period of time, and the failed units are removed. This is comparable to accelerating the burn-in procedure discussed in Chapter 6, for it tends to eliminate substandard units subject to infant mortality failures over a shorter period of time than simply burning them in under nominal conditions. The objective in environmental stress screening is to reach the flat portion of the bathtub curve in a minimum time and at minimum expense before a product is shipped.

In constructing programs for either environmental stress testing or screening, the selection of the stress levels and the choice of exposure times is a challenging task. Whereas theoretical models, such as those discussed in section 8.4 are helpful, the empirical knowledge gained from previous experience or industrial standards most often plays a larger role. Thermal cycling beyond the normal temperature limits is a frequent testing form. The test planner must decide on both a cycling rate and the number of cycles before proceeding to the next cycle magnitude. If too few cycles are used, the failures may not be precipitated; if too many are used, there is a diminishing return on the expenditure of time and equipment use. Often an important factor is that of using the same test for successive products to insure that reliability is being evaluated with a common standard. Figure 8.2 illustrates

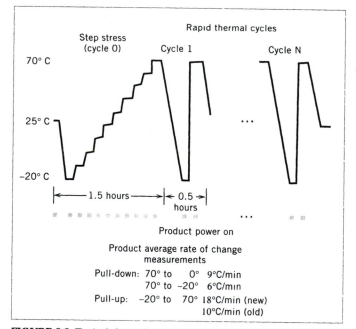

FIGURE 8.2 Typical thermal profiles used in environmental stress testing. (From Parker, T.P. and Harrison, G.L., *Quality Improvement Using Environmental Stress Testing*, pg. 17, AT&T Technical Journal, 71, #4, Aug. 1992. Reprinted by permissions.)

TABLE 8.2 Failure Times

i	ti	i	ti
0	0.00	5	1.50
1	0.62	6	1.62
2	0.87	7	1.76
3	1.13	8	1.88
4	1.25	9	2.03

one such thermal cycling prescription. Note that power on or off must be specified along with the temperature stress profile.

8.3   NONPARAMETRIC METHODS

We begin our treatment of life-testing with the use of nonparametric methods. Recall from Chapter 5.2 that these are methods in which the data are plotted directly, without an attempt to fit them to a particular distribution. Such analysis is valuable in allowing reliability behavior to be visualized and understood. It may also serve as a first step in making a decision whether to pursue parametric analysis, and in providing a visual indication of which class of distributions is most likely to be appropriate.

In either nonparametric or parametric analysis two classes of data may be encountered: ungrouped and grouped. Ungrouped data consists of a series of specific times at which the individual equipment failures occurred. Table 8.2 is an example of ungrouped data. Grouped data consist of the number of items failed within each of a number of time periods, with no information available on the specific times within the intervals at which failures took place. Table 8.3 is typical of grouped data. Both tables are examples of complete data; all the units are failed before the test is terminated.

Ungrouped data is more likely to be the result of laboratory tests in which the sample size is not large, but where instrumentation or personnel are available to record the exact times to failure. Larger sample sizes are often available for laboratory tests of less expensive equipment, such as electronic components. Then, however, it may not be economical to provide instrumenta-

TABLE 8.3 Grouped Failure Data

Time interval	Number of failures
$0 \leq t < 5$	21
$5 \leq t < 10$	10
$10 \leq t < 15$	7
$15 \leq t < 20$	9
$20 \leq t < 25$	2
$25 \leq t < 30$	1

tion for on-line recording of failure times. In such situations, the test is stopped at equal time increments, the components tested, and the number of failures recorded. The result is grouped data consisting of the number of failures during each time interval. Larger sample sizes are also likely to be obtained from field studies. But such data is often grouped in the form of monthly service reports or other consolidated data bases. Whether grouped or ungrouped, field data may require a fair amount of preliminary analysis to determine the appropriate times to failure. For example if the monthly service reports of failure for items that have been sold over several years are to be utilized, the time of sale must also be recorded to determine the time in use. Likewise, it may be necessary to include design or manufacturing modifications, unreported failures, and other complicating factors into the analysis to reduce the data to a usable form.

Ungrouped Data

Ungrouped data consists of a series of failure times $t_1\ t_2, \ldots, t_i, \ldots, t_N$ for the N units in the test. In statistical nomenclature the t_i are referred to as the rank statistics of the test. In Chapter 5 we discuss the utilization of such data to approximate the CDF in Eq. 5.12 as

$$\hat{F}(t_i) = i/(N+1). \tag{8.5}$$

Since the reliability and the CDF are related by $R = 1 - F$, we may make the estimate

$$\hat{R}(t_i) = \frac{N+1-i}{N+1}. \tag{8.6}$$

In addition to the reliability, we would also like to examine the behavior of the failure rate as a function of time. The use of Eqs. 6.10 and 6.14 to accomplish this is problematical since the required numerical differentiation amplifies the random behavior of the data. Instead we define the integral of the failure rate as

$$H(t) = \int_0^t \lambda(t')\, dt', \tag{8.7}$$

which is usually referred to as the cumulative hazard function since in some reliability literature $\lambda(t)$ is called the hazard function instead of the failure rate. Equation 6.18 may then be used to write the reliability as

$$R(t) = e^{-H(t)}, \tag{8.8}$$

which may be inverted to obtain

$$H(t) = -\ln R(t). \tag{8.9}$$

These equations reduce to $H(t) \to \lambda t$ in the case of a constant failure rate. In a hazard plot, $H(t)$ is graphed as a function of time. This provides some insight into the nature of the failure rate: a linear graph indicates a constant

TABLE 8.4 Ungrouped Data Computations

i	ti	R(ti)	H(ti)
0	0.00	1.00	0.0000
1	0.62	0.90	0.1054
2	0.87	0.80	0.2231
3	1.13	0.70	0.3567
4	1.25	0.60	0.5108
5	1.50	0.50	0.6931
6	1.62	0.40	0.9163
7	1.76	0.30	1.2040
8	1.88	0.20	1.6094
9	2.03	0.10	2.3026

failure rate, one whose curve is concave upward indicates a failure rate that is increasing with time, whereas a concave downward curve indicates a failure rate decreasing with time. To present $H(t)$ in a form suitable for plotting, we simply insert Eq. 8.6 into the right hand side of Eq. 8.9. Simplifying the algebra, we obtain

$$\hat{H}(t_i) = \ln(N + 1) - \ln(N + 1 - i) \qquad (8.10)$$

The use of these ungrouped data estimators for $R(t)$ and $H(t)$ are best understood with an example.

EXAMPLE 8.2

From the data in Table 8.2 construct graphs for the reliability and the cumulative hazard function as a function of time.

Solution The necessary calculations are carried out in Table 8.4. The results are plotted in Fig. 8.3. The concave upward behavior of $H(t)$ provides evidence of an increasing failure rate and therefore of wear or aging effects.

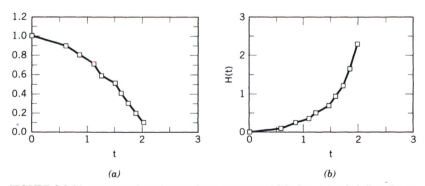

(a)   *(b)*

FIGURE 8.3 Nonparametric estimates from ungrouped life data (a) reliability, (b) cumulative hazard function

The estimate of the MTTF or variance of the failure distribution for ungrouped data is straightforward. We simply adopt the unbiased point estimators discussed in Chapter 5. The mean is given by Eq. 5.6,

$$\hat{\mu} = \frac{1}{N} \sum_{i=1}^{N} t_i, \tag{8.11}$$

and for the variance, Eq. 5.8, becomes

$$\hat{\sigma}^2 = \frac{1}{N-1} \sum_{i=1}^{N} (t_i - \hat{\mu})^2. \tag{8.12}$$

Equation 5.10 can likewise serve as a basis for calculating the skewness and the kurtosis of the time-to-failure distribution.

Grouped Data

Suppose that we want to estimate the reliability, failure rate, or cumulative hazard function of a failure distribution from data such as those given in Table 8.3. We begin with the reliability. The test is begun with N items. The number of surviving items is tabulated at the end of each of the M time intervals into which the data are grouped: $t_1, t_2, \ldots, t_i, \ldots t_M$. The number of surviving items at these times is found to be $n_1, n_2, \ldots, n_i, \ldots$. Since the reliability $R(t)$ is defined as the probability that a system will operate successfully for time t, we estimate the reliability at time t_i to be

$$\hat{R}(t_i) = \frac{n_i}{N}, \qquad i = 1, 2, \ldots, M, \tag{8.13}$$

which is a straightforward generalization of Eq. 5.11. Since the number of failures is generally significantly larger for grouped than for ungrouped data, it usually is not meaningful to derive more precise estimates. Knowing the values of the reliability at the t_i, we may combine Eqs. 8.9 and 8.13 to obtain an empirical plot of the hazard function:

$$\hat{H}(t_i) = \ln N - \ln n_i \tag{8.14}$$

These estimation procedures are illustrated in the following example.

EXAMPLE 8.3

From the data in Table 8.3 estimate the reliability and the cumulative hazard function. Is the failure rate increasing or decreasing?

Solution The necessary calculations, from Eqs. 8.12, 8.13 and 8.14 are indicated in Table 8.5. The resulting values for the quantities are plotted in Fig. 8.4. For $R(t)$ and $H(t)$. Since Fig 8.4*b* is nearly linear, the failure rate increases only slightly—if at all—with increasing time.

TABLE 8.5 Grouped Data Computations

i	ti	ni	R(ti)	H(ti)
0	0	50	1.00	0.0000
1	5	29	0.58	0.5447
2	10	19	0.38	0.9676
3	15	12	0.24	1.4271
4	20	3	0.06	2.8134
5	25	1	0.02	3.9120
6	30	0	0.00	

In addition to obtaining plots of the results for grouped data, we may estimate the mean, variance, or other properties of the failure distribution. We simply approximate $f(t)$ by a histogram. In the interval $t_{i-1} < t < t_i$ and set $f(t)$ equal to

$$f_i = \frac{n_{i-1} - n_i}{N\Delta_i}, \tag{8.15}$$

where the width of the interval is

$$\Delta_i = (t_i - t_{i-1}). \tag{8.16}$$

The integral of Eq. 3.15 is then estimated from

$$\hat{\mu} = \sum_{i=1}^{M} \bar{t}_i f_i \Delta_i, \tag{8.17}$$

where $\bar{t}_i = \frac{1}{2}(t_{i-1} + t_i)$. Likewise, the variance, given by Eq. 3.16, is estimated as

$$\hat{\sigma}^2 = \sum_{i=1}^{M} \bar{t}_i^2 f_i \Delta_i - \hat{\mu}^2. \tag{8.18}$$

8.4   CENSORED TESTING

Next we consider censored reliability tests. Censoring is said to occur if the data are incomplete, either because the test is not run to completion or

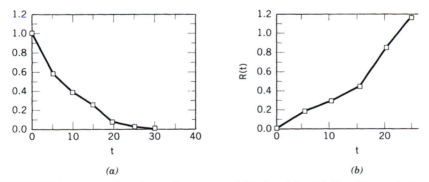

(a)   *(b)*

FIGURE 8.4 Nonparametric estimates from grouped life data (a) reliability, (b) cumulative hazard function

because specimens are removed during the test. Many reliability tests must either be stopped before all the specimens have failed, or intermediate results must be tabulated. The data are then said to be singly censored, or censored on the right, since most data are plotted with time on the horizontal axis. Data are said to be multiply censored if units are removed at various times during a life test. Such removals are usually required either because a mechanism that is not under study caused failure or because the unit is for some other reason no longer available for testing.

Singly-Censored Data

With single-censored grouped data we have available the number of failures for only some of the intervals, say for the first i ($<M$). For ungrouped data there are two types of single censoring. In type I the test is terminated after some fixed length of time; in type II the test is terminated after some fixed number of failures have taken place. This distinction becomes important when sampling for a particular distribution is considered. For the nonparametric methods used in this section, it is adequate to treat all singly-censored ungrouped data as failure-censored; we assume that of N units that begin a test, we are able to obtain the failure times for only the first n ($<N$) failures.

Censoring from the right of either grouped or ungrouped data simply removes that part of the curves in Figs. 8.3 or 8.4 to the right of the time at which the test is terminated. The graphical results still are very useful, for often the early part of the reliability curve is the most important for setting a warrantee period, for determining adequate safety, and for other purposes. Moreover, if early failures are under investigation, the first failures are of primary interest. Even when wearout is of concern, most engineering analysis can be completed without waiting until the very last test unit has failed.

Censoring from the right may be deliberately incorporated into a test plan in conjunction with specifying how many units are to be tested. The test engineer may require that a relatively large number of units be tested in order to obtain enough early failures in order to estimate better the failure rate curve for some specified period of time, say the warrantee period or the design life. If this is the case, many of the units will not fail until well after the time period of interest, and at least a few are likely to survive for very long periods. Thus terminating the test at the end of the period of interest is quite natural.

The standard formulas for the sample mean and variance, of course, can no longer be applied to singly-censored data. Likewise the methods discussed in Chapter 5.4 for estimating distribution parameters and their confidence intervals are no longer valid. Probability plotting methods, however, are applicable to censored data, and these are often particularly valuable in performing parametric analysis. If one of the standard PDFs, say the Weibull distribution, can be fitted to the data and the distribution's parameters estimated, the reliability can be extrapolated beyond the end of the test interval. Extreme care must be taken in employing such extrapolations, however, for if different failure modes appear after longer periods of time, the extrapolations may lead to serious errors.

Multiply-Censored Data

Multiply-censored data occurs in situations where some units are removed from the test before failure or because failure result from a mechanism not relevant to the test. Suppose, for example, that records are being kept on a fleet of trucks to determine the time-to-failure of the transmission. Trucks destroyed by severe accidents would be withdrawn from the test, assuming that a transmission failure was not the cause. Moreover, from time to time some of the trucks might be sold or for other reasons removed from the test population before failure occurs. When trucks are removed for such reasons, it is easy to pretend that the removed units were not part of the original sample. This would not bias the results, provided the censored units were representative of the total population, but it would amount to throwing away valuable data with a concomitant loss in precision of the life-testing results. It is preferable to include the effects of the removed but unfailed units in determining the reliability.

Multiple censoring may be called for even in situations in which all the test units are run to failure, for, in a complex piece of machinery, analysis may indicate two or more different failure modes. Thus, it may prove particularly advantageous to remove units that have not failed from the mode under study in order to describe a particular failure mode through the use of a specific distribution of times to failure. This requires, of course, that each piece of machinery be examined and a determination made of the failure mode.

In what follows, we examine the nonparametric analysis of multiply-censored data. These techniques have been developed the most extensively in the biomedical community, but they are also applicable to technological systems. Once the censoring is carried out and the reliability estimate is available, the substitution $\hat{F}(t_i) = 1 - \hat{R}(t_i)$ allows the probability plotting methods of Chapter 5 to be employed for parametric analysis.

Ungrouped Data Ungrouped censored data take the form shown in Table 8.6. They consist of a series of times, $t_1, t_2, \ldots, t_i, \ldots, t_N$. Each of these times represents the removal of a unit from the test. The removal may be due to failure, or it may be due to censoring (i.e., removal for any other reason). The convention is to indicate the times associated with censoring removals by placing a plus sign (+) after the number.

TABLE 8.6 Failure Times

27	39	40+	54	69
85+	93	102	135+	144

To estimate reliability, we begin by deriving a recursive relation for $R(t_i)$ in terms of $R(t_{i-1})$. Without censoring, it follows from Eq. 8.6 that

$$\hat{R}(t_{i-1}) = \frac{N + 2 - i}{N + 1}. \tag{8.19}$$

By taking the ratio

$$\frac{\hat{R}(t_i)}{\hat{R}(t_{i-1})} = \frac{N + 1 - i}{N + 2 - i}, \tag{8.20}$$

we obtain

$$\hat{R}(t_i) = \frac{N + 1 - i}{N + 2 - i} \hat{R}(t_{i-1}). \tag{8.21}$$

This expression may be interpreted in light of the definition of a conditional probability given by Eq. 2.4. The probability that a unit survives to t_i [i.e., $R(t_i)$] is just the product of the probability that it survives to t_{i-1} [i.e., $R(t_{i-1})$] multiplied by the conditional probability [i.e., $(N + 1 - i)/(N + 2 - i)$] that it will not fail between t_{i-1} and t_i, given that it is operating at t_{i-1}. Thus, for each t_i at which a failure takes place, we reduce the reliability by using Eq. 8.21.

In the event that a censoring action takes place at t_i, the reliability should not change. Therefore, we take

$$\hat{R}(t_i) = \hat{R}(t_{i-1}). \tag{8.22}$$

Equations 8.21 and 8.22 can be combined as an estimate of the conditional probability that a system that is operational at t_{i-1} will not fail until $t > t_i$.

$$\hat{R}(t_i \mid t_{i-1}) = \begin{cases} \left(\dfrac{N + 1 - i}{N + 2 - i} \right) & \textit{failure at } t_i \\[2mm] 1 & \textit{censor at } t_i \end{cases} \tag{8.23}$$

If both a failure and a censor take place at the same time, this formula may be applied unambiguously if the censor is assumed to follow immediately after the failure.

By analogy to Eq. 2.4, which defines conditional probability, we may write

$$R(t_i) = R(t_i \mid t_{i-1}) R(t_{i-1}). \tag{8.24}$$

Hence the reliability at any t_i can be determined by applying this relationship recursively

$$R(t_i) = R(t_i \mid t_{i-1}) R(t_{i-1} \mid t_{i-2}) R(t_{i-2} \mid t_{i-3}) \cdots R(t_1 \mid 0), \tag{8.25}$$

with $R(0) = 1$.

In practice, this estimate is used to calculate the values of the reliability only at the values of t_i at which failures occur. The time dependence of the reliability between these points may then be interpolated, for instance, by

TABLE 8.7 Spreadsheet for Multiply Censored Ungrouped Data Analysis in Example 8.4

	A	B	C	D
	i	ti	R(tilti-1)	R(ti)
1	i	ti	R(tilti-1)	R(ti)
2	1	27	0.90909	0.90909
3	2	39	0.90000	0.81818
4	3	40+	1.00000	
5	4	54	0.87500	0.71591
6	5	69	0.85714	0.61364
7	6	85+	1.00000	
8	7	93	0.80000	0.49091
9	8	102	0.75000	0.36818
10	9	135+	1.00000	
11	10	144	0.50000	0.18409

straight-line segments. Once the reliability has been calculated, Eq. 8.9 may be used to estimate the hazard function at the failure times.

Methods for treating multiply-censored data that are based on the use of the product of conditional reliabilities given in Eq. 8.25 are generally referred to as product limit methods. The foregoing procedure using Eq. 8.5 as a point of departure is due originally to Herd and Johnson. The Kaplan—Meier procedure, which is widely used in the biomedical community, is quite analogous; it begins with Eq. 5.11: $\hat{F}(t_i) = 1/N$ and yields the same results with the expectation that the factor in Eq. 8.23 is replaced by $(N - i)/(N + 1 - i)$. As N becomes larger, the differences between the two procedures become very small.*

EXAMPLE 8.4

Ten motors underwent life testing. Three of these motors were removed from the test and the remaining ones failed. The times in hours are given in Table 8.6. Use the Herd–Johnson method to plot the motor reliability versus time.

Solution The necessary calculations are indicated in Table 8.7. In columns A and B are the values of i and t_i. In column C $R(t_i \mid t_{i-1})$ is calculated from Eq. 8.23 and in D the values of $R(t_i)$ resulting from Eq. 8.24 are shown. The reliability is plotted in Fig. 8.5 for the values of t_i corresponding to failures.

Grouped Data The procedures for treating multiply-censored grouped data parallel those previously described for ungrouped data. Suppose that the number of failures and the number of non-failed items removed from the test is recorded for a number of intervals defined by t_0 (=0), t_1, t_2, t_3 ... t_i. We again use the recursive relationships given by Eqs. 8.24 and 8.25 to estimate the reliability, but now the t_i represent the time intervals over which the data

---

* W. Nelson, *Applied Life Data Analysis,* Chapt. 4, Wiley, New York, 1982.

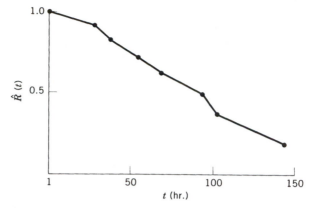

FIGURE 8.5 Reliability estimate from censored life data.

has been grouped. We must derive a new expression for $R(t_i | t_{i-1})$ which is applicable to grouped data.

Suppose that there are n_{i-1} items under test at the *beginning* of the ith interval for which $t_{i-1} < t < t_i$, and d_i failures occur during that interval. The conditional reliability may then be estimated from

$$\hat{R}(t_i | t_{i-1}) = 1 - \frac{d_i}{n_{i-1}}. \qquad (8.26)$$

If there were no censoring we would simply have

$$n_i = n_{i-1} - d_i, \qquad (8.27)$$

with $n_0 = N$, and Eq. 8.26 reduces to Eq. 8.13. Suppose, however, that during the ith interval c_i unfailed units are removed from the test. We then have

$$n_i = n_{i-1} - d_i - c_i. \qquad (8.28)$$

If c_i is a significant fraction of n_{i-1} Eq. 8.26 will tend to overestimate the reliability since for most of the interval there will be fewer than n_{i-1} units available for testing. If we assume that the c_i unfailed units are removed at random points throughout the interval, then a rough correction can be made to Eq. 8.26 by writing

$$\hat{R}(t_i | t_{i-1}) = 1 - \frac{d_i}{n_{i-1} - 0.5c_i}. \qquad (8.29)$$

In applying Eqs. 8.28 and 8.29 in conjunction with Eq. 8.25 to estimate reliability, the values of $\hat{R}(t_i | t_{i-1})$ and $\hat{R}(t_i)$ normally are only calculated at the end of those time intervals in which failure have occurred, for the value of the reliability would not change at intermediate times. The following example demonstrates the procedure.

EXAMPLE 8.5

Table 8.8 shows life data for 206 turbine disks at 100 hour intervals. Make a nonparametric estimate of the reliability versus time.

TABLE 8.8 Failure Data for 206 Turbine Disks*

Interval	Failures	Removals	Interval	Failures	Removals
0–200	0	4	1000–1200	0	18
200–300	1	2	1200–1300	2	5
300–400	1	11	1300–1400	1	13
400–500	3	10	1400–1500	0	14
500–700	0	32	1500–1600	1	14
700–800	1	10	1600–1700	1	14
800–900	0	11	1700–2000	0	5
900–1000	1	9	2000–2100	1	2

* Data from W. Nelson, *Applied Life Data Analysis*, Wiley, New York, 1982, p. 150.

Solution Since the censoring takes place randomly, we set up a spread sheet shown shown in Table 8.9. Columns A, B, and C are the values of i, t_i and n_i for those intervals in which failures take place. Columns F and G are calculated from Eqs. 8.28 and 8.29 respectively, and column H is calculated from Eq. 8.24.

Frequently field service records are tabulated over time intervals of equal length Δ, months, for instance. However only the time interval of purchase and the time interval during which failure occurs are recorded. Suppose at the end of some number of time intervals following the initiation of sales we want to use all of the available data to estimate the reliability. The recursive relations Eqs. 8.24 and 8.25 are still applicable, but care must be taken since inclusion of items of different ages in the reliability estimate is equivalent to multiple censoring from the right.

We retain the use of Eq. 8.28 to determine the number of items under test at the beginning of each interval. However, we now use Eq. 8.26 for the reliability since the censoring amounts to removal at the end of the ith time interval those operational items that are currently of age $i \cdot \Delta$ at the time the analysis is made. We must also make a correction to the time scale since the

TABLE 8.9 Spreadsheet for Multiply Censored Data Analysis in Example 8.5

	A	B	C	D	E	F	G	H
1	i	ti	n_{i-1}	di	ci	n_i	R(tilti-1)	R(ti)
2	2	200	206	0	4	202	1.0000	1.0000
3	3	300	202	1	2	199	0.9950	0.9950
4	4	400	199	1	11	187	0.9948	0.9899
5	5	500	187	3	10	174	0.9835	0.9736
6	8	800	142	1	10	131	0.9927	0.9665
7	10	1000	120	1	9	110	0.9913	0.9581
8	13	1300	92	2	5	85	0.9777	0.9367
9	14	1400	85	1	13	71	0.9873	0.9247
10	16	1600	57	1	14	42	0.9800	0.9063
11	17	1700	42	1	14	27	0.9714	0.8804
12	21	2100	9	1	2	6	0.8750	0.7703

items are sold throughout each time interval. If we assume that sales are approximately uniform during each time interval (since we have no basis for a more specific assumption) we estimate that the average age of the surviving items is $\Delta/2$ at the end of the first interval, $3\Delta/2$ at the end of the second, and in general $t_i = (i - 1/2)\Delta$. The procedure is made clearer with an example:

EXAMPLE 8.6

A new pager goes on sale beginning January 1. Monthly records are kept of the number sold, the number units returned and the month of sale for those returned. The first four months sales are Jan.–1430, Feb.–1657, March–1725, April–2198. For those sold in January, the returns during each month are J–31, F–71, M–56, A–53. For those sold in February the monthly returns are F–38, M–69, A–65, in March M–34, A–76, and in April A–43. Estimate the product reliability.

Solution We must first establish a time scale: In column B of Table 8.10 are the average ages in months at the end of each recording interval. In columns C–F are the monthly failures for those sold in January through April respectively, and column G contains the total number of failures during the first, second, third, and fourth months of operation. In columns H–K Eq. 8.28 is used to calculate the numbers in operation at the beginning of each monthly interval i for those sold in January through April respectively. Summing columns H–K in column L yields i n_{i-1} total number of units available at the beginning of each time interval. In columns M and N, the values of $\hat{R}(t_i | t_{i-1})$ and $\hat{R}(t_i)$ are calculated from Eqs. 8.26 and 8.24. The reliability is plotted in Fig. 8.6.

TABLE 8.10 Spreadsheet for Data Analysis in Example 8.6

	A	B	C	D	E	F	G
1					Failures		
2	i	ti	Jan.	Feb.	March	April	di
3	1	0.5	31	38	34	43	146
4	2	1.5	71	69	76		216
5	3	2.5	56	65			121
6	4	3.5	53				53

	H	I	J	K	L	M	N
1		#Test units					
2	Jan.	Feb.	March	April	n_{i-1}	R(tilti-1)	R(ti)
3	1430	1657	1725	2198	7010	0.9792	0.9792
4	1399	1619	1691		4709	0.9541	0.9342
5	1328	1550			2878	0.9580	0.8950
6	1272				1272	0.9583	0.8577

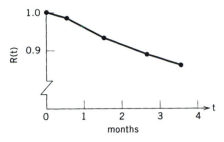

FIGURE 8.6 Reliability estimate for grouped censored life data.

8.5 ACCELERATED LIFE TESTING

Inadequate time to complete life testing is an ubiquitous problem in making reliability estimates. The censoring from the right discussed in the preceding section is a solution only if data from a sufficiently short time span is needed, or if that data can be confidently extrapolated to longer times. Fortunately, a number of acceleration methods may be used to counter the difficulties in performing life testing with time deadlines. Although none are without shortcomings, these procedures nevertheless contribute substantially to the timeliness with which reliability data are obtained. Accelerated tests can be divided roughly into two categories; compressed-time tests and advanced-stress tests.

Compressed-Time Testing

Unless the product is one that is expected to operate continuously, such as a wrist watch or an electric utility transformer, one can condense the component's lifetime by running it continuously to failure. Hence, many engines, motors, and other mechanical and electrical devices can be tested for durability in a small fraction of the calendar design life. Likewise, on-off cycles for many products can be accumulated over a condensed period of time compared to the calendar design life. Reliability tests are frequently performed in which appliance doors are opened and closed, consumer electronics is turned on and off, or pumps or motors are started and stopped to reach a design life target over a relatively short period of time. These are referred to as compressed-time tests, for the product is used more steadily or frequently in the test than in normal use, but the loads and environmental stresses are maintained at the level expected in normal use.

Precaution must be exercised in amassing data from compressed-time tests. In field use the appliance door may only be cycled (opened and closed) several times per day. But a compressed-time test can easily be performed in which the open-close cycle is performed a few times per minute. If the cycle is accelerated too much, however, the conditions of operation may change, increasing stress levels and thus artificially increasing failure rates. If the latch is worked several times per second, for example, the heat of friction may not have time to dissipate. This, in turn, would cause the latch to overheat; increasing the failure rate and perhaps activating failure mechanisms that would not plague ordinary operation. Conversely, tests in which engines, motors, or other systems, which normally operate for intermittent periods of time, are operated continually until failure occurs will not pick up the cyclical failure modes caused by starting and stopping. To detect these a separate cycling test is required, or the continuous operation must be interrupted by intervals long enough for ambient temperatures to be achieved. Compressed-time tests under the field conditions that a product will face may be more difficult to achieve. Nevertheless, some acceleration is possible. The field life of automobiles may be compressed by leasing them as taxicabs, that of a home kitchen appliances by testing them in restaurants. Differences, of course, will remain, but the data may be adequate for the design verification or other use for which it is needed.

EXAMPLE 8.7

Life testing was undertaken to examine the effect of operating time and number of on-off cycles on incandescent bulb life. Six volt flashlight bulbs were operated at 12.6 volts in order to increase the failure rates. The wall-clock failure times, in minutes, for 26 bulbs operated continually and 28 bulbs operated on a 30 sec. on-30 sec. off cycle are given in Table 8.11. Use probability plotting to fit the two sets of data to Weibull distributions, and determine the effect of on-off cycling on the life of the bulb.

Solution Recall from Chapter 5 that Weibull probability plots are made by plotting $y = \ln[\ln(1/(1 - F))]$ versus $\ln(t)$. The $F(t)$ is approximated at each failure by Eq. 5.12. The necessary calculations are performed in Table 8.12. In Figure 8.7, columns E and I are plotted versus columns G and C, respectively, and least-squares fits are

TABLE 8.11 Wall Clock Failure Times in Minutes

Steady State		Cyclic	
72	125	17	258
82	126	161	262
87	127	177	266
97	127	186	271
103	128	186	272
111	139	196	280
113	140	208	284
117	148	219	292
117	154	224	300
118	159	224	317
121	177	232	332
121	199	241	342
124	207	243	355
		243	376

TABLE 8.12 Spreadsheet for Weibull Analysis of Failure Data in Example 8.9

	A	B	C	D	E	F	G	H	I
				STEADY STATE:				CYCLIC:	
1									
2	i	t	x = ln(t)	F = i/27	y	t	x = ln(t)	F = i/29	y
3	1	72	4.2767	0.0370	−3.2770	17	2.8332	0.0345	−3.3498
4	2	82	4.4067	0.0741	−2.5645	161	5.0814	0.0690	−2.6386
5	3	87	4.4659	0.1111	−2.1389	177	5.1761	0.1034	−2.2146
6	4	97	4.5747	0.1481	−1.8304	186	5.2257	0.1379	−1.9077
7	5	103	4.6347	0.1852	−1.5857	186	5.2257	0.1724	−1.6647
8	6	111	4.7095	0.2222	−1.3811	196	5.2781	0.2069	−1.4619
9	7	113	4.7274	0.2593	−1.2036	208	5.3375	0.2414	−1.2864
10	8	117	4.7622	0.2963	−1.0458	219	5.3891	0.2759	−1.1308
11	9	117	4.7622	0.3333	−0.9027	224	5.4116	0.3103	−0.9900
12	10	118	4.7707	0.3704	−0.7708	224	5.4116	0.3448	−0.8607
13	11	121	4.7958	0.4074	−0.6477	232	5.4467	0.3793	−0.7404
14	12	121	4.7958	0.4444	−0.5314	241	5.4848	0.4138	−0.6272
15	13	124	4.8203	0.4815	−0.4204	243	5.4931	0.4483	−0.5197
16	14	125	4.8283	0.5185	−0.3135	243	5.4931	0.4828	−0.4167
17	15	126	4.8363	0.5556	−0.2096	258	5.5530	0.5172	−0.3171
18	16	127	4.8442	0.5926	−0.1077	262	5.5683	0.5517	−0.2202
19	17	127	4.8442	0.6296	−0.0068	266	5.5835	0.5862	−0.1251
20	18	128	4.8520	0.6667	0.0940	271	5.6021	0.6207	−0.0311
21	19	139	4.9345	0.7037	0.1959	272	5.6058	0.6552	0.0627
22	20	140	4.9416	0.7407	0.3001	280	5.6348	0.6897	0.1571
23	21	148	4.9972	0.7778	0.4082	284	5.6490	0.7241	0.2530
24	22	154	5.0370	0.8148	0.5226	292	5.6768	0.7586	0.3516
25	23	159	5.0689	0.8519	0.6469	300	5.7038	0.7931	0.4546
26	24	177	5.1761	0.8889	0.7872	317	5.7889	0.8276	0.5641
27	25	199	5.2933	0.9259	0.9565	332	5.8051	0.8621	0.6836
28	26	207	5.3327	0.9630	1.1927	342	5.8348	0.8966	0.8192
29	27					355	5.8721	0.9310	0.9836
30	28					376	5.9296	0.9655	1.2141

made. The first cyclic failure at 17 min. is an outlier, probably due to infant mortality, and would appear far to the left of the graph. Thus it is not included in the least-square fit. In terms of the slope a and the y intercept b, the Weibull shape and scale parameters are determined from Eqs. 5.33 and 5.34 to be

Steady St.: $\hat{m} = 4.41$, $\hat{\theta} = \exp(+21.8/4.41) = 140.2$ min. (clock time)

Cyclic: $\hat{m} = 4.51$, $\hat{\theta} = \exp(+25.3/4.51) = 273.1$ min. (clock time)

The shape factors are nearly identical, while the scale parameter for the cyclic case is approximately double that for steady-state operation. If we convert clock time to operating time and plot the results, the scale parameter would be 140 and (1/2) $273.1 = 137$. Thus the two sets of data give indistinguishable results when cast in terms of operating time. Therefore the effects of the on-off cycling on bulb lifetime are negligible.

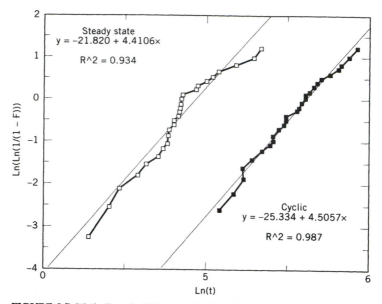

FIGURE 8.7 Weibull probability plot for light bulb accelerated life tests.

Advanced-Stress Testing

Systems that are normally in continuous operation or in which failures are caused by deterioration occurring, even though a unit is inactive, present some of the most difficult problems in accelerated testing. Failure mechanisms cannot be accelerated using the foregoing time compression techniques. Advanced-stress testing, however, may be employed to accelerate failures, since as increased loads or harsher environments are applied to a device, an increased failure rate may be observed. If a decrease in reliability can be quantitatively related to an increase in stress level, the life tests can be performed at high stress levels, and the reliability at normal levels inferred.

Both random failures and aging effects may be the subject of advanced stress tests. In the electronics industry, components are tested at elevated temperatures to increase the incidence of random failure. In the nuclear industry, pressure vessel steels are exposed to extreme levels of neutron irradiation to increase the rate of embrittlement. Similarly, placing equipment under a high-stress level for a short period of time in a proof test may be considered accelerated testing to reveal the early failures from defective manufacture.

The most elementary form of advances-stress test is the nonparametric estimate of the MTTF. Suppose that the MTTF is obtained at the number of different elevated-stress levels. The MTTF is then plotted versus some function of the stress level. Knowledge of either the stress effects or trial and error may be used to choose the function that will result in a linear graph. A curve is fitted to the data, and the MTTF is estimated at the stress level that the device is expected to experience during normal operation. This process is illustrated in the following example:

EXAMPLE 8.8

Accelerated life tests are run on four sets of 12 flashlight bulbs and the failure times in minutes are tabulated in Table 8.13. Estimate the MTTF at each voltage and extrapolate the results to the normal operating voltage of 6.0 volts.

Solution Using the spread sheet formula for the mean we have:

$$9.4 \text{ v: AVERAGE}(A3:A14) = 4{,}744 \text{ min.}$$

$$12.6 \text{ v: AVERAGE}(B3:B14) = 126. \text{ min}$$

$$14.3 \text{ v: AVERAGE}(C3:C14) = 29.0 \text{ min.}$$

$$16.0 \text{ v: AVERAGE}(D3:D14) = 10.3 \text{ min.}$$

In Fig. 8.8 ln(MTTF) is plotted versus volts, and the results fall nearly on a straight line as indicated by the .99 coefficient of determination. The least-squares fit indicates.

$$\ln(\text{MTTF}) = -1.14 \text{ v} + 19.3$$

Hence,

$$\text{MTTF} = \exp(19.3 - 1.14\,\text{v}) = 241 \times 10^6 \exp(-1.14\,\text{v}) \text{ min.}$$

$$= 167 \times 10^3 \exp(-1.14\,\text{v}) \text{ days}$$

At 6 volts:

$$\text{MTTF} = 167 \times 10^3 \exp(-1.14 \times 6) = 179 \text{ days} = 6 \text{ months}$$

The foregoing nonparametric process, while straightforward, has several drawbacks relative to the parametric methods to which we next turn. First, it requires that a complete set of life data be available at each stress level in

TABLE 8.13 Light Bulb Failure Times in Minutes

	A	B	C	D
1	9.4v	12.6v	14.3v	16.v
2				
3	63	87	9	7
4	3542	111	13	9
5	3782	117	23	9
6	4172	118	25	9
7	4412	121	28	9
8	4647	121	30	9
9	5610	124	32	10
10	5670	125	34	11
11	5902	128	37	12
12	6159	140	37	12
13	6202	148	39	13
14	6764	177	41	14

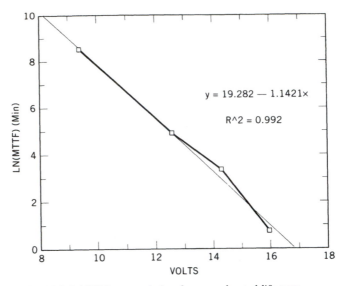

FIGURE 8.8 MTTF extrapolation from accelerated life tests.

order to use the sample mean to calculate the MTTF. Parametric methods can also utilize data that is censored as well as accelerated. Second, without attempting to fit the data to a distribution, one has no indication whether the shape, as well as the time scale of the distribution, is changing. Since changes in distribution shape are usually indications that a new failure mechanism is being activated by the higher-stress levels, there is a greater danger that the nonparametric estimate will be inappropriately extrapolated.

Parametric analysis may be applied to advanced-stress data as follows. As stress is increased above that encountered at normal operating levels, failures should occur at earlier times and therefore the CDF for failure should rise more rapidly. Let $F_a(t)$ be the failure CDF under accelerated-stress conditions and $F(t)$ be that obtained under ordinary operating conditions. Then, we would expect that at any time, $F_a(t) > F(t)$. True acceleration is said to take place if $F_a(t)$ and $F(t)$ are the same distribution and differ only by a scale factor in time. We then have

$$F_a(t) = F(\kappa t), \tag{8.30}$$

where $\kappa > 1$ is referred to as the acceleration factor.

The Weibull and lognormal distributions are particularly well suited for the analysis of advanced-stress tests, for in each case there is a scale parameter that is inversely proportional to the acceleration factor and a shape parameter that should be unaffected by acceleration. Thus, if the shape parameter remains relatively constant, some assurance is provided that no new failure mode has appeared.

The CDF for the Weibull distribution is given by Eq. 3.74. Thus at an advanced stress it will be given by

$$F_a(t) = 1 - e^{-(t/\theta')^m}, \tag{8.31}$$

where to satisfy Eq. 8.30 the scale parameter must be given by

$$\theta' = \theta/\kappa. \tag{8.32}$$

A special case of the Weibull distribution, of course, is the exponential distribution, where $m = 1$, is also used for accelerated testing. Likewise, the CDF for the lognormal distribution is given by Eq. 3.65. At corresponding advanced stress the distribution will be

$$F_a(t) = \Phi\left[\frac{1}{\omega}\ln\left(\frac{t}{t_0'}\right)\right], \tag{8.33}$$

where to satisfy Eq. 8.30 we must have

$$t_0' = t_0/\kappa. \tag{8.34}$$

The procedure for applying advanced-stress testing to determine the life of a device requires a good deal of care. One must be satisfied that the shape parameter is not changing, before making a statistical estimate of the scale parameter. This is often difficult, for at any one stress level the number of failures is not likely to be large enough to determine shape parameter within a narrow confidence interval, and moreover the estimates of these parameters will vary randomly from one stress level to the next. Thus, one must rely on other means to establish the shape parameter. Historical evidence from larger data bases may be used, or more advanced maximum likelihood methods may be used to combine the data under the assumption that there is a common shape parameter. Finally, additional data may be acquired at one or more of the stress levels to establish the parameter within a narrower bound. Some of these considerations are best illustrated by carrying through the analysis on a set of laboratory data. For this purpose we return to the light bulb data used in Examples 8.7 and 8.8:

EXAMPLE 8.9

Make Weibull plots of the accelerated-life test data in Table 8.13. Estimate the shape parameter and determine the acceleration factor as a function of voltage.

Solution For each of the four sets of data we make up a spread sheet analogous to Table 8.12. This is shown as Table 8.14. The first two columns contain the rank i, and the corresponding values of $y = \ln[\ln(1/(1-F))]$ with $F = i/(N+1)$. Columns C through F contain the failure times, copied from Table 8.13, and the corresponding values of $x = \ln(t)$ are calculated in columns G through J. The *x-y* curve for each voltage is shown in Fig 8.9. With the exception of one early failure at 63 min. in the 9.4 v data, the data sets appear to be reasonably represented by the Weibull distribution. Moreover the graphical representations appear to be of similar slope. To explore this further, we make least-squares fits of each of these data sets (deleting the one outlier) and obtain the slopes and the coefficients of determination:

9.4 v $a =$ SLOPE(B4:B14,G4:G14) $= 4.86$ $r^2 =$ RSQ(B4:B14,G4:G14) $= .891$

12.6 v $a =$ SLOPE(B3:B14,H3:H14) $= 2.10$ $r^2 =$ RSQ(B3:B14,H3:H14) $= .900$

TABLE 8.14 Spreadsheet for Weibull Analysis of Failure Data in Example 8.9

	A	B	C	D	E	F	G	H	I	J
			9 4v	12.6v	14.3v	16.v	9.4v	12.6v	14.3v	16.v
1										
2	i	y	t	t	t	t	x	x	x	x
3	1	−2.5252	63	87	9	7	4.143	4.466	2.197	1.94
4	2	−1.7894	3542	111	13	9	8.172	4.710	2.565	2.19
5	3	−1.3380	3782	117	23	9	8 238	4 762	3.135	2.19
6	4	−1.0004	4172	118	25	9	8.336	4.771	3.219	2.19
7	5	−0.7226	4412	121	28	9	8.392	4.796	3.332	2.19
8	6	−0.4796	4647	121	30	9	8.444	4.796	3.401	2.19
9	7	−0.2572	5610	124	32	10	8.632	4.820	3.466	2.30
10	8	−0.0455	5670	125	34	11	8.643	4.828	3.526	2.39
11	9	0.1644	5902	128	37	12	8.683	4.852	3.611	2.48
12	10	0.3828	6159	140	37	12	8.726	4.942	3 611	2.48
13	11	0.6269	6202	148	39	13	8.733	4.997	3.664	2.56
14	12	0.9419	6764	177	41	14	8 819	5.176	3.714	2.63
15										
16	ybar=	−0.5035				xbar=	8.529	4.8263	3.2868	2.31
17	m=	4.4				b=	−38.0	−21.7	−15.0	−10.7
18						theta=	5,672.6	139.9	30.0	11.4
19						ln(theta) =	8.643	4.941	3.401	2.4

14.3 v $a = $ SLOPE(B3:B14,I3:I14) $= 5.60$ $r^2 = $ RSQ(B3:B14,I3:I14) $= .862$

16.0 v $a = $ SLOPE(B3:B14,J3:J14) $= 3.79$ $r^2 = $ RSQ(B3:B14,J3:J14) $= .963$

These coefficients of determination reinforce the view that the data is reasonably fit by Weibull distributions. The varying values of the slopes reveals no systematic trend, and may well be due to large fluctuations caused by the small sample sizes. Thus the average over the four slopes, $a = m = 4.09$, may be a reasonable approximation to a

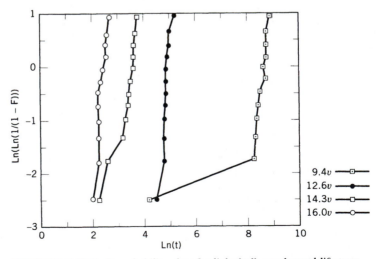

FIGURE 8.9 Weibull probability plots for light bulb accelerated life tests.

shape parameter for all of the data. We have an additional piece of evidence, however. The two larger data sets, $N = 24$, taken for steady state and cyclic operation at 12.6 v, shown in Fig. 8.7, yield values of 4.41 and 4.51. As a result we chose $m = 4.4$ as a reasonable estimate.

With the common shape factor, and therefore fixed slope, we may use Eq. 5.25 to make a least-squares fit for b, the y intercept, at each voltage: $b = \bar{y} - a\bar{x}$. The necessary calculations for b are carried out in Table 8.14. For each voltage the Weibull scale parameter $\hat{\theta}$ is then evaluated from Eq. 5.34. To estimate the acceleration factor as a function of voltage we first attempt a linear fit of the values given in Table 8.14 versus voltage. We obtain $r^2 = \text{RSQ(G18:J18,G1:J1)} = 0.77$, which is a poor fit. We next attempt a fit with $y = \ln(\hat{\theta})$ and obtain a coefficient of determination that is substantially closer to one: $r^2 = \text{RSQ(G19:J19,G1:J1)} = 0.98$. Therefore we make a least-square fit of $\ln(\hat{\theta})$ versus voltage and find $a = \text{SLOPE(G19:J19,G1:J1)} = -0.96$ and $\text{INTERCEPT(G19:J19,G1:J1)} = 17.4$. Thus we may write $\ln(\hat{\theta}') = -0.96\nu + 17.4$ or $\hat{\theta}' = 36.0 \cdot 10^6 \exp(-0.96\nu)$. From Eq. 8.32 we find the acceleration factor to be

$$\kappa = \hat{\theta}/\hat{\theta}' = \exp[0.96(\nu - 6)]$$

Other distributions, such as the normal and extreme value, may also be used in advanced-stress testing. In these cases, however, the analysis is more complex since both distribution parameters change if Eq. 8.30 remains valid. For example in the normal distribution, we have $\mu' = \mu/\kappa$ and $\sigma' = \sigma/\kappa$. Thus lines drawn on probability plots at different stress levels will no longer be parallel with the time scaling. The normal distribution is more useful in modeling phenomena in which stress levels have additive instead of multiplicative effects on the times to failure. For μ is a displacement rather than a scale parameter, and thus in such situations only μ and not σ will be effected. A similar behavior is observed if the extreme value distribution is employed.

Acceleration Models

As in compressed-time testing, the extrapolations involved in advanced-stress testing may be problematical in situations where it is feasible to run accelerated tests at only one or two stress levels. Then it is impossible to define an empirical relationship between stress and reliability from which the extrapolation to normal operating conditions can be made. In such situations the existence of a well-understood acceleration model can replace the empirical extrapolation. For example, the rate at which a wide variety of chemical reactions take place, whether they be corrosion of metals, breakdown of lubricants, or diffusion of semiconductor materials, obeys the Arrhenius equation.

$$rate \sim e^{-\Delta H/kT}, \tag{8.35}$$

where ΔH is the activation energy, k is the Boltzmann constant, and T is the absolute temperature. Thus, for systems in which chemical reactions are responsible for failure, an increase in temperature increases the failure rate in a prescribed manner.

Since the times to failure will increase as the rate decreases, we may equate the scale parameter for the Weibull distribution to the inverse of the rate

$$\theta = Ae^{\Delta H/kT}, \tag{8.36}$$

where A is a proportionality constant. The Arrhenius equation may also be used for lognormal fitting simply by substituting the scale parameter t_0 for θ in the following equations. Suppose that T_0 is the nominal temperature at which the device is designed to operate. The acceleration factor, defined in Eq. 8.30 may then be determined simply by taking the ratio θ_0/θ_1 of scale parameters at the nominal and elevated temperatures, T_0 and T_1.

$$\kappa(T_1) = \exp\left\{(\Delta H/k)\left[\frac{1}{T_0} - \frac{1}{T_1}\right]\right\}. \tag{8.37}$$

Before this expression may be used for accelerated testing, however, the activity energy ΔH must be determined. This can be accomplished by taking the ratio between θ_1 and θ_2 at two elevated temperatures and solving Eq. 8.36 for ΔH:

$$\Delta H = k\left(\frac{1}{T_1} - \frac{1}{T_2}\right)^{-1} \ln\left(\frac{\theta_1}{\theta_2}\right). \tag{8.38}$$

Thus tests must first be run at two reference temperatures T_1 and T_2 to determine the Weibull parameters θ_1 and θ_2. Then, once ΔH has been determined, the acceleration factor can be calculated as a function of temperature.

Other time-scaling laws are also available. Empirical relations are often applied to voltage, humidity and other environmental factors. Accelerated testing is useful, but it must be carried out with great care to ensure that results are not erroneous. We must be certain that the phenomena for which the acceleration factor κ has been calculated are the failure mechanisms. Experience gained with similar products and a careful comparison of the failure mechanisms occurring in accelerated and real-time tests will help determine whether we are testing the correct phenomena.

8.6 CONSTANT FAILURE RATE ESTIMATES

In this section we examine in more detail the testing procedures for determining the MTTF when the data are exponentially distributed. This is justified both because the exponential distribution (i.e., the constant failure rate model) is the most widely applied in reliability engineering, and because it provides insight into the problems of parameter estimation that are indicative of those encountered with other distributions.

We must, of course, determine whether the constant failure rate model is applicable to the test at hand. At least four approaches to this problem may be taken. The exponential distribution may be assumed, based on experience with equipment of similar design. It may be identified by using one of the standard statistical goodness-of-fit criteria or by probability plotting, and examining the results visually for the required straight-line behavior. Finally, it may be argued from the failure mode whether the failures are random, as opposed to early or aging failures. If defective products or aging effects are identified as causing some of the failures, the data must be censored appropriately.

The exponential distribution has only a single parameter to be estimated, the failure rate λ. Rather than estimate the failure rate directly, most sampling schemes are cast in terms of the MTTF, denoted by MTTF $\equiv \mu = 1/\lambda$. For uncensored data the value of μ may be estimated from Eq. 8.11. Moreover, when N, the number of test specimens, is sufficiently large, the central limit theorem, which was discussed in Chapter 5, may be used to estimate a confidence interval. In particular, the 69% confidence interval is given by $\hat{\mu} \pm \sigma/\sqrt{N}$, where σ^2 is the variance of the distribution. Since for the exponential distribution $\sigma = \mu$, we may estimate the 69% confidence interval from $\hat{\mu} \pm \hat{\mu}/\sqrt{N}$.

Censoring on the Right

It is clear from the foregoing expressions that for a precise estimate a large sampling size is required. Using many test specimens is expensive, but, more important, a very long time is required to complete the test. As N becomes large, the last failure is likely to occur only after several MTTFs have elapsed. Moreover, the analysis of the failures that occur after long periods of time is problematic for two reasons. First, a design life is normally less than the MTTF, and it is often not possible to hold up final design, production, or operation while tests are carried out over many design lives. Equally important, many of the last failures are likely to be caused by aging effects. Thus they must be removed from the data by censoring if a true picture of the random failures is to be gained.

Type I and type II censoring from the right are attractive alternatives to uncensored sampling. By limiting the period of the test while increasing the number of units tested, we can eliminate most of the aging failures, and estimate more precisely the time-independent failure rate. Within this framework four different test plans may be used. With the assumption that the test is begun with N test units, these plans may be distinguished as follows. If the test is terminated at some specified time, say t_*, then type I censoring is said to take place. If the test is terminated immediately after a particular number of failures, say n, then type II censoring is said to take place. With either type I or type II censoring, we may run the test in either of two ways. In the nonreplacement method each unit is removed from the test at the time of failure. In the replacement method each unit is immediately repaired or replaced following failure so that there are always N units operating until the test is terminated.

The choice between type I and type II censoring involves the following trade-off. Type I censoring is more convenient because the duration of the test t_* can be specified when the test is planned. The time t_n of the nth failure, at which a test with type II censoring is terminated, however, cannot be predicted with precision at the time the test is planned, for t_n is a random variable. Conversely, the precision of the measurement of the MTTF for the exponential distribution is a function of the number of failures rather than of the test time. Therefore, it is often considered advisable to wait until some specified number of failures have occurred before concluding the test.

A number of factors also come into play in determining whether nonreplacement or replacement tests are to be used. In laboratory tests the cost of the test units compared with the cost of the apparatus required to perform the test may be the most significant factor. Consider two extreme examples. First, if jet engines are being tested, nonreplacement is the likely choice. When a specified number of engines are available, more will fail within a given length of time if they are all started at the same time than if some of them are held in reserve to replace those that fail. The same is true of any other expensive piece of equipment that is to be tested as a whole.

Conversely, suppose that we are testing fuel injectors for large internal-combustion engines. The supply of fuel injectors may be much larger than the number of engines upon which to test them. Therefore, it would make sense to keep all the engines running for the entire length of the test by immediately replacing each fuel injector following failure, provided that the replacement can be carried out swiftly and at minimum cost. Minimizing cost is an important provision, for generally the personnel costs are larger with replacement tests; in nonreplacement tests personnel or instrumentation is required only to record the failure times. In replacement tests personnel and equipment must be available for carrying out the repairs or replacements within a short period of time.

The situation is likely to be quite different when the data are to be accumulated from actual field experience with breakdowns. Here, in the normal course of events, equipment is likely to be repaired or replaced over a time span that is short compared to the MTTF. Conversely, records may indicate only the number of breakdowns, not when they occurred. The number of breakdowns might be inferred, for example, from spare parts orders or from numbers of service calls. In these circumstances replacement testing describes the situation. Moreover, unlike nonreplacement testing, the MTTF estimation does not require that the times of failures be recorded.

One last class of test remains to be mentioned. Sometimes referred to as percentage survival, it is a simple count of the fraction (or percentage) of failed units. From the properties of the exponential distribution, we infer the MTTF. This test procedure requires no surveillance, for failed equipment does not need to be replaced or times of failure recorded. Not surprisingly, the estimate obtained is less precise. The method is normally not recommended, unless failures are not apparent at the time they take place and can only be determined by destructive testing or other invasive techniques following the conclusion of the test.

MTTF Estimates

With the exception of the percentage survival technique, the same estimator may be shown to be valid for all the test procedures described:*

$$\hat{\mu} = \frac{T}{n},$$

$$T = \text{total operational time of all test units}, \tag{8.39}$$

$$n = \text{number of failures}.$$

For each class of test, however, the total operating time T is calculated differently.

Consider first nonreplacement testing with type I censoring (i.e., the test is terminated at some predetermined time t_*). If $t_1, t_2, \ldots, t_n$ are the times of the n failures, the total operational time T for the N units tested is

$$T = \sum_{i=1}^{n} t_i + (N - n) t_*, \tag{8.40}$$

since $N - n$ units operate for the full time t_*.

EXAMPLE 8.10

A 30-day nonreplacement test is carried out on 20 rate gyroscopes. During this period of time 9 units fail; examination of the failed units indicates that none of the failures is due to defective manufacture or to wear mechanisms. The failure times (in days) are 27.4, 13.5, 10.5, 20.0, 23.6, 29.1, 27.7, 5.1, and 14.4. Estimate the MTTF.

Solution From Eq. 8.40 with $N = 20$ and $n = 9$,

$$T = \sum_{i=1}^{9} t_i + (20 - 9) \times 30$$

$$= 171.3 + 11 \times 30 = 501.3$$

$$\hat{\mu} = \frac{T}{n} = \frac{501.3}{9} = 55.7 \text{ days}.$$

For type II censoring the test is stopped at t_n, the time of the nth failure. Thus, if there is no replacement of test units, the total operating time is

* I. Bazovsky, *Reliability Theory and Practice*, Prentice-Hall, Englewood Cliffs, NJ, 1961.

calculated from

$$T = \sum_{i=1}^{n} t_i + (N - n) t_n, \tag{8.41}$$

since the unfailed $(N - n)$ units are taken out of service at the time of the nth failure. Note that in the event that some of the units, say k of them, are removed from the test because they fail from another mechanism, such as aging, then T is still calculated by Eq. 8.40 or Eq. 8.41. Now, however, the estimate is obtained by dividing only by the number $n - k$ of random failures:

$$\hat{\mu} = \frac{T}{n - k}. \tag{8.42}$$

EXAMPLE 8.11

The engineer in charge of the test in the preceding problem decides to continue to test until 10 of the 20 rate gyroscopes have failed. The tenth failure occurs at 41.2 days, at which time the test is terminated. Estimate the MTTF.

Solution   From Eq. 8.41 with $N = 20$ and $n = 10$,

$$T = \sum_{i=1}^{10} t_n + (20 - 10)41.2$$

$$T = (171.3 + 41.2) + 10 \times 41.2 = 624.5$$

$$\hat{\mu} = \frac{T}{n} = \frac{624.5}{10} = 62.4 \text{ days.}$$

In replacement testing all N units are operated for the entire length of the test. Thus, for type I censoring, we have $T = Nt_*$, where t_* is the specified test time. Hence

$$\hat{\mu} = \frac{Nt_*}{n}. \tag{8.43}$$

For type II censoring, we have $T = Nt_n$, where t_n is the time at which the nth unit fails. Thus $T = Nt_n$ or

$$\hat{\mu} = \frac{Nt_n}{n}. \tag{8.44}$$

EXAMPLE 8.12

A chemical plant has 24 process control circuits. During 5000 hr of plant operation the circuits experience 14 failures. After each failure the unit is immediately replaced. What is the MTTF for the control circuits?

Solution From Eq. 8.43

$$T = Nt_* = 24 \times 5000 = 120,000$$

$$\hat{\mu} = \frac{T}{n} = \frac{120,000}{14} = 8571 \text{ hr.}$$

EXAMPLE 8.13

Six units of a new high-precision pressure monitor are placed on an industrial furnace. After each failure the monitor is immediately replaced. However, the eighth failure occurs after only 840 hours of service. It is decided that the high-temperature environment is too severe for the instruments to function reliably, and the furnace is shut down to replace the pressure monitors with a more reliable, and expensive, design. Assuming that the failures are random, estimate the MTTF of the monitors.

Solution From Eq. 8.44

$$T = Nt_8 = 6 \times 840 = 5040 \text{ hr}$$

$$\hat{\mu} = \frac{T}{n} = \frac{5040}{8} = 630 \text{ hr.}$$

As alluded to earlier, the MTTF may also be estimated from the percentage survival method. We begin by first estimating the reliability at the end of the test, time t_0 as $R(t_0) = 1 - n/N$. With an exponential distribution however, the reliability is given by

$$R(t_0) = \exp(-t_0/\mu). \tag{8.45}$$

Thus, combining these equations, we estimate MTTF from

$$\mu = \frac{t_0}{\ln[1/(1 - n/N)]}. \tag{8.46}$$

EXAMPLE 8.14

A National Guard unit is supplied with 20,000 rounds of ammunition for a new model rifle. After 5 years, 18,200 rounds remain unused. From these 200 rounds are chosen randomly and test-fired. Twelve of them misfire. Assuming that the misfires are random failures of the ammunition caused by storage conditions, estimate the MTTF.

Solution In Eq. 8.46 take $n = 12$, $N = 200$, and $t_0 = 5$ years. We have

$$\hat{\mu} = \frac{5}{\ln\{1/[1 - 12/200]\}} = 81 \text{ years.}$$

Confidence Intervals

We next consider the precision of the MTTF estimates made with Eq. 8.39. The confidence limits for both replacement and nonreplacement tests may be expressed in terms of $\hat{\mu}$ and the number of failures by using the χ^2 distribution. The results are given conveniently by the curves shown in Fig. 8.10. We consider type II censoring first.

Let $U_{\alpha/2,n}$ and $L_{\alpha/2,n}$ be the upper and lower limits for the $100 \times (1 - \alpha)$ percent confidence interval for type II censoring. The two-sided confidence

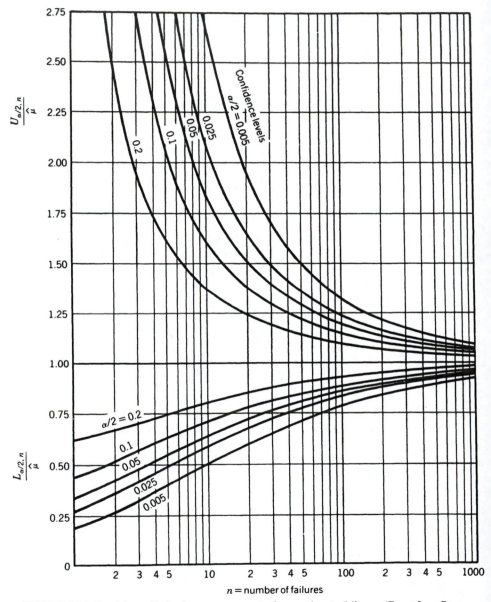

FIGURE 8.10 Confidence limits for measurement of mean-time-to-failures. (From Igor Bazovsky, *Reliability Theory and Practice,* © 1961, p. 241, with permission from Prentice-Hall, Englewood Cliffs, NJ.)

interval states that if the test is stopped after the nth failure, there is a $1 - \alpha$ probability that the true value of n lies between $L_{\alpha/2,n}$ and $U_{\alpha/2,n}$:

$$P\{L_{\alpha/2,n} \leq \mu < U_{\alpha/2,n}\} = 1 - \alpha. \tag{8.47}$$

It turns out that the ratios $L_{\alpha/2,n}/\hat{\mu}$ and $U_{\alpha/2,n}/\hat{\mu}$ are independent of the operating time T. Therefore, they can be plotted as functions of α and n, the number of failures. The plot is shown in Fig. 8.10. Thus, if $\hat{\mu}$ has been estimated from one of the forms of Eq. 8.39, the confidence interval can be read from Fig. 8.10. This is best illustrated by examples.

EXAMPLE 8.15

What is the 90% confidence interval for the rate gyroscopes tested in Example 8.11 taking the failure at 41.2 days into account?

Solution For a 90% confidence interval we have $100(1 - \alpha) = 90$, or $\alpha = 0.1$ and $\alpha/2 = 0.05$. For $n = 10$ failures we find from Fig. 8.10 that

$$\frac{L_{0.05,10}}{\hat{\mu}} \approx 0.65, \qquad \frac{U_{0.05,10}}{\hat{\mu}} \approx 1.82.$$

Therefore, using $\hat{\mu} = 62.4$ days from Example 8.11:

$$L_{0.05,10} \approx 0.65 \times 62.4 = 41 \text{ days},$$
$$U_{0.05,10} \approx 1.82 \times 62.4 = 114 \text{ days},$$
$$41 < \mu < 114 \text{ days with 90\% confidence.}$$

With slight modifications the results of Fig. 8.10 may also be applied to type I censoring, where the test is ended at some time t_*. Using the properties of the χ^2 distribution, it may be shown that the upper confidence limit and $\hat{\mu}$ remain the same. The lower confidence limit, in general, decreases. It may be related to the results in Fig. 8.10 by

$$\frac{L^*_{\alpha/2,n}}{\hat{\mu}} = \frac{n}{n+1} \frac{L_{\alpha/2,(n+1)}}{\hat{\mu}}, \tag{8.48}$$

where L^* is the value for type I censoring, and L is the plotted value for type II censoring. Again, the confidence limits are applicable to both nonreplacement and replacement testing.

EXAMPLE 8.16

During the first year of operation a demineralizer suffers seven shutdowns. Estimate the MTBF and the 95% confidence interval.

Solution   From Eq. 8.39

$$\hat{\mu} = \text{MTBF} = \frac{T}{n} = \frac{12 \text{ months}}{7} = 1.71 \text{ months.}$$

For a 95% confidence interval $\alpha = 0.05$ and $\alpha/2 = 0.025$. From Fig. 8.10,

$$\frac{L^*_{0.025, n}}{\hat{\mu}} = \frac{n}{n+1} \frac{L_{0.025, n+1}}{\hat{\mu}} = \frac{7}{8} \frac{L_{0.025,8}}{\hat{\mu}} \approx \frac{7}{8} \times 0.57 = 0.50$$

$$L_{0.025,7} = 0.50 \times 1.71 = 0.86 \text{ month,}$$

$$U_{0.025,7} = 2.5 \times 1.71 = 4.27 \text{ months.}$$

Thus

$$0.86 \text{ months} < \mu < 4.27 \text{ months}$$

with 95% confidence.

In some situations, particularly in setting specifications, we are not interested in the MTBF, but only in assuring that it be greater than some specified value. If the MTBF must be greater than the specified value at a confidence level of $\alpha/2$, we estimate $L_{\alpha/2,n}/\hat{\mu}$ or $L^*_{\alpha/2,n}/\hat{\mu}$ from Fig. 8.10 and determines the value of $\hat{\mu}$ with an appropriate form of Eq. 8.39.

EXAMPLE 8.17

A computer specification calls for an MTBF of at least 100 hr with 90% confidence. If a prototype fails for the first time at 210 hr, can these test data be used to demonstrate that the specification has been met?

Solution $\hat{\mu} = T/n = 210/1 = 210$ hr. For the 90% one-sided confidence interval $\alpha/2 = 0.1$. From Fig. 8.10,

$$L_{0.1,1}/\hat{\mu} \approx 0.44,$$

$$L_{0.1,1} = 0.44 \times 210 = 93 \text{ hr.}$$

The test is inadequate, since the lower confidence limit is smaller than the specified value of 100 hr.

A word is in order concerning the percentage survival test discussed earlier. It is a form of binomial sampling, with the ratio n/N being the estimate of the failure probability of failure. Consequently, the method discussed in Chapter 2 can be used to estimate the confidence interval of the failure probability, and from this the confidence interval on the MTTF can be estimated. The uncertainty is greater than that obtained from testing in which the actual failure times are recorded.

EXAMPLE 8.18

Estimate the 90% confidence interval for the National Guard ammunition problem, Example 8.14.

Solution Since, in 5 years, 12 of 200 rounds fail, the 5-year failure probability may be calculated from Eq. 2.66 to be

$$\hat{p} = \frac{n}{N} = \frac{12}{200} = 0.06 = 1 - \hat{R}.$$

Since this test is a form of binomial sampling, we can look up the 90% confidence interval on p from Appendix B. We obtain for $n = 12$, $0.01 < p < 0.31$. For a constant failure rate we have

$$p = 1 - e^{-t/\mu} \quad \text{or} \quad \mu = -t/\ln(1 - p).$$

Therefore, with $t = 25$ years,

$$\frac{-25}{\ln(1 - 0.31)} < \mu < \frac{-25}{\ln(1 - 0.01)}$$

$$67 \text{ years} < \mu < 2487 \text{ years}.$$

with 90% confidence.

Bibliography

Bazovsky, I., *Reliability Theory and Practice*, Prentice-Hall, Englewood Cliffs, NJ, 1961.

Crowder, M. J., A. C. Kimber, R. L. Smith, and T. J. Sweeting, *Statistical Analysis of Reliability Data*, Chapman & Hall, London, 1991.

Kapur, K. C., and L. R. Lamberson, *Reliability in Engineering* Wiley, NY, 1977.

Kececioglu, D., *Reliability and Life Testing Handbook*, Vol I & II, Prentice-Hall, Englewood Cliffs, NJ, 1993.

Lawless, J. F., *Statistical Models and Methods for Lifetime Data*, Wiley, NY, 1982.

Mann, N. R., R. E. Schafer, and N. D. Singpurwalla, *Methods for Statistical Analysis of Reliability and Life Data*, Wiley, NY, 1974.

Nelson, W., *Accelerated Testing*, Wiley, NY, 1990.

———., *Applied Life Data Analysis*, Wiley, NY, 1982.

Tobias, P. A., and D. C. Trindade, *Applied Reliability*, Van Nostrand–Reinhold, NY, 1986.

Exercises

8.1 Suppose that "bugs" are detected and corrected in developmental software at 1.4, 8.9, 24.3, 68.1, 117.2, and 229.3 hrs.

    (a) Estimate the reliability growth coefficient, α.

    (b) Calculate the coefficient of determination for α.

8.2 The wearout times of 10 emergency flares in minutes are 17.0, 20.6, 21.3, 21.4, 22.7, 25.6, 26.5, 27.0, 27.7, and 29.7. Use the nonparametric method to make plots of the reliability and cumulative hazard function.

8.3 Determine the MTTF of the data in Example 5.7.

8.4 For the data in Example 5.7, make a nonparametric graph of the reliability and cumulative hazard function.

8.5 The L_{10} life is defined at the time at which 10% of a product has failed.

(a) Estimate L_{10} for the failure data in Example 5.2.

(b) Estimate the MTTF for that data.

8.6 For the flashlight bulb data in Example 5.2 make nonparametric plots of the reliability and cumulative hazard function.

8.7 A new robot system undergoes test-fix-test-fix development testing. The number of failures during each 100-hr interval in the first 700 hr of operation are recorded. They are 14, 7, 6, 4, 3, 1, and 1.

(a) Plot the cumulative MTBF $\equiv T/n$ on log-log paper and approximate the data by a straight line.

(b) Estimate α from the slope of the line.

8.8 Data for the failure times of 318 radio transmitter receivers are given in the following table.*

Time interval, hr	Failures	Time interval, hr	Failures
0–50	41	300–350	18
50–100	44	350–400	16
100–150	50	400–450	15
150–200	48	450–500	11
200–250	28	500–550	7
250–300	29	550–600	11

At 600 hr, 51 of the receiver–transmitters remained in operation. Use the nonparametric method described in the text to plot the reliability and cumulative hazard function versus time.

8.9 Fifteen components undergo a 100 hour life-test. Failures occur a 31.4, 45.9, 50.2, 58.4, 70.7, 73.2, 86.6 and 96.3 hours. From previous experience the data is expected to obey a lognormal distribution. Make a probability plot and estimate the lognormal parameters; then estimate the MTTF.

---

* From W. Mendenhall and R. J. Hader, "Estimation of Parameters of Mixed Exponential Distribution Failure Times from Censored Life Test Data," *Biometrika*, **63**, 449–464 (1958).

8.10 The following uncensored grouped data were collected on the failure time of feedwater pumps, in units of 1000 hr:

Interval	Number of failures
$0 \leqslant t \leqslant 6$	5
$6 \leqslant t \leqslant 12$	19
$12 \leqslant t \leqslant 18$	61
$18 \leqslant t \leqslant 24$	27
$24 \leqslant t \leqslant 30$	20
$30 \leqslant t \leqslant 36$	17

Make a nonparametric plot of the reliability and of the cumulative hazard function versus time.

8.11 The test started in Exercise 8.9 is run to completion. The remaining samples fail at 100.6, 117.9, 124.8, 148.7, 159.5, 205.2, and 232.5 hours. Redo the analysis and compare the lognormal parameters and the MTTF to the values obtained in Exercise 8.9

8.12 The following numbers of bends to failure were recorded for 20 paper clips: 11, 29, 15, 20, 19, 11, 12, 9, 9, 8, 13, 20, 11, 22, 20, 9, 25, 19, 11, and 10.

(a) Make a nonparametric plot of $R(t)$, the reliability.
(b) Attempt to fit your data to Weibull, lognormal and/or normal distributions and determine the parameters.
(c) Briefly discuss your results.

8.13 Repeat Exercise 8.9 but fit the data to a two-parameter Weibull distribution.

8.14 Consider the following multiply censored data* for the field windings for 16 generators. The times to failure and removal times (in months) are 31.7, 39.2, 57.5, 65.0+, 65.8, 70.0, 75.0+, 75.0+, 87.5+, 88.3+, 94.2+, 101.7+, 105.8, 109.2+, 110.0, and 130.0+. Make a nonparametric plot of the reliability.

8.15 Suppose that a device undergoing accelerated testing can be described by a Weibull distribution with a shape factor of $m = 2.0$. Under accelerated test conditions, with an acceleration factor of $\kappa = 5.0$, 50% of the devices are found to fail during the first month. Under *normal operating conditions,* estimate how long the device will last before the failure probability reaches 10%. (This is referred to as the L_{10} life of the device).

---

* From Nelson, *Applied Life Data Analysis,* Wiley, New York, 1982

8.16 The data that follows is obtained for the time to failure of 128 appliance motors

(a) Make a histogram of the PDF.

(b) Plot the reliability.

(c) Plot the cumulative hazard function.

hours	# failures	hours	# failures
0–10	4	50–60	31
10–20	8	60–70	22
20–30	11	70–80	10
30–40	16	80–90	2
40–50	23	90–100	1

8.17 Estimate the mean and variance of the data in Exercise 8.16

8.18 Make a Weibull plot and a normal plot of the grouped data in Exercise 8.16. Determine which is the better fit and estimate the parameters for that distribution.

8.19 Make a two-parameter Weibull plot of the multiply-censored winding data from Exercise 8.14 and estimate m and θ.

8.20 A wear test is run on 20 specimens and the following failure times in hours are obtained: 81, 91, 95+, 97, 100+, 106, 109, 110+, 112, 114+, 117+, 120, 126, 128, 130, 132+, 139, 144, 154, and 163. Using the product-limit technique to account for the censoring:

(a) Make a nonparametric plot of the reliability.

(b) Fit the data to a normal distribution and estimate the parameters.

8.21 Of a group of 180 transformers, 20 of them fail within the first 4000 hr of operation. The times to failure in hours are as follows:*

10	1046	2096	3200
314	1570	2110	3360
730	1870	2177	3444
740	2020	2306	3508
990	2040	2690	3770

(a) Make a normal probability plot.

(b) Estimate μ and σ for the transformers.

(c) Estimate how many transformers will fail between 4000 and 8000 hr.

8.22 Plot the data from the Exercise 8.21 on exponential paper to estimate whether the failure rate increases or decreases with time.

---

* Data from Nelson, op cit.

8.23 Twenty units of a catalytic converter are tested to failure without censoring. The times-to-failure (in days) are the following:

2.6	3.2	3.4	3.9	5.6
7.1	8.4	8.8	8.9	9.5
9.8	11.3	11.8	11.9	12.7
12.3	16.0	21.9	22.4	24.2

Make an exponential probability plot, and determine whether the failure rate is increasing or decreasing with time.

8.24 A producer of consumer products offers a three year double-your-money back guarantee over a limited marketing area and collects the failure data tabulated below.

(a) Make a nonparametric plot of $R(t)$.

(b) Fit the data to a Weibull distribution and estimate the parameters.

(c) Fit the data to a lognormal distribution and estimate the parameters.

(d) Does the Weibull or the lognormal distribution yield the better fit?

Quarter sold:	W 92	S 92	S 92	F 92	W 93	S 93	S 93	F 93	W 94	S 94	S 94	F94
Number sold:	842	972	1061	1293	939	1014	1036	1185	979	1125	1205	1300
Number failed:												
W92	18											
S92	42	22										
S92	33	42	21									
F92	32	39	45	26								
W93	32	37	43	54	19							
S93	27	35	38	51	38	22						
S93	34	31	42	50	39	43	20					
F93	42	35	37	46	34	39	43	23				
W94	27	32	35	46	37	39	40	50	19			
S94	26	26	29	40	32	36	38	48	44	26		
S94	21	31	36	43	33	37	41	42	41	44	28	
F94	25	27	31	41	29	33	35	45	35	46	49	24

8.25 Make a Weibull plot of Exercise 8.23 and estimate the parameters m and θ.

8.26 The following multiply-censored times-to-failure (in hours) have been obtained from a battery powered motor used in inexpensive consumer products: 22, 37, 41, 43, 56, 57+, 58, 61, 62+, 63+, 64, 64, 65+, 69, 69, 69+, 70, 76+, 78, 87, 88+, 89, 94, 100, and 119. Using the product-limit technique to account for the censoring:

(a) Make a nonparametric plot of the reliability and cumulative hazard function.

(b) Fit the data to a Weibull distribution and estimate the parameters.

8.27 Suppose that instead of Eq. 5.12, we use Eq. 5.13 as a starting point for nonparametric analysis. Derive the expressions for $\hat{R}(t_i)$ and $\hat{H}(t_i)$, that should be used in place of Eqs. 8.6 and 8.10

8.28 Microcircuits undergo accelerated life testing. The analysis is to be carried out using nonparametric methods for ungrouped data.

  (a) The first test series on six prototype microcircuits results in the following times to failure (in hours): 1.6, 2.6, 5.7, 9.3, 18.2, and 39.6. Plot a graph of the estimated reliability.

  (b) The second test series of six prototype microcircuits results in the following times to failure (in hours): 2.5, 2.8, 3.5, 5.7, 10.3, and 23.5. Combine these data with the data from *a* and plot the reliability estimate on the same graph used for *a*.

8.29 At rated voltage a microcircuit has been estimated to have an MTTF of 20,000 hr. An accelerated life test is to be carried out to verify this number. It is known that the microcircuit life is inversely proportional to the cube of the voltage. At least 10% of the test circuits must fail before the test is terminated if we are to have confidence in the result. If the test must be completed in 30 days, at what percentage of the rated voltage should the circuits be tested?

8.30 A life test with type II censoring is performed on 50 servomechanisms that are thought to have a constant failure rate. The test is terminated after the twentieth failure. The times to failure (in months) are as follows:

0.10	0.29	0.49	0.51	0.55
0.63	0.68	1.16	1.40	2.24
2.25	2.64	2.99	3.01	3.06
3.15	3.51	3.53	3.99	4.05

The failed servomechanisms are not replaced.

  (a) Make an exponential probability plot and estimate whether the failure rate is constant.

  (b) Make a point estimate of the MTTF from the appropriate form of Eq. 8.39.

  (c) Using the MTTF from *b*, draw a straight line through the data plotted for *a*.

  (d) What is the 90% confidence interval on the MTTF?

  (e) Draw the straight lines on your plot in *a* corresponding to the confidence limits on the MTTF.

8.31 Suppose that in Exercise 8.30 the life test had to be stopped at 3 months because of a production deadline. Based on a 3-month test, estimate the MTTF and the corresponding 90% confidence interval.

8.32 Sets of electronic components are tested at 100°F and 120°F and the MTTFs are found to be 80 hr and 35 hr, respectively. Assuming that the Arrhenius equation is applicable, estimate the MTTF at 70°F.

8.33 A nonreplacement reliability test is carried out on 20 high-speed pumps to estimate the value of the failure rate. In order to eliminate wear failures, it is decided to terminate the test after half of the pumps have failed. The times of the first 10 failures (in hours) are 33.7, 36.9, 46.8, 56.6, 62.1, 63.6, 78.4, 79.0, 101.5, and 110.2.

   (a) Estimate the MTTF.

   (b) Determine the 90% confidence interval for the MTTF.

8.34 A nonreplacement test with type I censoring is run for 50 hours on 30 microprocessors. Five failures occur at 12, 19, 28, 39, and 47 hours. Estimate the value of the constant failure rate.

8.35 A replacement test is run for 30 days using 18 test setups. During the test there are 16 failures. Assuming an exponential distribution, estimate the MTTF.

CHAPTER 9

Redundancy

"If it doesn't have at least two engines and two pilots, I don't get on it."

S. Hanauer

9.1 INTRODUCTION

It is a fundamental tenet of reliability engineering that as the complexity of a system increases, the reliability will decrease, unless compensatory measures are taken. Since a frequently used measure of complexity is the number of components in a system, the decrease in reliability may then be expressed in terms of the product rule derived in Chapter 6. To recapitulate, if the component failures are mutually independent, the reliability of a system with N nonredundant components is

$$R = R_1 R_2 \ldots R_n \ldots R_N \tag{9.1}$$

where R_n is the reliability of the nth component. The dramatic deterioration of system reliability that takes place with increasing numbers of components is illustrated graphically by considering systems with components of identical reliabilities. In Fig. 9.1, system reliability versus component reliability is plotted, each curve representing a system with a different number of components. It is seen, for example, that as the number of components is increased from 10 to 50, the component reliability must be increased from 0.978 to 0.996 to maintain a system reliability of 0.80.

An alternative to the requirements for increased component reliability is to provide redundancy in part or all of a system. In what follows, we examine a number of different redundant configurations and calculate the effect on system reliability and failure rates. We also discuss specifically several of the trade-offs between different redundant configurations as well as the increased problem of common-mode failures in highly redundant systems.

The graphical presentation of systems provided by reliability block diagrams adds clarity to the discussion of redundancy. In these diagrams, which

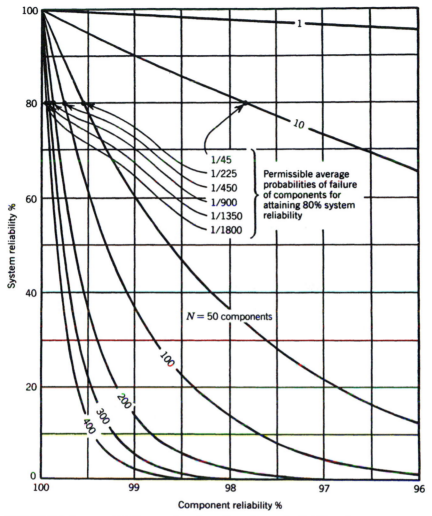

100

80

System reliability %

60

1/45
1/225
1/450
1/900
1/1350
1/1800

Permissible average
probabilities of failure
of components for
attaining 80% system
reliability

1

10

40

$N = 50$ components

20

100

200

300

400

0

100   99   98   97   96

Component reliability %

FIGURE 9.1 System reliability as a function of number and reliability of components. (From Norman H. Roberts, *Mathematical Methods of Reliability Engineering*, p. 112, McGraw-Hill, New York, 1964. Reprinted by permission.)

have their origin in electric circuitry, a signal enters from the left, passes through the system, and exits on the right. Each component is represented as a block in the system; when enough blocks fail so that all the paths by which the signal may pass from left (input) to right (output) are cut, the system is said to fail. The reliability block diagram of a nonredundant system is the series configuration shown in Fig. 9.2a; the failure of either block (unit) clearly causes system failure. The simplest redundant configurations are the parallel systems shown in Fig. 9.2b and c. In the active parallel system shown in 9.2b both blocks (units) must fail to cut the signal path and thus cause system failure. In the standby parallel system shown in Fig. 9.2c the arrow

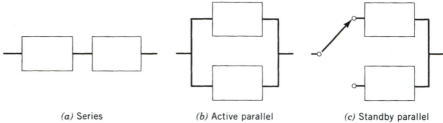

(a) Series   (b) Active parallel   (c) Standby parallel

FIGURE 9.2 Reliability block diagrams: (a) series, (b) active parallel, (c) standby parallel.

switches from the upper block (the primary unit) to the lower block (the standby unit) upon failure of the primary unit. Thus, both units must fail for the system to fail. More general redundant configurations may also be represented as reliability block diagrams. Figures 9.9. and 9.11 are examples of redundant configurations considered in the following sections.

9.2   Active and Standby Redundancy

We begin our examination of redundant systems with a detailed look at the two-unit parallel configurations pictured in Fig. 9.2. They differ in that both units in active parallel are employed and therefore subject to failure from the onset of operation, whereas in a standby parallel the second unit is not brought into operation until the first fails, and therefore cannot fail until a later time. In this section we derive the reliabilities for the idealized configurations, and then in Section 9.3 we discuss some of the limitations encountered in practice. Similar considerations also arise in treating multiple redundancy with three or more parallel units and in the more complex redundant configurations considered the subsequent sections.

Active Parallel

The reliability $R_a(t)$ of a two-unit active parallel system is the probability that either unit 1 or unit 2 will not fail until a time greater than t. Designating random variables t_1 and t_2 to represent the failure times we have

$$R_a(t) = P\{t_1 > t \cup t_2 > t\}. \qquad (9.2)$$

Thus Eq. 2.10 yields

$$R_a(t) = P\{t_1 > t\} + P\{t_2 > t\} - P\{t_1 > t \cap t_2 > t\}. \qquad (9.3)$$

Next we make an important assumption. Assume that the failures are independent events and thus replace the last term in Eq. 9.3 by $P\{t_1 > t\}P\{t_2 > t\}$. Denoting the reliabilities of the units as

$$R_i(t) = P\{t_i > t\}, \qquad (9.4)$$

we may then write

$$R_a(t) = R_1(t) + R_2(t) - R_1(t) R_2(t). \tag{9.5}$$

Standby Parallel

The derivation of the standby parallel reliability $R_s(t)$ is somewhat more lengthy since the failure time t_2 or the standby unit is dependent on the failure time t_1 of the primary unit. Only the second unit must survive to time t for the system to survive, but with the condition that it cannot fail until after the first unit fails. Hence we may write

$$R_s(t) = P\{t_2 > t \mid t_2 > t_1\}. \tag{9.6}$$

There are two possibilities. Either the first unit doesn't fail, $t_1 > t$, or the first unit fails, but the standby unit does not, $t_1 < t \cap t_2 > t$. Since these two possibilities are mutually exclusive, according to Eq. 2.12 we may just add the probabilities,

$$R_s(t) = P\{t_1 > t\} + P\{t_1 < t \cap t_2 > t\}. \tag{9.7}$$

The first term is just $R_1(t)$, the reliability of the primary unit. The second term requires more careful attention. Suppose that the PDF for the primary unit is $f_1(t)$. Then the probability of unit 1 failing between t' and $t' + dt'$ is $f_1(t')\, dt'$. Since the standby unit is put into operation at t', the probability that it will survive to time t is $R_2(t - t')$. Thus the system reliability, given that the first failure takes place between t' and $t' + dt'$ is $R_2(t - t')f_1(t')\, dt'$. To obtain the second term in Eq. 9.7 we integrate primary failure time t' between zero and t:

$$P\{t_1 < t \cap t_2 > t\} = \int_0^t R_2(t - t') f_1(t')\, dt'. \tag{9.8}$$

The standby system reliability then becomes

$$R_s(t) = R_1(t) + \int_0^t R_2(t - t') f_1(t')\, dt', \tag{9.9}$$

or using Eq. 6.10 to express the PDF in terms of reliability we obtain

$$R_s(t) = R_1(t) - \int_0^t R_2(t - t') \frac{d}{dt'} R_1(t')\, dt'. \tag{9.10}$$

Constant Failure Rate Models

General expressions for active or standby systems reliability can be obtained by inserting Eq. 6.18 for the reliability with time-dependent failure rates into Eqs. 9.5 or 9.10. Comparisons are simplest, however, if we employ a constant failure rate model. Assume that the units are identical, each with a failure rate λ. Equation 6.25, $R = \exp(-\lambda t)$, may then be inserted to obtain

$$R_a(t) = 2e^{-\lambda t} - e^{-2\lambda t} \tag{9.11}$$

for active parallel, and

$$R_s(t) = (1 + \lambda t)e^{-\lambda t} \tag{9.12}$$

for standby parallel.

The system failure rate can be determined for each of these cases using Eq. 6.15. For the active system we have

$$\lambda_a(t) = -\frac{1}{R_a}\frac{d}{dt}R_a = \lambda\left(\frac{1 - e^{-\lambda t}}{1 - 0.5e^{-\lambda t}}\right), \tag{9.13}$$

while for the standby system

$$\lambda_s(t) = -\frac{1}{R_s}\frac{d}{dt}R_s = \lambda\left(\frac{\lambda t}{1 + \lambda t}\right). \tag{9.14}$$

Figure 9.3 shows both the reliability and the failure rate for the two parallel systems, along with the results for a system consisting of a single unit. The results for the failure rates are instructive. For even though the units' failure rates are constants, the failure rates of the redundant systems as a whole are functions of time. Characteristic of systems with redundancy, they have zero failure rates at $t = 0$. The failure rates then increase to an asymptotic value of λ, the value for a single unit. At intermediate times the failure rate for the standby system is smaller than for the active parallel system. This is reflected in a larger reliability for the standby system.

Two additional measures are useful in assessing the increased reliability that results from redundant configurations. These are the mean-time-to-failure or MTTF and the rare event estimate for reliability at times which are small compared to the MTTF of single units. The values of the MTTF for active and standby parallel systems of two identical units are obtained by substituting Eqs. 9.11 and 9.12 into Eq. 6.22. We have

$$\text{MTTF}_a = \tfrac{3}{2}\,\text{MTTF} \tag{9.15}$$

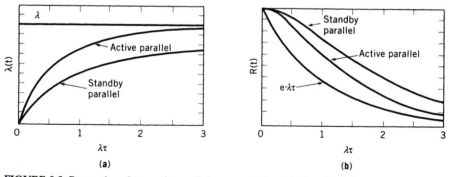

FIGURE 9.3 Properties of two-unit parallel systems: (a) reliability, (b) failure rate.

and

$$\text{MTTF}_s = 2 \text{ MTTF}, \tag{9.16}$$

where MTTF $= 1/\lambda$ for each of the two units. Thus, there is a greater gain in MTTF for the standby than for the active system.

Frequently, the reliability is of most interest for times that are small compared to the MTTF, since it is within the small-time domain where the design life of most products fall. If the single unit reliability, $R = \exp(-\lambda t)$, is expanded in a power series of λt, we have

$$R(t) = 1 - \lambda t + \tfrac{1}{2}(\lambda t)^2 - \tfrac{1}{6}(\lambda t)^3 + \cdots \tag{9.17}$$

The rare event approximation has the form of one minus the leading term in λt. Thus

$$R(t) \approx 1 - \lambda t, \qquad \lambda t \ll 1 \tag{9.18}$$

for a single unit. Employing the same exponential expansion for the redundant configurations we obtain

$$R_a(t) \approx 1 - (\lambda t)^2, \qquad \lambda t \ll 1, \tag{9.19}$$

from Eq. 9.11 and

$$R_s(t) \approx 1 - \tfrac{1}{2}(\lambda t)^2, \qquad \lambda t \ll 1. \tag{9.20}$$

from Eq. 9.12. Hence, for short times the failure probability, $1 - R$, for a standby system is only one-half of that for an active parallel system.

EXAMPLE 9.1

The MTTF of a system with a constant failure rate has been determined. An engineer is to set the design life so that the end-of-life reliability is 0.9.

(a) Determine the design life in terms of the MTTF.

(b) If two of the systems are placed in active parallel, to what value may the design life be increased without causing a decrease in the end-of-life reliability?

Solution Let the failure rate be $\lambda \equiv 1/\text{MTTF}$.

(a) $R = e^{-\lambda T}$. Therefore, $T = (1/\lambda) \ln(1/R)$.

$$T = \ln\left(\frac{1}{R}\right) \times \text{MTTF} = \ln\left(\frac{1}{0.9}\right) \text{MTTF} = 0.105 \text{ MTTF}.$$

(b) From Eq. 9.11, $R = 2e^{-\lambda T} - e^{-2\lambda T}$. Let $x \equiv e^{-\lambda T}$. Therefore, $x^2 - 2x + R = 0$. Solve the quadratic equation:

$$x = \frac{+2 \pm \sqrt{4 - 4R}}{2} = 1 - \sqrt{1 - R}.$$

The "+" solution is eliminated, since x cannot be greater than one. Since $x = e^{-\lambda T} = 1 - \sqrt{1 - R}$, then with $\lambda = 1/\text{MTTF}$,

$$T = \ln \left[\frac{1}{(1 - \sqrt{1 - R})} \right] \times \text{MTTF},$$

$$= \ln \left[\frac{1}{(1 - \sqrt{1 - 0.9})} \right] \times \text{MTTF} = 0.380 \, \text{MTTF}.$$

Thus the redundant system may have nearly four times the design life of the single system, even though it may be seen from Eq. 9.15 that the MTTF of the redundant system is only 50% longer.

9.3 REDUNDANCY LIMITATIONS

The results for active and standby reliability presented thus far are highly idealized. In practice, a number of factors can significantly reduce the reliability of redundant systems. In reality, these factors and their mitigation often are dominant in determining the level of reliability which can be achieved. For active parallel systems, common mode failures and load sharing phenomena tend to be of most concern. For standby systems, switching failures and failure of the standby unit before switching are important considerations.

Common-Mode Failures

Common-mode failures are caused by phenomena that create dependencies between two or more redundant components which cause them to fail simultaneously. Such failures have the potential for negating much of the benefit gained with redundant configurations. Common-mode failures may be caused by common electric connections, shared environmental stresses such as dust or vibration, common maintenance problems, or a host of other factors. In commercial aviation, for example, a great deal of redundancy is employed, allowing high levels of safety to be achieved. Thus when problems do occur frequently they may be attributed to common-mode failures: the dust rising from a volcanic eruption in Alaska that caused simultaneous malfunctioning of all of a commercial airliner's engines, or the pieces of a fractured jet engine turbine blade that cut all of the redundant hydraulic control lines and caused the crash of a DC10.

Viewed in terms of the reliability block diagrams in Fig. 9.2, common-mode failure mechanisms have the same effect as putting in an additional component in series with the parallel configuration. For identical units with reliability R, the active parallel reliability given by Eq. 9.5 becomes

$$R'_a = (2R - R^2) R', \qquad (9.21)$$

where R' is the contribution to decreased reliability from common mode failures. The effects are illuminated if we recast this equation in terms of the failure probability $p = 1 - R$, $p' = 1 - R'$ and $p'_a = 1 - R'_a$ corresponding to each of the reliability's. Equation 9.21 may be written as

$$p'_a = p' + p^2 - p'p^2. \qquad (9.22)$$

Suppose we have an aircraft engine with a failure probability per flight of $p = 10^{-6}$ and a common mode failure probability a thousand times smaller: $p' = 10^{-9}$. For a two engine aircraft in the absence of common-mode failures the failure probability would be $p^2 = 10^{-12}$, but from Eq. 9.22 we see that

$$p'_a = 10^{-9} + 10^{-12} - 10^{-21}. \tag{9.23}$$

Thus the system failure probability, $p'_a \approx 10^{-9}$ is totally dominated by common mode failure, although it is still far more reliable than if a single engine had been used.

A great deal of the engineering of redundant systems is expended on identifying possible common mode mechanisms and eliminating them. Nevertheless, some possibilities may be impossible to eliminate entirely, and therefore reliability modeling must take them into account. Most commonly, such phenomena are modeled through the following constant failure rate model.* Suppose that λ is the total failure rate of a single unit. We divide λ into two contributions

$$\lambda = \lambda_I + \lambda_c, \tag{9.24}$$

where λ_I is the rate of independent failure and λ_c is the common-mode failure rate. These partial failure rates may be used to express common-mode failure rates in active parallel systems as follows. Define the factor β as the ratio

$$\beta = \lambda_c/\lambda. \tag{9.25}$$

Each of the units then has an failure mode reliability of

$$R_I = e^{-\lambda_I t}, \tag{9.26}$$

which accounts only for independent failures. Therefore the system reliability for independent failure is determined by using λ_I in Eq. 9.11. We multiply this system reliability by $\exp(-\lambda_c t)$ to account for common-mode failures. Thus, for the two units in parallel.

$$R_a(t) = (2e^{-\lambda_I t} - e^{-2\lambda_I t})e^{-\lambda_c t}, \tag{9.27}$$

or using $\lambda_c = \beta\lambda$ and $\lambda_I = (1 - \beta)\lambda$ we may write

$$R_a(t) = [2 - e^{-(1-\beta)\lambda t}]e^{-\lambda t}. \tag{9.28}$$

The loss of reliability with the increase in the β factor is clearly seen by looking at the rare event approximation at small λt, for we now have a term which is linear in λt:

$$R_a(t) \approx 1 - \beta\lambda t - (1 - 2\beta + \beta^2/2)(\lambda t)^2 + \cdots, \tag{9.29}$$

as opposed to $1 - (\lambda t)^2$ as in Eq. 9.19. The effect of common-mode failures can also be seen in the reduction in the mean-time-to-failure:

$$\text{MTTF}_a = \left[2 - \frac{1}{2 - \beta}\right]\text{MTTF}. \tag{9.30}$$

* K. L. Flemming and P. H. Raabe, "A Comparison of Three Methods for the Quantitative Analysis of Common Cause Failures," *General Atomic Report*, GA-A14568, 1978.

EXAMPLE 9.2

Suppose that a unit has a design-life reliability of 0.95.

(*a*) Estimate the reliability if two of these units are put in active parallel and there are no common-mode failures.

(*b*) Estimate the maximum fraction β of common failures that is acceptable if the parallel units in *a* are to retain a system reliability of at least 0.99.

Solution   From Eq. 9.18 take $\lambda t = 0.05$.

(*a*) $R \approx 1 - (\lambda T)^2$, $R = 0.9975$.

(*b*) From Eq. 9.29,

$$\tilde{R} = 1 - R = 0.01 \approx \beta \lambda T + \left(1 - 2\beta + \frac{\beta^2}{2}\right)(\lambda T)^2.$$

Thus, with $\lambda T \approx 0.05$, we have

$$0.00125\beta^2 + 0.045\beta - 0.0075 = 0.$$

Therefore,

$$\beta = \frac{-0.045 \pm (2.0625 \times 10^{-3})^{1/2}}{0.0025}.$$

For β to be positive, we must take the positive root. Therefore, $\beta \leq 0.166$.

Load Sharing

Load sharing is a second cause of reliability degradation in active parallel systems. For redundant engines, motors, pumps, structures and many other devices and systems, the failure of one unit will increase the stress level on the other and therefore increase its failure rate. A simple example is two flashlight batteries placed in parallel to provide a fixed voltage. Assume the circuit is designed so that if either fails the other will supply adequate voltage. Nevertheless, the current through the remaining battery will be higher, and this will cause greater heating in the internal resistance. The net result is that the remaining battery will operate at a higher temperature and thus tend to deteriorate faster.

Fortunately, in a redundant system with sufficient capacity, the increased failure rate should not lead to unacceptable failure probabilities. If the first failure is detected, the system may be required to operate for only a short period of time before repairs are made. Thus if one engine fails in a multi-engine aircraft, it is only necessary that the flight continue to the nearest airfield without incurring a significant probability of a second engine failure. From this standpoint, the degradation is less serious than the potential for common-mode failures.

In Chapter 11, Markov methods are used to develop the following model for shared load redundancy with time-independent failure rates. Suppose that

$\lambda^* > \lambda$ is the increased failure rate of the remaining unit after the first has failed. Then, in the absence of common-mode failures,

$$R_a(t) = 2e^{-\lambda^* t} + e^{-2\lambda t} - 2e^{-(\lambda+\lambda^*)t}. \tag{9.31}$$

This may be seen to reduce to Eq. 9.11 in the limiting case that $\lambda^* = \lambda$. A conservative design procedure, which always gives an underestimate of the reliability, is to replace λ by λ^* in Eq. 9.31, thereby assuming that each unit is carrying the entire load of the system.

If λ^* becomes too large, all of the benefit of the redundancy may be lost, and in fact the system may be less reliable than a single unit with failure rate λ. For example, it may be shown that if $\lambda^* > 1.56\,\lambda$, the MTTF will be less than for a single unit. In the limit as $\lambda^* \to \infty$ Eq. 9.31 reduces to the reliability for the two units placed in series. This may be understood as follows. If either unit failing gives rise to the second unit failing almost instantaneously then indeed the system failure rate will be twice that of a single unit. For in doubling the number of units, one increases the possibility of a first failure.

EXAMPLE 9.3

In an active parallel system each unit has a failure rate of 0.002 hr^{-1}.

(a) What is the MTTF$_a$ if there is no load sharing?

(b) What is the MTTF$_a$ if the failure rate increases by 20% as a result of increased load?

(c) What is the MTTF$_a$ if one simply (and conservatively) increased both unit failure rates by 20%?

Solution

(a) $\text{MTTF}_a = \dfrac{3}{2}\,\text{MTTF} = \dfrac{3}{2\lambda} = \dfrac{3}{2 \times 0.002} = 750$ hr

(b) $\text{MTTF}_a = \displaystyle\int_0^\infty R_a(t)\,dt = \int_0^\infty [2e^{-\lambda^* t} + e^{-2\lambda t} - 2e^{-(\lambda+\lambda^*)t}]\,dt$

or

$$\text{MTTF}_a = \frac{2}{\lambda^*} + \frac{1}{2\lambda} - \frac{2}{\lambda+\lambda^*}.$$

Thus with

$$\lambda^* = 1.2 \times 0.002 = 0.0024 \ hr^{-1}$$

we have

$$\text{MTTF}_a = \frac{2}{0.0024} + \frac{1}{2 \times 0.002} - \frac{2}{0.0044} = 629 \text{ hrs}$$

(c) $\text{MTTF}_a = \dfrac{3}{2\lambda^*} = \dfrac{3}{2 \times 0.0024} = 625$ hr

Switching and Standby Failures

Common-mode failures are less likely for standby than for active parallel configurations because the secondary system may be quite different from the primary. For example, the causes of the failure of electric power are likely to be quite different than those that may cause the diesel backup generator to fail. Nevertheless, care must also be exercised in the design and operation of systems with standby redundancy. Some smaller possibility of common-mode failure incapacitating both primary and secondary units may remain. In addition, two new failure modes, unique to standby configurations, must be addressed: switching failures and secondary unit failure while in the standby mode. The following illustration may be helpful in understanding these modes.

Suppose power is supplied by a diesel generator. A second identical generator is used for backup. If there is some probability, p, that a switch can not be made to the second generator upon failure of the primary unit, as derived in Chapter 11, the reliability of the system is obtain by multiplying the second term in Eq. 9.12 by $(1 - p)$:

$$R_s(t) = [1 + (1 - p)\lambda t]e^{-\lambda t}. \tag{9.32}$$

One cause of switching failures is the failure of the control mechanism in sensing the primary unit failure and turning on the secondary unit. Time is also an important consideration, for in certain situations some delay can be tolerated before the backup unit takes over. For example, if a pump supplying coolant to a reservoir fails, it may only be necessary for the backup system to come on before the reservoir drains. On a shorter time scale, if a process control computer fails there may be a period of seconds or less before the backup is required. If some time delay is tolerable, repeated attempts to switch the system may be made, or parts replaced.

Failure of the secondary unit to function may result not only from switching failures. The secondary system may also have failed in the standby mode before the primary system failure. Such failures are most prone to happen in situations where the secondary unit is called upon very infrequently and therefore may have been allowed to deteriorate while in the standby mode. In Chapter 11 an expression for reliability in which both failure modes are present is developed. The result is equivalent to affixing the multiplicitive factor $(\lambda^+ t)^{-1}(1 - e^{-\lambda^+ t})$ to the second term in Eq. 9.32

$$R_s(t) = \left[1 + (1 - p)\frac{\lambda}{\lambda^+}(1 - e^{-\lambda^+ t})\right]e^{-\lambda t}, \tag{9.33}$$

where λ^+ is the failure rate of the secondary unit while in standby.

EXAMPLE 9.4

An engineer designs a standby system with two identical units to have an idealized MTTF, of 1000 days. To be conservative, she then assumes a switching failure probability of 10% and the failure rate of the unit in standby of 10% of the unit in operation.

Assuming constant failure rates, estimate the reduced $MTTF_s$ of the system with switching and standby failures included.

Solution For the idealized MTTF, we have $MTTF_s = 2/\lambda$ or

$$\lambda = 2/1000 \text{ days} = 0.002 \text{ day}^{-1}.$$

For the reduced MTTF, we have

$$MTTF_s = \int_0^\infty R_s(t) \, dt = \int_0^\infty \left\{ \left[1 + (1 - p) \frac{\lambda}{\lambda^+} (1 - e^{-\lambda^+ t}) \right] e^{-\lambda t} \right\} dt$$

or

$$MTTF_s = \frac{1}{\lambda} [1 + (1 - p)(1 + \lambda^+/\lambda)^{-1}].$$

Thus with $p = 0.1$ and $\lambda^+/\lambda = 0.1$ we have:

$$MTTF_s = \frac{1}{0.002} [1 + (1 - 0.1)(1 + 0.1)^{-1}] = 909 \text{ days}$$

Cold, Warm, and Hot Standby

The trade-off between switching failures and failure in standby must be considered in the design of standby redundancy; it is the primary consideration in determining whether cold, warm, or hot standby is to be used. In cold standby the secondary unit is shut down until needed. This typically reduces the value of λ^+ to a minimum. However, it tends to result in the largest values of p. Thus in our example of the diesel generator, it is most likely not to have failed if it has not been operating. However, coming from cold startup to a fully loaded operation on short notice may cause sufficient transient stress to result in a significant demand failure probability. In warm standby the transient stresses are reduced by having the secondary unit continuously in operation, but in an idling or unloaded state. In this case p may be expected to be smaller, at the expense of a moderately increased value of λ^+. Even smaller values of p are achieved by having the secondary unit in hot standby, that is, continuously operating at a full load. In this case—for identical units—the failure rate will equal that of the primary system, $\lambda^+ = \lambda$, causing Eq. 9.33 to reduce to

$$R_s(t) = (2 - p) e^{-\lambda t} - (1 - p) e^{-2\lambda t}. \tag{9.34}$$

We see from this equation that if the switching failure can be made very small, which is the object of hot standby, the equation is equivalent to an active parallel system. Thus the reliability is markedly less than for an idealized standby system. In many instances of warm or hot standby, however, secondary unit failures in standby can be detected and repaired fairly rapidly. The modeling of such repairable systems is taken up in Chapters 10 and 11.

Redundant computer control systems present a somewhat different situation than that encountered with motors, engines, pumps, or other energy or mass delivery systems. In order to start from cold standby not only must the computer be powered, but the current data must be loaded to memory. Hot standby is particularly advantages in these cases where switching the output from the primary to the secondary computer is a relatively simple matter. There is, however, one difficulty. A means must be established for detecting which computer is wrong. This is straightforward if the computer stops functioning altogether. However, if the failure mode is a type that caused the computer to give incorrect but plausible output, then a means for knowing where the incorrect information is being produced is a necessity. For these situations the 2/3 voting systems discussed in the following section are widely used.

9.4   MULTIPLY REDUNDANT SYSTEMS

The reliability of a system can be further enhanced by placing increased numbers of components in parallel. Such redundancy can take either active or standby form. In $1/N$ and m/N redundancy, respectively, one or m of the N units must function for the system to function. Consider $1/N$ redundancy first for active and then for standby parallel. In either of these configurations the probability of system malfunction becomes increasingly small, and as a result increased attention must be given to the complications discussed in Section 9.3.

1/N Active Redundancy

Suppose that we have N components in parallel; if any one of them functions, the system will function successfully. Thus, in order for the system to fail, all the components must fail. This may be written as follows. Let X_i denote the event of the ith component failure and X the system failure. Thus, for a system of N parallel components, we have

$$X = X_1 \cap X_2 \cap \ldots \cap X_N, \qquad (9.35)$$

and the system reliability is

$$R_a = 1 - P\{X_1 \cap X_2 \cap \ldots \cap X_N\}. \qquad (9.36)$$

If the failures are mutually independent, we may use the definition of independence to write

$$R_a = 1 - P\{X_1\}P\{X_2\} \ldots P\{X_N\}. \qquad (9.37)$$

The $P\{X_i\}$ are the component failure probabilities; therefore, they are related to the reliabilities by

$$P\{X_i\} = 1 - R_i. \qquad (9.38)$$

Consequently, we have for $1/N$ active redundancy

$$R_a = 1 - \prod_i (1 - R_i). \qquad (9.39)$$

For identical components this may be simplified. Suppose that all the R_i have the same value, $R_i = R$. Equation 9.39 then reduces to

$$R_a = 1 - (1 - R)^N. \tag{9.40}$$

The degree of improvement in system reliability brought about by multiple redundancy is indicated in Fig. 9.4, where system reliability is plotted versus component reliability for different numbers of parallel components. Two other characterizations of the increased reliability are given by the rare event approximation and the MTTF. The expansion of Eq. 9.18 yields $1 - R \approx \lambda t$ for small λt and results in the reduction of Eq. 9.40 to

$$R_a(t) \approx 1 - (\lambda t)^N, \qquad \lambda t \ll 1. \tag{9.41}$$

We may use the binomial expansion, introduced in Chapter 2, to express the reliability in a form that is more convenient for evaluating the MTTF. The binomial coefficients allow us to write in general

$$(p + q)^N = \sum_{n=0}^{N} C_n^N p^{N-n} q^n, \tag{9.42}$$

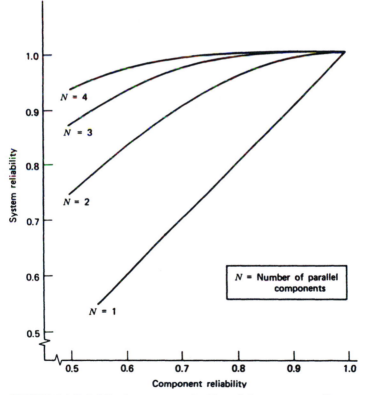

FIGURE 9.4 Reliability improvement by N parallel components. (From K. C. Kapur and L. R. Lamberson, *Reliability in Engineering Design*. Copyright © 1977, by John Wiley and Sons. Reprinted by permission.)

where the C_n^N coefficients are given by Eq. 2.43. Taking $p = 1$ and $q = -R$, we obtain

$$(1 - R)^N = \sum_{n=0}^{N} C_n^N (-1)^n R^n. \tag{9.43}$$

Therefore, since $C_0^N = 1$, we may write Eq. 9.40 as

$$R_a = \sum_{n=1}^{N} (-1)^{n-1} C_n^N R^n. \tag{9.44}$$

We next assume a constant failure rate for each component and replace R with $e^{-\lambda t}$. Applying Eq. 6.22, to express the MTTF in terms of $R_a(t)$, we obtain

$$\text{MTTF}_a = \sum_{n=1}^{N} (-1)^{n-1} \frac{C_n^N}{n\lambda}. \tag{9.45}$$

While the forgoing relationships indicate that in principle, reliabilities very close to one are obtainable, common-mode failures become an increasingly overriding factor when N is taken to be three or more. If the β factor method is applied, for example, the loss of reliability may be dominated not by the $(\lambda t)^N$ of Eq. 9.41 but by a $\beta \lambda t$ term as in Eq. 9.29. Likewise, the load sharing phenomena becomes increasingly serious as additional units fail. A four engine aircraft, flying on one engine may be expected to be under higher stress than a two engine aircraft flying on one.

EXAMPLE 9.5

A temperature sensor is to have a design-life reliability of no less than 0.98. Since a single sensor is known to have a reliability of only 0.90, the design engineer decides to put two of them in parallel. From Eq. 9.5 the reliability should then be 0.99, meeting the criterion. Upon reliability testing, however, the reliability is estimated to be only 0.97. The engineer first deduces that the degradation is due to common-mode failures and then considers two options: (1) putting a third sensor in parallel, and (2) reducing the probability of common-mode failures.

(a) Assuming that the sensors have constant failure rates, find the value of β that characterizes the common-mode failures.

(b) Will adding a third sensor in parallel meet the reliability criterion if nothing is done about common-mode failures?

(c) By how much must β be reduced if the two sensors in parallel are to meet the criterion?

Solution   If the design-life reliability of a sensor is $R_1 = e^{-\lambda T} = 0.9$, then $\lambda T = \ln(1/R_1) = \ln(1/0.9) = 0.10536$.

(a) Let $R_2 = 0.97$ be the system reliability for two sensors in parallel. Then β is found in terms of R_2 from Eq. 9.28 to be

$$\beta = 1 + \frac{1}{\lambda T}\ln(2 - R_2 e^{\lambda T}) = 1 + \frac{1}{0.10536}\ln\left(2 - \frac{0.97}{0.9}\right),$$

$$= 0.2315.$$

(*b*) The reliability for three sensors in parallel is given by Eq. 9.40 with $N = 3$. Using $\lambda_l = (1 - \beta)\lambda$ and $\lambda_c = \beta\lambda$, we may expand the bracketed term to obtain

$$R_3 = [3 - 3e^{-(1-\beta)\lambda T} + e^{-2(1-\beta)\lambda T}]e^{-\lambda T}.$$

From *a* we have $(1 - \beta)\lambda T = (1 - 0.2315) \times 0.10536 = 0.08097$, and thus $e^{-(1 - \beta)\lambda T} = 0.92222$. Thus the reliability is

$$R_3 = [3 - 3 \times 0.92222 + (0.92222)^2] \times 0.9 = 0.975$$

Therefore, the criterion is not met by putting a third sensor in parallel.

(*c*) To meet the criterion with two sensors in parallel, we must reduce β enough so that the equation in part *a* is satisfied with $R_2 = 0.98$. Thus

$$\beta = 1 + \frac{1}{0.10536} \ln \left(2 - \frac{0.97}{0.9} \right) = 0.1165.$$

Therefore, β must be reduced by at least

$$1 - \frac{0.1165}{0.2315} \approx 50\%.$$

1/N Standby Redundancy

We may derive expressions for $1/N$ standby reliability by noting that the derivation of the recursive equation, Eq. 9.10, is valid even if $R_1(t)$ represents a standby system. Thus we may derive the reliability of a standby system of N identical units in terms of a system of $N - 1$ units. Suppose we denote the reliability of the n unit system as R_n, and thus of the $n - 1$ system as R_{n-1}, where the reliability of a single unit is $R_1 = R$. We may now rewrite Eq. 9.10 as

$$R_n(t) = R_{n-1}(t) - \int_0^t R(t - t') \frac{d}{dt'} R_{n-1}(t') \, dt'. \tag{9.46}$$

Thus R_2, in the constant failure rate approximation given by Eq. 9.12, may be shown to result from inserting $R = R_1 = e^{-\lambda t}$ into the right hand side of this expression. Likewise if Eq. 9.12 is inserted into the right hand side of this expression we obtain

$$R_3(t) = [1 + \lambda t + \tfrac{1}{2}(\lambda t)^2]e^{-\lambda t}. \tag{9.47}$$

This expression can be inserted into the right of Eq. 9.46 to obtain R_4 and so on. In general, for N units in standby redundancy we obtain

$$R_s(t) = \sum_{n=0}^{N-1} \frac{1}{n!} (\lambda t)^n e^{-\lambda t}. \tag{9.48}$$

Equation 6.22 then yields a standby MTTF of

$$\text{MTTF}_s = N/\lambda. \tag{9.49}$$

To calculate the rare event approximation we first note that the exponential expansion can be written as two sums:

$$e^{\lambda t} = \sum_{n=0}^{N-1} \frac{1}{n!} (\lambda t)^n + \sum_{n=N}^{\infty} \frac{1}{n!} (\lambda t)^n. \tag{9.50}$$

Solving for the first sum, and inserting the result into Eq. 9.48, we obtain after simplification

$$R_s(t) = 1 - \sum_{n=N}^{\infty} \frac{1}{n!} (\lambda t)^n e^{-\lambda t}. \tag{9.51}$$

Thus taking the lowest order terms, we find for small λt that

$$R_s(t) \approx 1 - \frac{1}{N!} (\lambda t)^N. \tag{9.52}$$

We see that the $1/N$ standby configuration comes closer to one in the rare event approximation than does Eq. 9.41 for the active parallel system. Of course switching failures and failures in the standby state must be included to make more realistic comparisons.

m/N Active Redundancy

In the $1/N$ systems considered thus far, if any one of the two or more units functions, the system operates successfully. We now turn to the m/N system in which m is the minimum number that must function for successful system operation. The m/N is popular for relief valves, pumps, motors, and other equipment that must have a specified capacity to meet design criteria. In such systems it is often possible to increase reliability without a commensurate cost increase, for components of off-the-shelf sizes may meet capacity requirements while at the same time allowing for some degree of redundancy. In instrumentation and control systems m/N configurations are popular for two reasons. The spurious fail-safe operation of a single unit is prevented from causing undesirable consequences. Likewise, voting can be applied to the output of redundant instruments or computers.

An m/N system may be represented in a reliability block diagram, as shown for a 2/3 system in Figure 9.5. Now, however, the block representing

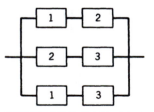

FIGURE 9.5 Reliability block diagram for a $\frac{2}{3}$ system.

each component must be repeated in the diagram. Thus the system reliability cannot be calculated as in earlier $1/N$ cases because the three parallel chains contain some of the same components and therefore cannot be independent of one another.

For identical components, the reliability of an m/N system may be determined by again returning to the binomial distribution. Suppose that p is the probability of failure over some period of time for one unit. That is,

$$p = 1 - R, \tag{9.53}$$

where R is the component reliability. From the binomial distribution the probability that n units will fail is just

$$P\{\mathbf{n} = n\} = C_n^N p^n (1 - p)^{N-n}. \tag{9.54}$$

The m/N system will function if there are no more than $N - m$ failures. Thus

$$P\{\mathbf{n} \le N - m\} = \sum_{n=0}^{N-m} C_n^N p^n (1 - p)^{N-n} \tag{9.55}$$

is the reliability. Combining Eqs. 9.53 and 9.55 then yields

$$R_a = \sum_{n=0}^{N-m} C_n^N (1 - R)^n R^{N-n}. \tag{9.56}$$

Alternately, since

$$P\{\mathbf{n} > N - m\} = \sum_{n=N-m+1}^{N} C_n^N p^n (1 - p)^{N-n} \tag{9.57}$$

is the probability that the system will fail, we may also write the system reliability as

$$R_a = 1 - \sum_{n=N-m+1}^{N} C_n^N (1 - R)^n R^{N-n}. \tag{9.58}$$

Equations 9.56 and 9.58 are identical in value. Depending on the ratio of m to N, one may be more convenient than the other to evaluate. For example, in a $1/N$ system Eq. 9.58 is simpler to evaluate, since the sum on the right-hand side has only one term, $n = N$, yielding Eq. 9.40.

In dealing with redundant configurations, whether of the $1/N$ or m/N variety, we can simplify the calculations substantially with little loss of accuracy if the component failure probabilities are small (i.e., when the component's reliability approaches one). In these situations a reasonable approximation includes only the leading term in the summation of Eq. 9.58. To illustrate, suppose that R is very close to one; we may replace it by one in the R^{N-n} term to yield

$$R_a \approx 1 - \sum_{n=N-m+1}^{N} C_n^N (1 - R)^n. \tag{9.59}$$

We note, however, that the terms in the $(1 - R)^n$ series decrease very rapidly in magnitude as the exponent is increased. Consequently, we need include only the term with the lowest power of $1 - R$. Thus the reliability is approximately

$$R_a \approx 1 - C_{N-m+1}^N (1 - R)^{N-m+1}. \tag{9.60}$$

If the rare event approximation, $1 - R \approx \lambda t$, is employed, then

$$R_a \approx 1 - C_{N-m+1}^N (\lambda t)^{N-m+1}. \tag{9.61}$$

EXAMPLE 9.6

A pressure vessel is equipped with six relief valves. Pressure transients can be controlled successfully by any three of these valves. If the probability that any one of these valves will fail to operate on demand is 0.04, what is the probability on demand that the relief valve system will fail to control a pressure transient? Assume that the failures are independent.

Solution   In this situation, the foregoing equations are valid if unreliability, $\bar{R}_a = 1 - R_a$, is defined as demand failure probability. Using the rare-event approximation, we have from Eq. 9.60, with $N = 6$ and $m = 3$, $0.04 = 1 - R$:

$$\bar{R}_a \approx C_4^6 (0.04)^4 = \frac{6!}{2!4!} (0.04)^4 = 15 \times 256 \times 10^{-8}$$

$$\bar{R}_a \approx 0.38 \times 10^{-4}.$$

9.5   REDUNDANCY ALLOCATION

High reliability can be achieved in a variety of ways; the choice will depend on the nature of the equipment, its cost, and its mission. If we were to provide an emergency power supply for a hospital, an air traffic control system, or a nuclear power plant, for example, the most cost-effective solution might well be to use commercially available diesel generators as the components in a redundant configuration. On the other hand, the use of redundancy may not be the optimal solution in systems in which the minimum size and weight are overriding considerations: for example, in satellites or other space applications, in well-logging equipment, and in pacemakers and similar biomedical applications. In such applications space or weight limitations may dictate an increase in component reliability rather than redundancy. Then more emphasis must be placed on robust design, manufacturing quality control, and on controlling the operating environment.

Once a decision is made to include redundancy, a number of design trade-offs must be examined to determine how redundancy is to be deployed. If the entire system is not to be duplicated, then which components should be duplicated? Consider, for example, the simple two-component system shown in Fig. 9.6*a*. If the reliability $R_a = R_1 R_2$ is not large enough, which component should be made redundant? Depending on the choice, the system Fig. 9.6*b*

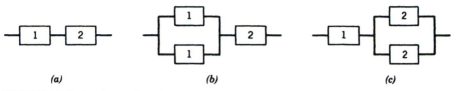

FIGURE 9.6 Redundancy allocation.

or *c* will result. It immediately follows that

$$R_b = (2R_1 - R_1^2)R_2, \tag{9.62}$$

$$R_c = R_1(2R_2 - R_2^2). \tag{9.63}$$

Or taking the differences of the results, we have

$$R_b - R_c = R_1 R_2 (R_2 - R_1). \tag{9.64}$$

Not surprisingly, this expression indicates that the greatest reliability is achieved in the redundant configuration if we duplicate the component that is least reliable; if $R_2 > R_1$, then system R_b is preferable, and conversely. This rule of thumb can be generalized to systems with any number of nonredundant components; the largest gains are to be achieved by making the least reliable components redundant. In reality, the relative costs of the components also must be considered. Since component costs are normally available, the greatest impediment to making an informed choice is lack of reliability data for the components involved. Trade-offs in the allocation of redundancy often involve additional considerations. Two examples are those between high- and low-level redundancy, and those between fail-safe and fail-to-danger consequences.

EXAMPLE 9.7

Suppose that in the system shown in Fig. 9.6 the two components have the same cost, and $R_1 = 0.7$, $R_2 = 0.95$. If it is permissible to add two components to the system, would it be preferable to replace component 1 by three components in parallel or to replace components 1 and 2 each by simple parallel systems?

Solution   If component 1 is replaced by three components in parallel, then from Eq. 9.40

$$R_a = [1 - (1 - R_1)^3]R_2 = 0.973 \times 0.95 = 0.92435.$$

If each of the two components is replaced by a simple parallel system,

$$R_b = [1 - (1 - R_1)^2][1 - (1 - R_2)^2] = 0.91 \times 0.9975 = 0.9077.$$

In this problem the reliability R_1 is so low that even the reliability of a simple parallel system, $2R_1 - R_1^2$, is smaller than that of R_2. Thus replacing component 1 by three parallel components yields the higher reliability.

High- and Low-Level Redundancy

One of the most fundamental determinants of component configuration concerns the level at which redundancy is to be provided. Consider, for example, the system consisting of three subsystems, as shown in Fig. 9.7. In high-level redundancy, the entire system is duplicated, as indicated in Fig. 9.7a, whereas in low-level redundancy the duplication takes place at the subsystem or component level indicated in Fig. 9.7b. Indeed, the concept of the level at which redundancy is applied can be further generalized to lower and lower levels. If each of the blocks in the diagram is a subsystem, each consisting of components, we might place the redundancy at a still lower component level. For example, computer redundancy might be provided at the highest level by having redundant computers, at an intermediate level by having redundant circuit boards within a single computer, or at the lowest level by having redundant chips on the circuit boards.

Suppose that we determine the reliability of each of the systems in Fig. 9.7 with the component failures assumed to be mutually independent. The reliability of the system without redundancy is then

$$R_0 = R_a R_b R_c. \tag{9.65}$$

The reliability of the two redundant configurations may be determined by considering them as composites of series and parallel configurations.

For the high-level redundancy shown in Fig. 9.7a, we simply take the parallel combination of the two series systems. Since the reliability of each series subsystem is given by Eq. 9.65, the high-level redundant reliability is given by

$$R_{HL} = 2R_0 - R_0^2, \tag{9.66}$$

or equivalently,

$$R_{HL} = 2R_a R_b R_c - R_a^2 R_b^2 R_c^2. \tag{9.67}$$

Conversely, to calculate the reliability of the low-level redundant system, we first consider the parallel combinations of component types a, b, and c separately. Thus the two components of type a in parallel yield

$$R_A = 2R_a - R_a^2, \tag{9.68}$$

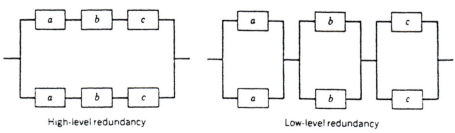

High-level redundancy   Low-level redundancy

FIGURE 9.7 High- and low-level redundancy.

and similarly,

$$R_B = 2R_b - R_b^2, \qquad R_C = 2R_c - R_c^2. \tag{9.69}$$

The low-level redundant system then consists of a series combination of the three redundant subsystems. Hence

$$R_{LL} = R_A R_B R_C, \tag{9.70}$$

or, inserting Eqs. 9.68 and 9.69 into this expression, we have

$$R_{LL} = (2R_a - R_a^2)(2R_b - R_b^2)(2R_c - R_c^2). \tag{9.71}$$

Both the high- and the low-level redundant systems have the same number of components. They do not result, however, in the same reliability. This may be demonstrated by calculating the quantity $R_{LL} - R_{HL}$. For simplicity we examine systems in which all the components have the same reliability, R. Then

$$R_{HL} = 2R^3 - R^6 \tag{9.72}$$

and

$$R_{LL} = (2R - R^2)^3. \tag{9.73}$$

After some algebra we have

$$R_{LL} - R_{HL} = 6R^3(1 - R)^2. \tag{9.74}$$

Consequently, $R_{LL} > R_{HL}$.

Regardless of how many components the original system has in series, and regardless of whether two or more components are put in parallel, low-level redundancy yields higher reliability, but only if a very important condition is met. The failures must be truly independent in both configurations. In reality, common-mode failures are more likely to occur with low-level than with high-level redundancy. In high-level redundancy similar components are likely to be more isolated physically and therefore less susceptible to common local stresses. For example, a faulty connector may cause a circuit board to overheat and then the two redundant chips on that board to fail. But if the redundant chips are on different circuit boards in a high-level redundant system, this common-mode failure mechanism will not exist. Physical isolation, in general, may eliminate many causes of common-mode failures, such as local flooding and overheating.

Some insight into common-mode failures may be gained as follows. Consider the same high- and low-level redundant systems for which the results are given by Eqs. 9.72 and 9.73, and let the component reliability be represented by $R = e^{-\lambda t}$. Suppose that because components in the high-level system are physically isolated, there are no significant common-mode failures. Then we may write simply

$$R_{HL} = e^{-3\lambda t}(2 - e^{-3\lambda t}). \tag{9.75}$$

In the low-level system, however, we specify that some fraction, β, of the failure rate λ is due to common-mode failures. In this case the quantities R_a, R_b, and

R_c will no longer reduce to Eq. 7.73, or

$$R_{LL} = (2e^{-\lambda t} - e^{-2\lambda t})^3, \tag{9.76}$$

where there are no common-mode failures. Rather, the β-factor model replaces Eqs. 9.68 and 9.69 by Eq. 9.28 to yield

$$R_A = R_B = R_C = 2e^{-\lambda t} - e^{-2\lambda t}e^{\beta\lambda t}. \tag{9.77}$$

Then, from Eq. 9.70, we find the low-level redundant system reliability is reduced to

$$R_{LL} = (2e^{-\lambda t} - e^{-2\lambda t}e^{\beta\lambda t})^3. \tag{9.78}$$

This must be compared to Eq. 9.75 to determine how large β can become before the advantage of low-level is lost. Consider the following example.

EXAMPLE 9.8

Suppose that the design-life reliability of each of the components in the high- and low-level redundant systems pictured in Fig. 9.7 is 0.99. What fraction of the failure rate in the low-level system may be due to common-mode failures, without the advantage of low-level redundancy being lost?

Solution   Set $R_{HL} = R_{LL}$, using Eqs. 9.75 and 9.78 at the end of the design life:

$$e^{-3\lambda T}(2 - e^{-3\lambda T}) = (2e^{-\lambda T} - e^{-2\lambda T + \beta\lambda T})^3.$$

Solving for β yields

$$\beta = \frac{1}{\lambda T}\ln[2 - (2 - e^{-3\lambda T})^{1/3}] + 1.$$

Since $e^{-\lambda T} = 0.99$, $\lambda T = 0.01005$. Thus

$$\beta = \frac{1}{0.01005}\ln[2 - (2 - 0.99^3)^{1/3}] + 1 = 0.0197.$$

Fail-Safe and Fail-to-Danger

Thus far we have lumped all failures together. There are situations, however, in which different failure modes can have quite different consequences. Judgment must then be exercised in allocating redundancy between modes. One of the most common examples occurs in the trade-off between fail-safe and fail-to-danger encountered in the design of m/N alarm and safety systems.

Consider an alarm system. The alarm may fail in one of two ways. It may fail to function even though a dangerous situation exists, or it may give a spurious or false alarm even though no danger is present. The first of these is referred to as fail-to-danger and the second as fail-safe. Generally, the fail-to-danger probability is made much smaller than the fail-safe probability. Even then, small fail-safe probabilities are also required. If too many spurious alarms are sounded, they will tend to be ignored. Then, when the real danger is present, the alarm is also likely to be ignored.

Two factors are central to the trade-offs between fail-safe and fail-to-danger modes. First, many design alterations that decrease the fail-to-danger probability are likely to increase the fail-safe probability. Power supply failures, which are often a primary cause of failure of crudely designed safety systems, are an obvious example. Often, the system can be redesigned so that power supply failure will cause the system to fail-safe instead of to-danger. Specifically, instead of leaving the system unprotected following the failure, the power supply failure will cause the system to function spuriously. Of course, if no change is made in the probability of power supply failure, the amelioration of system fail-to-danger will result in an increased number of spurious operations.

Second, as increased redundancy is used to reduce the probability of fail-to-danger, more fail-safe incidents are likely to occur. To demonstrate this, consider a $1/N$ parallel system with which are associated two failure probabilities p_d and p_s for fail-to-danger and fail-safe, respectively. The system fail-to-danger unreliability $\tilde{R}_{dg}$ is found by noting that all units must fail. Hence

$$\tilde{R}_{dg} = p_d^N \qquad (9.79)$$

However, the system fail-safe reliability is calculated by noting that any one-unit failure with probability p_s will cause the system to fail-safe. Thus

$$\tilde{R}_{sf} = 1 - (1 - p_s)^N. \qquad (9.80)$$

If $p_s \ll 1$, then $(1 - p_s)^N \approx N p_s$, and we see that the fail-safe probability grows linearly with the number of units in parallel,

$$\tilde{R}_{sf} \approx N p_s \qquad (9.81)$$

The m/N configuration has been extensively used in electronic and other protection systems to limit the number of spurious operations at the same time that the redundancy provides high reliability. In such systems the fail-to-danger unreliability is obtained from Eq. 9.57:

$$\tilde{R}_{dg} = P\{\mathbf{n} \geq N - m\} = \sum_{n=N-m+1}^{N} C_n^N p_d^n (1 - p_d)^{N-n}. \qquad (9.82)$$

With the approximation that $p_d \ll 1$ this reduces to a form analogous to Eq. 9.61:

$$\tilde{R}_{dg} \approx C_{N-m+1}^N p_d^{N-m+1}. \qquad (9.83)$$

Conversely, at least m spurious signals must be generated for the system to fail-safe. Assuming independent failures with probability p_s, we have

$$\tilde{R}_{sf} = P\{\mathbf{n} \geq m\} = \sum_{n=m}^{N} C_n^N p_s^n (1 - p_s)^{N-n}. \qquad (9.84)$$

Now, assuming that $p_s \ll 1$, we may approximate this expression by

$$\tilde{R}_{sf} \approx C_m^N p_s^m. \qquad (9.85)$$

From Eqs. 9.83 and 9.85 the trade-off between fail-to-danger and spurious operation is seen. The fail-safe probability is decreased by increasing m, and the fail-to-danger probability is decreased by increasing $N - m$. Of course, as N becomes large, common-mode failures may severely limit further improvement.

EXAMPLE 9.9

You are to design an m/N detection system. The number of components, N, must be as small as possible to minimize cost. The fail-to-danger and the fail-safe probabilities for the identical components are

$$p_d = 10^{-2}, \qquad p_s = 10^{-2}.$$

Your design must meet the following criteria:

1. Probability of system fail-to-danger $< 10^{-4}$.
2. Probability of system fail-safe $< 10^{-2}$.

What values of m and N should be used?

Solution   Make a table of unreliabilities (i.e., the failure probabilities) for fail-safe and fail-to-danger using the rare-event approximations given by Eqs. 9.85 and 9.83.

m/N	$\bar{R}_{sf}$ Eq. 9.85	$\bar{R}_{dg}$ Eq. 9.83
1/1	$p_s = 10^{-2}$	$p_d = 10^{-2}$
1/2	$2p_s = 2 \times 10^{-2}$	$p_d^2 = 10^{-4}$
2/2	$p_s^2 = 10^{-4}$	$2p_d = 2 \times 10^{-2}$
1/3	$3p_s = 3 \times 10^{-2}$	$p_d^3 = 10^{-6}$
2/3	$3p_s^2 = 3 \times 10^{-4}$	$3p_d^2 = 3 \times 10^{-4}$
3/3	$p_s^3 = 10^{-6}$	$3p_d = 3 \times 10^{-2}$
1/4	$4p_s = 4 \times 10^{-2}$	$p_d^4 = 10^{-8}$
2/4	$6p_s^2 = 6 \times 10^{-4}$	$4p_d^3 = 4 \times 10^{-6}$
3/4	$4p_s^3 = 4 \times 10^{-6}$	$6p_d^2 = 6 \times 10^{-6}$
4/4	$p_s^4 = 10^{-8}$	$4p_d = 4 \times 10^{-2}$

At least four components are required to meet both criteria. They are met by a 2/4 system.

Voting Systems

In addition to the use of m/N redundancy to reduce the spurious operation of safety and alarm systems, it plays an important role in the design of computer control systems that must feed continuous streams of highly reliable output to guarantee safe operations. Temperature controllers in chemical plants, automated avionics controls, controls for respirators and other biomedical devices offer a few examples where accurate sensing and control often requires the use of redundancy.

In these situations the most frequent configuration is a 2/3 voting system. Three process computers or other instruments operate in parallel. A voter then compares the outputs of the three units, and if one differs from the other two, its output is ignored. The configuration reliability is then obtained by putting the voter reliability in series with the 2/3 result obtained from Eq. 9.56:

$$R_{sys} = (3R^2 - 2R^3)R_v, \tag{9.86}$$

where R and R_v are the computer and voter reliabilities, respectively. Clearly the voter must have a very small failure probability if the system is to operate satisfactorily. Fortunately, the voter is typically a very simple device compared to the computer, and therefore may be expected to have a much smaller failure probability.

In some situations the electronic voter may be replaced by an operator decision. Suppose, for example that three computers are used to calculate the pitch and yawl of an aircraft. The pilot and copilot might have the displays from two of the computers in front of them with a third placed to be readily visible by both of them. Therefore comparisons can be made readily, and the malfunctioning computer switched out of the system. Of course this system also creates an additional opportunity for pilot error.

More extensive voting systems may be required to achieve exceedingly small failure probabilities in computer controlled systems. In one such configuration each of the computers has a spare, which may be kept in hot standby and switched into the circuit upon detection of a failure by the voter. An alternative configuration is a 3/5 majority vote system. In each of these configurations at least three computers must fail before the system fails, but each requires that additional computers be purchased.

EXAMPLE 9.10

Derive the MTTF and the rare-event approximation for

(*a*) a 2/3 voting system,

(*b*) a 3/5 voting system.

Assume the failure probability of the voter can be neglected. How do the results compare to those for a single unit?

    Solution (2/3) From Eq. 9.86 we have

$$R = e^{-\lambda t} : R_{2/3} = 3e^{-2\lambda t} - 2e^{-3\lambda t}.$$

Using the definition of MTTF given by Eq. 6.22 and evaluating the integrals we have

$$\text{MTTF}_{2/3} = \frac{3}{2\lambda} - \frac{2}{3\lambda} = \frac{5}{6}\text{MTTF}.$$

For the rare-event approximation Eq. 9.61 yields

$$R_{2/3} \approx 1 - C_2^3(\lambda t)^2 = 1 - 3(\lambda t)^2.$$

(3/5) From Eq. 9.56 we have

$$R_{3/5} = \sum_{n=0}^{2} C_n^5 (1-R)^n R^{5-n} = R^5 + 5(1-R)R^4 + 10(1-R)^2 R^3.$$

Thus,

$$R_{3/5} = 10R^3 - 15R^4 + 6R^5 = 10e^{-3\lambda t} - 15e^{-4\lambda t} + 6e^{-5\lambda t}$$

and we can again apply Eq. 6.22 to obtain

$$\text{MTTF}_{3/5} = \frac{10}{3\lambda} - \frac{15}{4\lambda} + \frac{6}{5\lambda} = \frac{47}{60}\text{MTTF}.$$

For the rare-event approximation Eq. 9.61 yields

$$R_{3/5} \approx 1 - C_3^5 (\lambda t)^3 = 1 - 10(\lambda t)^3$$

Increased number of voting components decreases the system MTTF. However, at short times the rare-event approximations indicated that the reliability is increasingly close to one. For example with $\lambda t = 0.1$ we have

$$R_{1/1} \approx 0.90, \quad R_{2/3} \approx 0.97 \quad \text{and} \quad R_{3/5} \approx 0.99.$$

Finally, it should be noted that in an electronic system, transient faults, which may last only a fraction of a second, are expected to occur more frequently than "hard" irrecoverable failure. Thus in voting systems, software is often included to test for transient faults and restart the computer once the fault is corrected. If this is not done the failure probability may be too large even if three or more faults must occur before the system will fail. In this case the failure mode is referred to as "exhaustion of spares." Conversely if the testing to determine whether a correctable fault or an irreparable failure has taken place takes a significant length of time, there is a small possibility that a fault will cause a second computer to malfunction before the spare can be switched in. The system is then said to have a fault handling or switching failure. The achievement of very small failure probabilities in systems such as shown in Fig. 9.8 often hinges on balancing the gains and losses incurred with the use of such sophisticated fault handling systems.

9.6 REDUNDANCY IN COMPLEX CONFIGURATIONS

Systems may take on a variety of complex configurations. In what follows we examine the analysis of redundancy in two classes of systems: those that may be analyzed in terms of series and parallel configurations, and those in which the components are linked in such a way that they cannot. For brevity, we primarily treat configurations involving only active parallel units. However, with proper care the analysis can be extended to systems containing standby configurations.

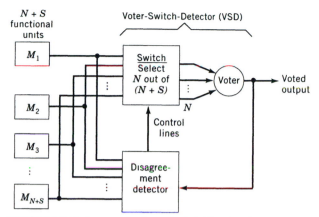

FIGURE 9.8 Basic organization of a hybrid redundant system. From S. A. Elkind, "Reliability and Availability Techniques," *The Theory and Practice of Reliable System Design*, D. P. Siewiorek and R. S. Swarz (eds.) Digital Press, Bedford, MA 1982.

Series–Parallel Configurations

As long as a system can be decomposed into series and parallel subsystem configurations, the techniques of the preceding sections can be employed repeatedly to derive expressions for system reliability. As an example consider the reliability block diagram shown for a system in Fig. 9.9. Components a_1 through a_4 have reliability R_a and components b_1 and b_2 have reliability R_b. For the following analysis to be valid, the failures of the components must be independent of one another.

We begin by noting that there are two sets of subsystems with type a components, consisting of a simple parallel configuration as shown in Fig. 9.10a. Thus we define the reliability of these configurations as

$$R_A = 2R_a - R_a^2. \qquad (9.87)$$

The system configuration then appears as the reduced block diagram shown in Fig. 9.10b. We next note that each newly defined subsystem A is in series

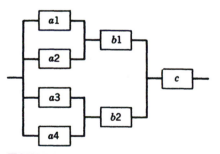

FIGURE 9.9 Reliability block diagram of a series–parallel configuration.

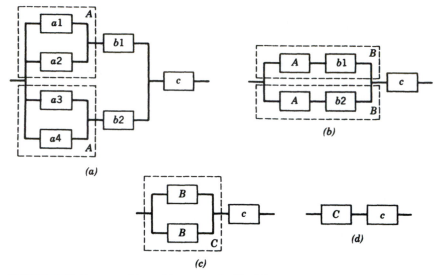

FIGURE 9.10 Decomposition of the system in Fig. 9.9.

with a component of type b. We may therefore define a subsystem B by

$$R_B = R_A R_b, \tag{9.88}$$

and the reduced block diagram then appears as in Fig. 9.10c. Since the two subsystems B are in parallel, we may write

$$R_C = 2R_B - R_B^2 \tag{9.89}$$

to yield the simplified configuration shown in Fig. 9.10d. Finally, the total system consists of the series of subsystems C and component c. Thus

$$R = R_C R_c. \tag{9.90}$$

Having derived an expression for the system reliability, we may combine Eqs. 9.87 through 9.90 to obtain the system reliability in terms of that of R_a, R_b, and R_c

$$R = (2R_a - R_a^2) R_b [2 - (2R_a - R_a^2) R_b] R_c. \tag{9.91}$$

Standby configurations can also be included within series–parallel configurations. Suppose components a_1 and a_2 are in a 1/2 standby configuration, and that components a_3 and a_4 are in the same configuration. In the constant failure rate approximation we would simply replace R_A by R_S, given by Eq. 9.12, and proceed as before. We would obtain, instead of Eq. 9.91,

$$R = R_s R_b (2 - R_s R_b) R_c \tag{9.92}$$

EXAMPLE 9.11

Suppose that in Fig. 9.9, $R_a = R_b = e^{-\lambda t} \equiv R_*$ and $R_c = 1$. Find R in the rare-event approximation.

Solution We simplify Eq. 9.91,

$$R = R_*^2(2 - R_*)[2 - (2 - R_*)R_*^2]$$

and write it as a polynomial in R_*:

$$R = 4R_*^2 - 2R_*^3 - 4R_*^4 + 4R_*^5 - R_*^6.$$

Then we expand $R_*^N = e^{-N\lambda t} \approx 1 - N\lambda t + \frac{1}{2}N^2(\lambda t)^2 - \cdots$ to obtain for small λt

$$R \approx 4[1 - 2\lambda t + 2(\lambda t)^2] - 2[1 - 3\lambda t + \frac{9}{2}(\lambda t)^2] - 4[1 - 4\lambda t + 8(\lambda t)^2]$$

$$+ 4[1 - 5\lambda t + \frac{1}{2}25(\lambda t)^2] - 1 + 6\lambda t - 18(\lambda t)^2$$

$$R \approx (4 - 2 - 4 + 4 - 1) - (8 - 6 - 16 + 20 - 6)(\lambda t)$$

$$- (-8 + 9 + 32 - 50 + 18)(\lambda t)^2 + \cdots$$

$$R \approx 1 - (\lambda t)^2.$$

Had the coefficient of the $(\lambda t)^2$ term also been zero, we would have needed to carry terms in $(\lambda t)^3$.

Linked Configurations

In some situations the linkage of the components or subsystems is such that the foregoing technique of decomposing into parallel and series configurations cannot be applied directly. Such is the case for the system configuration shown in Fig. 9.11, consisting of subsystem types 1, 2, and 3, with reliabilities R_1, R_2, and R_3.

To analyze this and similar systems, we decompose the problem into a combination of series–parallels by utilizing the total probability rule given in Eq. 2.20.

$$P\{Y\} = P\{Y | X\}P\{X\} + P\{Y | \tilde{X}\}P\{\tilde{X}\}. \tag{9.93}$$

Suppose we let X be the event that subsystem 2a fails. Then $P\{X\} = 1 - R_2$ and $P\{\tilde{X}\} = R_2$. If we then let Y denote successful system operation, the system reliability is defined as $R = P\{Y\}$. Now suppose we define the conditional reliabilities that the system function with subsystem 2a failed as

$$R^- = P\{Y | X\} \tag{9.94}$$

and with 2a operational as

$$R^+ = P\{Y | \tilde{X}\}. \tag{9.95}$$

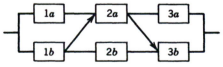

FIGURE 9.11 Reliability block diagram of a cross-linked system.

Inserting these probabilities into Eq. 9.93, we may write the system reliability as

$$R = R^-(1 - R_2) + R^+R_2. \tag{9.96}$$

We must now evaluate the conditional reliabilities R^+ and R^-. For R^- in which 2a has failed, we disconnect all the paths leading through 2a in Fig. 9.11; the result appears in Fig. 9.12a. Conversely, for R^+ in which 2a is functioning, we pass a path through 2a, thereby bypassing 2b with the result shown in Fig. 9.12b.

We see that when 2a is failed, the reduced system consists of a series of three subsystems, 1b, 2b, and 3b; subsystems 1a and 3a no longer make any contribution to the value of R^-. We obtain

$$R^- = R_1R_2R_3. \tag{9.97}$$

When 2a is operating, we have a series combination of two parallel configurations, 1a and 1b in the first and 3a and 3b in the second; since component 2b is always bypassed, it has no effect on R^+. Therefore, we have

$$R^+ = (2R_1 - R_1^2)(2R_3 - R_3^2). \tag{9.98}$$

Finally, substituting these expressions into Eq. 9.96, we find the system reliability to be

$$R = R_1R_2R_3(1 - R_2) + (2R_1 - R_1^2)(2R_3 - R_3^2)R_2 \tag{9.99}$$

EXAMPLE 9.12

Evaluate Eq. 9.99 in the rare-event approximation with $R_n = e^{-\lambda t}$ for all n.

Solution Let $R_* = R_n$. Then Eq. 9.99 becomes $R = R_*^3(1 - R_*) + (2R_* - R_*^2)^2R_*$. Writing this expression as a polynomial in R_*, we have $R = 5R_*^3 - 5R_*^4 + R_*^5$. Now we expand $R_*^N = e^{-\lambda t} = 1 - N\lambda t + \frac{1}{2}N^2(\lambda t)^2 - \cdots$ to obtain:

$$R = 5 - 15\lambda t + \frac{1}{2}45(\lambda t)^2 - \cdots$$

$$-5 + 20\lambda t - \frac{1}{2}80(\lambda t)^2 + \cdots$$

$$+ 1 - 5\lambda t + \frac{1}{2}25(\lambda t)^2 - \cdots$$

Hence,

$$R = 1 - 5(\lambda t)^2 + \cdots$$

If the $(\lambda t)^2$ term were zero, we would need to carry the $(\lambda t)^3$ term in the expansion.

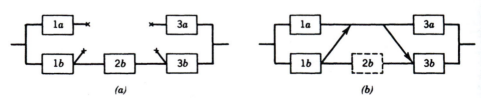

(a)   (b)

FIGURE 9.12 Decomposition of the system in Fig. 9.11.

Bibliography

Barlow, R. E., and F. Proschan, *Mathematical Theory of Reliability*, Wiley, NY, 1965.

Henley, E. J., and H. Kumamoto, *Reliability Engineering and Risk Assessment*, Prentice-Hall, Englewood Cliffs, NJ, 1981.

Roberts, N. H., *Mathematical Methods in Reliability Engineering*, McGraw-Hill, NY, 1964.

Sandler, G. H., *System Reliability Engineering*, Prentice-Hall, Englewood Cliffs, NJ, 1963.

Siewiorek, D. P., and R. S. Swarz, *Reliable Computer Systems*, 2nd ed. Digital Press, 1992.

Exercises

9.1 A nonredundant system with 100 components has a design-life reliability of 0.90. The system is redesigned so that it has only 70 components. Estimate the design life of the redesigned systems, assuming that all the components have constant failure rates of the same value.

9.2 At the end of one year of service the reliability of a component with a constant failure rate is 0.95.

   (a) What is the failure rate (include units)?
   (b) If two of the components are put in active parallel, what is the one year reliability? (Assume no dependencies.)
   (c) If 10% of the component failure rate may be attributed to common-mode failures, what will the one-year reliability be of the two components in active parallel?

9.3 Thermocouples of a particular design have a failure rate of $\lambda = 0.008/$ hr. How many thermocouples must be placed in active parallel if the system is to run for 100 hrs with a system failure probability of no more than 0.05? Assume that all failures are independent.

9.4 In an attempt to increase the MTTF, an engineer puts two devices in parallel and tests the resulting parallel system. The MTTF increases by only 40%. Assuming the device failure rate is a constant, what fraction of it, β, is due to common-mode failures of the parallel system?

9.5 A disk drive has a constant failure rate and an MTTF of 5000 hr.

   (a) What will the probability of failure be for one year of operation?
   (b) What will the probability of failure be for one year of operation if two of the drives are placed in active parallel and the failures are independent?
   (c) What will the probability of failure be for one year of operation if the common-mode errors are characterized by $\beta = 0.2$?

9.6 Suppose the design life reliability of a standby system consisting of two identical units must be at least 0.95. If the MTTF for each unit is 3 months, determine the design life. (Assume constant failure rates and neglect switching failures, etc.)

9.7 Find the variance in the time to failure, assuming a constant failure rate λ:

(a) For two units in series.

(b) For two units in active parallel.

(c) Which is larger?

9.8 Suppose that the reliability of a single unit is given by a Weibull distribution with $m = 2$. Use Eq. 9.10 to show that a standby system consisting of two such units has a reliability of

$$R_s(t) = e^{-(t/\theta)^2} + \sqrt{2\pi}(t/\theta)\,\mathrm{erf}(\sqrt{1/2}t/\theta)\,e^{-\frac{1}{2}(t/\theta)^2}$$

where the error function is defined by

$$\mathrm{erf}(y) = \frac{1}{\sqrt{\pi}}\int_0^y e^{-x^2}dx.$$

9.9 Suppose that two identical units are placed in active parallel. Each has a Weibull distribution with known θ and $m > 1$.

(a) Determine the system reliability.

(b) Find a rare-event approximation for a.

9.10 Suppose that the units in Exercise 9.9 each have a Weibull distribution with $m = 2$. By how much is the MTTF increased by putting them in parallel?

9.11 A component has a one-year design-life reliability of 0.9; two such components are placed in active parallel. What is the one-year reliability of the resulting system:

(a) In the absence of common-mode failures?

(b) If 20% of the failures are common-mode failures?

9.12 Suppose that the PDF for time-to-failure for a single unit is uniform:

$$f(t) = \begin{cases} 1/T, & 0 < t < T \\ 0, & otherwise \end{cases}.$$

(a) Find and plot $R(t)$ for a single unit.

(b) Find and plot $R(t)$ for two units in active parallel.

(c) Find and plot $R(t)$ for two units in standby parallel.

(d) Find the MTTF for parts a, b, and c.

9.13 An amplifier with constant failure rate has a reliability of 0.90 at the end of one month of operation. If an identical amplifier is placed in standby parallel and there is a 3% switching failure probability, what will the reliability of the parallel system be at the end of one year?

9.14 Consider the standby system described by Eq. 9.33:

(a) Find the MTTF.

(b) Show that your result from *a* reduces to Eq. 9.15 as $p \to 0$ and $\lambda^+ \to \lambda$.

(c) Show that your result from *a* reduces to a single unit MTTF as $p \to 1$.

(d) Find the rare-event approximation for Eq. 9.33.

9.15 Consider a system with three identical components with failure rate λ_1. Find the system failure rate:

(a) For all three components in series.

(b) For all three components in active parallel.

(c) For two components in parallel and the third in series.

(d) Plot the results for *a*, *b*, and *c* on the same scale for $0 \leq t \leq 5/\lambda$.

9.16 For a 1/2 parallel system with load sharing:

(a) Show that for $\lambda^*/\lambda > 1.56$ will have a smaller MTTF than a single unit.

(b) Find the rare-event approximation for the case where $\lambda^*/\lambda = 1.56$.

(c) Using rare-event approximations, compare reliabilities at $\lambda t = 0.05$ for a single unit, for $\lambda^*/\lambda = 1.56$ and for $\lambda^*/\lambda = 1.0$.

(d) Discuss your results.

9.17 In a 1/2 active parallel system each unit has a failure rate of 0.05 day^{-1}.

(a) What is the system MTTF with no load sharing?

(b) What is the system MTTF if the failure rate increases by 10% as a result of increased load?

(c) What is the system MTTF if one increases both unit failure rates by 10%?

9.18 An engineer running a 1/2 identical unit system in cold standby finds the switching failure probability is 0.2 while the failure rate in standby is negligible. He converts to hot standby and eliminates the switching failure probability, but discovers that now the failure rate of the unit in standby is 30% of the active unit. As measured by system MTTF, has going from cold to hot standby improved or degraded the system? By how much?

9.19 Suppose that a system consists of two subsystems in active parallel. The reliability of each subsystem is given by the Rayleigh distribution

$$R(t) = e^{-(t/\theta)^2}.$$

Assuming that common-mode failures may be neglected, determine the system MTTF.

9.20 Repeat exercise 9.18 assuming that the failure rate of the unit in standby is only 20% of the active unit.

9.21 The design criterion for the ac power system for a reactor is that its failure probability be less than 2×10^{-5}/year. Off-site power failures may be expected to occur about once in 5 years. If the on-site ac power system consists of two independent diesel generators, each of which is capable of meeting the ac power requirements, what is the maximum failure probability per year that each diesel generator can have if the design criterion is to be met? If three independent diesel generators are used in active parallel, what is the value of the maximum failure probability? (Neglect common-mode failures.)

9.22 Consider a 1/3 system in active parallel, each unit of which has a constant failure rate λ.

(a) Plot the system failure rate $\lambda(t)$ in units of λ versus λt from $\lambda t = 0$, to large enough λt to approach an asymptotic system failure rate.

(b) What is the asymptotic value $\lambda(\infty)$?

(c) At what interval should the system be shut down and failed components replaced if there is a criterion that $\lambda(t)$ should not exceed 1/3 of the asymptotic value?

9.23 An engineer designs a system consisting of two subsystems in series. The reliabilities are $R_1 = 0.98$ and $R_2 = 0.94$. The cost of the two subsystems is about equal. The engineer decides to add two redundant components. Which of the following would it be better to do?

(a) Duplicate subsystems 1 and 2 in high-level redundance.

(b) Duplicate subsystems 1 and 2 in low-level redundance.

(c) Replace the second subsystem with 1/3 redundance.

Justify your answer.

9.24 For a 2/3 system:

(a) Express $R(t)$ in terms of the constant failure rates.

(b) Find the system MTTF.

(c) Calculate the reliability y when $\lambda t = 1.0$ and compare the result to a single unit and to a 1/2 system with the same unit failure rate.

9.25 Suppose that a system consists of two components, each with a failure rate λ, placed in series. A redundant system is built consisting of four components. Derive expressions for the system failure rates

(a) for high-level redundancy,

(b) for low-level redundancy.

(c) Plot the results of a and b along with the failure rate of the nonredundant system for $0 \leq t \leq 2/\lambda$.

9.26 Suppose that in Exercise 9.21 one-fourth of the diesel generator failures are caused by common-mode effects and therefore incapacitate all the active parallel systems. Under these conditions what is the maximum failure probability (i.e., random and common-mode) that is allowable if two diesel generators are used? If three diesel generators are used?

9.27 The failure rate on a jet engine is $\lambda = 10^{-3}/\text{hr}$. What is the probability that more than two engines on a four-engine aircraft will fail during a 2-hr flight? Assume that the failures are independent.

9.28 The shutdown system on a nuclear reactor consists of four independent subsystems, each consisting of a control rod bank and its associated drives and actuators. Insertion of any three banks will shut down the reactor. The probability that a subsystem will fail is 0.2×10^{-4} per demand. What is the probability per demand that the shutdown system will fail, assuming that common-mode failures can be neglected?

9.29 Two identical components, each with a constant failure rate, are in series. To improve the reliability two configurations are considered:

(a) for high-level redundancy,

(b) for low-level redundancy.

Calculate the system MTTF in terms of MTTF of the system mean-time-to-failure without redundancy.

9.30 Consider two components with the same MTTF. One has an exponential distribution, the other a Rayleigh distribution (see Exercise 9.19). If they are placed in active parallel, find the system MTTF in terms of the component MTTF.

9.31 A radiation-monitoring system consists of a detector, an amplifier, and an annunciator. Their lifetime reliabilities and costs are, respectively, 0.83 ($1200), 0.58 ($2400), and 0.69 ($1600).

(a) How would you allocate active redundancy to achieve a system lifetime reliability of 0.995?

(b) What is the cost of the system?

9.32 For constant failure rates evaluate R_{HL} and R_{LL} for high- and low-level redundancy in the rare-event approximation beginning with Eqs. 9.72 and 9.73.

9.33 A system consists of three components in series, each with a reliability of 0.96. A second set of three components is purchased and a redundant system is built. What is the reliability of the redundant system (*a*) with high-level redundancy, (*b*) with low-level redundancy?

9.34 The identical components of the system below have fail-to-danger probabilities of $p_d = 10^{-2}$ and fail-safe probabilities of $p_s = 10^{-1}$.

(a) What is the system fail-to-danger probability?

(b) What is the system fail-safe probability?

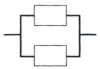

9.35 Calculate the reliabilities of the following systems:

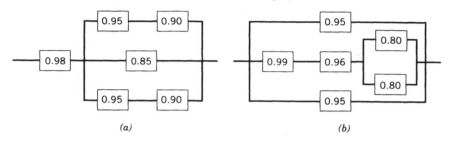

(a)   (b)

9.36 A device consist of two components in series with a (1/2) standby system as shown. Each component has the same constant failure rate.

(a) What is $R(t)$?

(b) What is the rare-event approximation for $R(t)$?

(c) What is the MTTF?

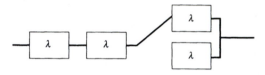

9.37 Calculate the reliability for the following system, assuming that all the component failure rates are equal. Then use the rare-event approximation to simplify your result.

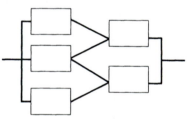

9.38 Calculate the reliability, $R(t)$, for the following systems, assuming that all the components have failure rate λ. Then use the rare-event approximation to simplify the result.

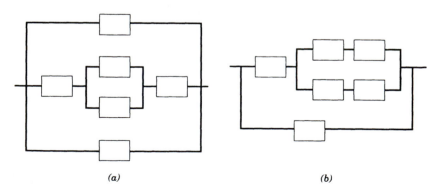

(a)   (b)

9.39 Given the following component reliabilities, calculate the reliability of the two systems.

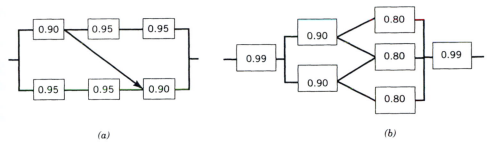

(a)   (b)

9.40 Calculate the reliabilities of the following two systems, assuming that all the component reliabilities are equal. Then determine which system has the higher reliability.

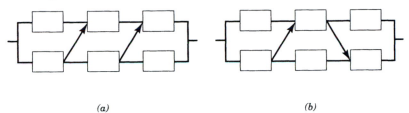

(a)   (b)

CHAPTER 10

Maintained Systems

"A little neglect may breed great mischief ...
for want of a nail the shoe was lost;
for want of a shoe the horse was lost;
and for want of a horse the rider was lost."

Benjamin Franklin
Poor Richard's Almanac 1758

10.1 INTRODUCTION

Relatively few systems are designed to operate without maintenance of any kind, and for the most part they must operate in environments where access is very difficult, in outer space or high-radiation fields, for example, or where replacement is more economical than maintenance. For most systems there are two classes of maintenance, one or both of which may be applied. In preventive maintenance, parts are replaced, lubricants changed, or adjustments made before failure occurs. The objective is to increase the reliability of the system over the long term by staving off the aging effects of wear, corrosion, fatigue, and related phenomena. In contrast, repair or corrective maintenance is performed after failure has occurred in order to return the system to service as soon as possible. Although the primary criteria for judging preventive-maintenance procedures is the resulting increase in reliability, a different criterion is needed for judging the effectiveness of corrective maintenance. The criterion most often used is the system availability, which is defined roughly as the probability that the system will be operational when needed.

The amount and type of maintenance that is applied depends strongly on its costs as well as the cost and safety implications of system failure. Thus, for example, in determining the maintenance for an electric motor used in a manufacturing plant, we would weigh the costs of preventive maintenance against the money saved from the decreased number of failures. The failure costs would need to include, of course, both those incurred in repairing or replacing the motor, and those from the loss of production during the unscheduled downtime for repair. For an aircraft engine the trade-off would be much different: the potentially disastrous consequences of engine failure would eliminate repair maintenance as a primary consideration. Concern would be with how much preventive maintenance can be afforded and with the possibility of failures induced by faculty maintenance.

In both preventive and corrective maintenance, human factors play a very strong role. It is for this reason that laboratory data are often not representative of field data. In field service the quality of preventive maintenance is not likely to be as high. Moreover, repairs carried out in the field are likely to take longer and to be less than perfect. The measurement of maintenance quantities thus depends strongly on human reliability so that there is great difficulty in obtaining reproducible data. The numbers depend not only on the physical state of the hardware, but also on the training, vigilance, and judgment of the maintenance personnel. These quantities in turn depend on many social and psychological factors that vary to such an extent that the probabilities of maintenance failures and repair times are generally more variable than the failure rates of the hardware.

In this chapter we first examine preventive maintenance. Then we define and discuss availability and other quantities needed to treat corrective maintenance. Subsequently, we examine the repair of two types of failure: those that are revealed (i.e., immediately obvious) and those that are unrevealed (i.e., are unknown until tests are run to detect them). Finally, we examine the relation of a system to its components from the point of view of corrective maintenance.

10.2 PREVENTIVE MAINTENANCE

In this section we examine the effects of preventive maintenance on the reliability of a system or component. We first consider ideal maintenance in which the system is restored to an as-good-as-new condition each time maintenance is applied. We then examine more realistic situations in which the improvement in reliability brought about by maintenance must be weighed against the possibility that faulty maintenance will lead to system failure. Finally, the effects of preventive maintenance on redundant systems are examined.

Idealized Maintenance

Suppose that we denote the reliability of a system without maintenance as $R(t)$, where t is the operation time of the system; it includes only the intervals when the system is actually operating, and not the time intervals during which it is shut down. If we perform maintenance on the system at time intervals T, then, as indicated in Fig. 10.1, for $t < T$ maintenance will have no effect on

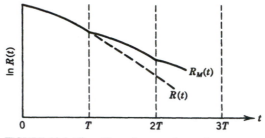

FIGURE 10.1 The effect of preventive maintenance on reliability.

reliability. That is, if $R_M(t)$ is the reliability of the maintained system,

$$R_M(t) = R(t), \qquad 0 \le t < T. \tag{10.1}$$

Now suppose that we perform maintenance at T, restoring the system to an as-good-as-new condition. This implies that the maintained system at $t > T$ has no memory of accumulated wear effects for times before T. Thus, in the interval $T < t \le 2T$, the reliability is the product of the probability $R(T)$ that the system survived to T, and the probability $R(t - T)$ that a system as good as new at T will survive for a time $t - T$ without failure:

$$R_M(t) = R(T)R(t - T), \qquad T \le t < 2T. \tag{10.2}$$

Similarly, the probability that the system will survive to time t, $2T \le t < 3T$, is just the reliability $R_M(2T)$ multiplied by the probability that the newly restored system will survive for a time $t - 2T$:

$$R_M(t) = R(T)^2 R(t - 2T), \qquad 2T \le t < 3T. \tag{10.3}$$

The same argument may be used repeatedly to obtain the general expression

$$R_M(t) = R(T)^N R(t - NT), \qquad NT \le t < (N + 1)T,$$
$$N = 0, 1, 2, \dots . \tag{10.4}$$

The MTTF for a system with preventive maintenance can be determined by replacing $R(t)$ by $R_M(t)$ in Eq. 6.22:

$$\text{MTTF} = \int_0^\infty R_M(t) \, dt. \tag{10.5}$$

To evaluate this expression, we first divide the integral into time intervals of length T:

$$\text{MTTF} = \sum_{N=0}^\infty \int_{NT}^{(N+1)T} R_M(t) \, dt. \tag{10.6}$$

Then, inserting Eq. 10.4, we have

$$\text{MTTF} = \sum_{N=0}^\infty \int_{NT}^{(N+1)T} R(T)^N R(t - NT) \, dt. \tag{10.7}$$

Setting $t' = t - NT$ then yields

$$\text{MTTF} = \sum_{N=0}^{\infty} R(T)^N \int_0^T R(t')\, dt'. \tag{10.8}$$

Then, evaluating the infinite series,

$$\sum_{n=0}^{\infty} R(T)^N = \frac{1}{1 - R(T)}, \tag{10.9}$$

we have

$$\text{MTTF} = \frac{\int_0^T R(t)\, dt}{1 - R(T)}. \tag{10.10}$$

We would now like to estimate how much improvement, if any, in reliability we derive from the preventive maintenance. The first point to be made is that in random or chance failures (i.e., those represented by a constant failure rate λ), idealized maintenance has no effect. This is easily proved by putting $R(t) = e^{-\lambda t}$ on the right-hand side of Eq. 10.4. We obtain

$$R_M(t) = (e^{-\lambda t})^N e^{-\lambda(t - NT)} = e^{-N\lambda t} e^{-\lambda(t - NT)} = e^{-\lambda t} \tag{10.11}$$

or simply

$$R_M(t) = R(t), \qquad 0 \le t \le \infty. \tag{10.12}$$

Preventive maintenance has a quite definite effect, however, when aging or wear causes the failure rate to become time-dependent. To illustrate this effect, suppose that the reliability can be represented by the two-parameter Weibull distribution described in Chapter 3. For the system without maintenance we have

$$R(t) = \exp\left[-\left(\frac{t}{\theta}\right)^m \right]. \tag{10.13}$$

Equation 10.4 then yields for the maintained system

$$R_M(t) = \exp\left[-N\left(\frac{T}{\theta}\right)^m \right] \exp\left[-\left(\frac{t - NT}{\theta}\right)^m \right], \quad NT \le t < (N+1)T, \tag{10.14}$$

$$N = 0, 1, 2, \ldots.$$

To examine the effect of maintenance, we calculate the ratio $R_M(t)/R(t)$. The relationship is simplified if we calculate this ratio at the time of maintenance $t = NT$:

$$\frac{R_M(NT)}{R(NT)} = \exp\left[-N\left(\frac{T}{\theta}\right)^m + \left(\frac{NT}{\theta}\right)^m \right]. \tag{10.15}$$

Thus there will be a gain in reliability from maintenance only if the argument of the exponential is positive, that is, if $(NT/\theta)^m > N(T/\theta)^m$. This reduces to the condition

$$N^{m-1} - 1 > 0. \tag{10.16}$$

This states simply that m must be greater than one for maintenance to have a positive effect on reliability; it corresponds to a failure rate that is increasing with time through aging. Conversely, for $m < 1$, preventive maintenance decreases reliability. This corresponds to a failure rate that is decreasing with time through early failure. Specifically, if new defective parts are introduced into a system that has already been "worn in," increased rates of failure may be expected. These effects on reliability are illustrated in Fig. 10.2 where Eq. 10.14 is plotted for both increasing ($m > 1$) and decreasing ($m < 1$) failure rates, along with random failures ($m = 1$).

Naturally, a system may have several modes of failure corresponding to increasing and decreasing failure rates. For example, in Chapter 6 we note that the bathtub curve for a device may be expressed as the sum of Weibull distributions

$$\int_0^t \lambda(t')\, dt' = \left(\frac{t}{\theta_1}\right)^{m_1} + \left(\frac{t}{\theta_2}\right)^{m_2} + \left(\frac{t}{\theta_3}\right)^{m_3}, \tag{10.17}$$

For this system we must choose the maintenance interval for which the positive effect on wearout time is greater than the negative effect on wearin time. In practice, the terms in Eq. 10.17 may be due to different components of the system. Thus we would perform preventive maintenance only on the components for which the wearout effect dominates. For example, we may replace worn spark plugs in an engine without even considering replacing a fuel injection system with a new one, which might itself be defective.

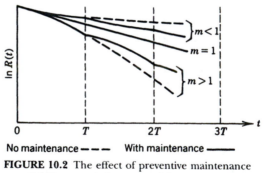

FIGURE 10.2 The effect of preventive maintenance on reliability: $m > 1$, increasing failure rate; $m < 1$, decreasing failure rate; $m = 1$, constant failure rate.

EXAMPLE 10.1

A compressor is designed for 5 years of operation. There are two significant contributions to the failure rate. The first is due to wear of the thrust bearing and is described by a Weibull distribution with $\theta = 7.5$ year and $m = 2.5$. The second, which includes all other causes, is a constant failure rate of $\lambda_0 = 0.013/\text{year}$.

(*a*) What is the reliability if no preventive maintenance is performed over the 5-year design life?

(*b*) If the reliability of the 5-year design life is to be increased to at least 0.9 by periodically replacing the thrust bearing, how frequently must it be replaced?

Solution Let $T_d = 5$ be the design life.

(*a*) The system reliability may be written as

$$R(T_d) = R_0(T_d) R_M(T_d),$$

where

$$R_0(T_d) = e^{-\lambda_0 T_d} = e^{-0.013 \times 5} = 0.9371,$$

is the reliability if only the constant failure rate is considered. Similarly,

$$R_M(T_d) = e^{-(T_d/\theta)^m} = e^{-(5/7.5)^{2.5}} = 0.6957$$

is the reliability if only the thrust bearing wear is considered. Thus,

$$R(T_d) = 0.9371 \times 0.6957 = 0.6519.$$

(*b*) Suppose that we divide the design life into N equal intervals; the time interval, T, at which maintenance is carried out is then $T = T_d/N$. Correspondingly, $T_d = NT$. For bearing replacement at time interval T, we have from Eq. 10.14,

$$R_M(T_d) = \exp\left[-N\left(\frac{T_d}{N\theta}\right)^m\right] = \exp\left[-N^{1-m}\left(\frac{T_d}{\theta}\right)^m\right].$$

For the criterion to be met, we must have

$$R_M(T_d) = \frac{R(T_d)}{R_0(T_d)} \geq \frac{0.9}{0.9371}, \qquad R_M(T_d) \geq 0.9604.$$

With $(T_d/\theta)^m = (5/7.5)^{2.5} = 0.36289$, we calculate

$$R_M(T_d) = \exp(-0.36289 N^{-1.5}).$$

N	1	2	3	4	5
$R_M(T_d)$	0.696	0.880	0.933	0.956	0.968

Thus the criterion is met for $N = 5$, and the time interval for bearing replacement is $T = T_d/N = \frac{5}{5} = 1$ year.

In Chapter 6 we state that even when wear is present, a constant failure rate model may be a reasonable approximation, provided that preventive maintenance is carried out, with timely replacement of wearing parts. Although this may be intuitively clear, it is worthwhile to demonstrate it with our present model. Suppose that we have a system for which wearin effects can be neglected, allowing us to ignore the first term in Eq. 10.17 and write

$$R(t) = \exp\left[-\frac{t}{\theta_2} - \left(\frac{t}{\theta_3}\right)^{m_3}\right]. \tag{10.18}$$

The corresponding expression for the maintained system given by Eq. 10.4 becomes

$$R_M(t) = \exp\left[-N\left(\frac{T}{\theta_3}\right)^{m_3}\right]\exp\left[-\frac{t}{\theta_2} - \left(\frac{t-NT}{\theta_3}\right)^{m_3}\right], \quad NT \le t \le (N+1)T.$$

$$(10.19)$$

For a maintained system the failure rate may be calculated by replacing R by R_M in Eq. 6.15:

$$\lambda_M(t) = -\frac{1}{R_M(t)}\frac{d}{dt}R_M(t). \tag{10.20}$$

Thus, taking the derivative, we obtain

$$\lambda_M(t) = \frac{1}{\theta_2} + \frac{m_3}{\theta_3}\left(\frac{t-NT}{\theta_3}\right)^{m_3-1}, \quad NT \le t < (N+1)T. \tag{10.21}$$

Provided that the second term, the wear term, is never allowed to become substantial compared to the first, the random-failure term, the overall failure rate may be approximated as a constant by averaging over the interval T. This is illustrated for a typical set of parameters in Fig. 10.3.

Imperfect Maintenance

Next consider the effect of a less-than-perfect human reliability on the overall reliability of a maintained system. This enters through a finite probability p that the maintenance is carried out unsatisfactorily, in such a way that the faulty maintenance causes a system failure immediately thereafter. To take this into account in a simple way, we multiply the reliability by the maintenance nonfailure probability, $1 - p$, each time that maintenance is performed. Thus Eq. 10.4 is replaced by

$$R_M(t) = R(T)^N(1-p)^N R(t-NT), \quad NT < t < (N+1)T,$$

$$N = 0, 1, 2, \ldots. \tag{10.22}$$

The trade-off between the improved reliability from the replacement of wearing parts and the degradation that can come about because of mainte-

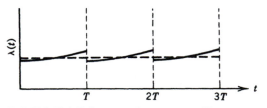

FIGURE 10.3 Failure rate for a system with preventive maintenance.

nance error may now be considered. Since random failures are not affected by preventive maintenance, we consider the system in which only aging is present, by using Eq. 10.13 with $m > 1$. Once again the ratio R_M/R after the Nth preventive maintenance is a useful indication of performance. Note that for $p \ll 1$, we may approximate

$$(1 - p)^N \approx e^{-Np} \tag{10.23}$$

to obtain

$$\frac{R_M(NT)}{R(NT)} = \exp\left[-N\left(\frac{T}{\theta}\right)^m - Np + \left(\frac{NT}{\theta}\right)^m\right]. \tag{10.24}$$

For there to be an improvement from the imperfect maintenance, the argument of the exponential in this expression must be positive. This reduces to the condition

$$p < (N^{m-1} - 1)\left(\frac{T}{\theta}\right)^m. \tag{10.25}$$

Consequently, the benefits from imperfect maintenance are not seen until a long time, when either N or T is large. This is plausible because after a long time wear effects degrade the reliability enough that the positive effect of maintenance compensates for the probability of maintenance failure. This is illustrated in Fig. 10.4.

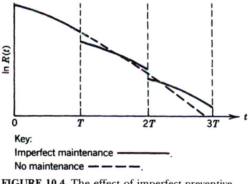

Key:
Imperfect maintenance ————.
No maintenance — — — —.

FIGURE 10.4 The effect of imperfect preventive maintenance on reliability.

EXAMPLE 10.2

Suppose that in Example 10.1 the probability of faulty bearing replacement causing failure of the compressor is $p = 0.02$. What will the design-life reliability be with the annual replacement program?

Solution At the end of the design life ($T_d = 5$ years) maintenance will have been performed four times. From the preceding problem we take the perfect maintenance result to be

$$R(T_d) = R_0 R_M = 0.937 \times 0.968 = 0.907.$$

With imperfect maintenance,

$$R(T_d) = R_0 R_M (1 - p)^4 = 0.907 \times 0.98^4 = 0.907 \times 0.922 = 0.836.$$

In evaluating the trade-off between maintenance and aging, we must examine the failure mode very closely. Suppose, for example, that we consider the maintenance of an engine. If after maintenance the engine fails to start, but no damage is done, the failure may be corrected by redoing the maintenance. In this case p may be set equal to zero in the model just given, with the understanding that preventive maintenance includes a checkout and a repair of maintenance errors.

The situation is potentially more serious if the maintenance failure damages the system or is delayed because it is an induced early-failure. We consider each of these problems separately. Suppose first that after maintenance the engine is started and is irreparably damaged by the maintenance error. Whether maintenance is desirable in these circumstances strongly depends on the failure mode that the maintenance is meant to prevent. If the engine's normal mode of failure is simply to stop running because a component is worn, with no damage to the remainder of the engine, it is unlikely that even the increased reliability provided by the preventive maintenance is economically worthwhile. Provided that there are no safety issues at stake, it may be more expedient to wait for failure, and then repair, rather than to chance damage to the system through faulty maintenance. If we are concerned about servicing an aircraft engine, however, the situation is entirely different. Damaging or destroying an occasional engine on the ground following faulty maintenance may be entirely justified in order to decrease the probability that wear will cause an engine to fail in flight.

Consider, finally, the situation in which the maintenance does not cause immediate failure but adds a wearin failure rate. This may be due to the replacement of worn components with defective new ones. However, it is equally likely to be due to improper installation or reassembly of the system, thereby placing excessive stress on one or more of the components. After the first repair, we then have a failure rate described by a bathtub curve, as in Eq. 10.17, with the first term stemming at least in part from imperfect maintenance. The reliability is then determined by inserting Eq. 10.17 into Eq. 10.4. If we assume that the early failure term is due to faulty maintenance, it may be shown by again calculating $R_M(NT)/R(NT)$ that the reliability is improved only if

$$(1 - N^{m_1-1}) \left(\frac{T}{\theta_1}\right)^{m_1} < (N^{m_3-1} - 1) \left(\frac{T}{\theta_3}\right)^{m_3}, \qquad m_1 < 1, m_3 > 1. \quad (10.26)$$

Whether or not an increase in overall reliability is the only criterion to be used once again depends on whether the failure modes are comparable in the system damage that is done. If no safety questions are involved, it is primarily a question of weighing the costs of repairing the failures caused by aging against those induced by maintenance errors. This might be the case, for example, with an automobile engine. With an aircraft engine, however, prevention of failure in flight must be the overriding criterion; the cost of repairing the engine following failure, of course, is not relevant if the plane crashes. In this, and similar situations, the more important consideration is often the effect of maintenance errors on redundant systems because maintenance is one of the primary causes of common-mode failures. We examine these next.

Redundant Components

The foregoing expressions for $R_M(t)$ may be used in calculating the reliability of redundant systems as in Chapter 9, but only if the maintenance failures on different components are independent of one another. This stipulation is frequently difficult to justify. Although some maintenance failures are independent, such as the random neglect to tighten a bolt, they are more likely to be systematic; if the wrong lubricant is put in one engine, it is likely to be put in a second one also.

The common-mode failure model introduced in Chapter 9 may be applied with some modification to treat such dependent maintenance failures. As an example we consider a parallel system consisting of two identical components. If the maintenance is imperfect but independent, we may insert Eq. 10.22 into Eq. 9.5 to obtain

$$R_I(t) = 2R(T)^N(1 - p)^N R(t - NT) - R(T)^{2N}(1 - p)^{2N}R(t - NT)^2,$$

$$NT \le t < (N + 1)T, \quad (10.27)$$

$$N = 0, 1, 2, \ldots .$$

Suppose that a maintenance failure on one component implies that the same failure occurs simultaneously in the other. We account for this by separating out the maintenance failures into a series component, much as we did with the common-mode failure rate λ_c in Chapter 9. Thus the system failure is modeled by taking the reliability for perfect maintenance (i.e., $p = 0$) and multiplying by $1 - p$ for each time that maintenance is performed. Thus, for dependent maintenance failures,

$$R_D(t) = \{2R(T)^N R(t - NT) - R(T)^{2N}R(t - NT)^2\}(1 - p)^N,$$

$$NT \le t < (N + 1)T \quad (10.28)$$

$$N = 0, 1, 2, \ldots .$$

The degradation from maintenance induced common-mode failures is indicated by the ratio of Eqs. 10.28 to 10.27. We find

$$\frac{R_D(NT)}{R_I(NT)} = \frac{1 - \frac{1}{2}R(T)^N}{1 - \frac{1}{2}(1-p)^N R(T)^N}. \tag{10.29}$$

The value of this ratio is less than one, and it decreases each time imperfect preventive maintenance is performed.

10.3   CORRECTIVE MAINTENANCE

With or without preventive maintenance, the definition of reliability has been central to all our deliberations. This is no longer the case, however, when we consider the many classes of systems in which corrective maintenance plays a substantial role. Now we are interested not only in the probability of failure, but also in the number of failures and, in particular, in the times required to make repairs. For such considerations two new reliability parameters become the focus of attention. Availability is the probability that a system is available for use at a given time. Roughly, it may be viewed as a fraction of time that a system is in an operational state. Maintainability is a measure of how fast a system may be repaired following failure. Both availability and maintainability, however, require more formal definitions if they are to serve as a quantitative basis for the analysis of repairable systems.

Availability

For repairable systems a fundamental quantity of interest is the availability. It is defined as follows:

$$A(t) = \text{probability that a system is performing} \atop \text{satisfactorily at time } t. \tag{10.30}$$

This is referred to as the point availability. Often it is necessary to determine the interval or mission availability. The interval availability is defined by

$$A^*(T) = \frac{1}{T}\int_0^T A(t)\, dt. \tag{10.31}$$

It is just the value of the point availability averaged over some interval of time, T. This interval may be the design life of the system or the time to accomplish some particular mission. Finally, it is often found that after some initial transient effects the point availability assumes a time-independent value. In these cases the steady-state or asymptotic availability is defined as

$$A^*(\infty) = \lim_{T\to\infty} \frac{1}{T}\int_0^T A(t)\, dt. \tag{10.32}$$

If a system or its components cannot be repaired, the point availability is just equal to the reliability. The probability that it is available at t is just equal to the probability that it has not failed between 0 and t:

$$A(t) = R(t) \tag{10.33}$$

Combining Eqs. 10.31 and 10.33, we obtain

$$A^*(T) = \frac{1}{T}\int_0^T R(t)\ dt. \tag{10.34}$$

Thus, as T goes to infinity, the numerator, according to Eq. 6.22, becomes the MTTF, a finite quantity. The denominator, T, however, becomes infinite. Thus the steady-state availability of a nonrepairable system is

$$A^*(\infty) = 0. \tag{10.35}$$

Since all systems eventually fail, and there is no repair, the availability averaged over an infinitely long time span is zero.

EXAMPLE 10.3

A nonrepairable system has a known MTTF and is characterized by a constant failure rate. The system mission availability must be 0.95. Find the maximum design life that can be tolerated in terms of the MTTF.

Solution For a constant failure rate the reliability is $R = e^{-\lambda t}$. Insert this into Eq. 10.34 to obtain

$$A^*(T) = \frac{1}{\lambda T}(1 - e^{-\lambda T}).$$

Expanding the exponential then yields

$$A(T) = \frac{1}{\lambda T}(1 - 1 + \lambda T - \tfrac{1}{2}(\lambda T)^2 + \cdots).$$

Thus $A(T) \approx 1 - \tfrac{1}{2}\lambda T$, for $\lambda T \ll 1$ or $0.95 = 1 - \tfrac{1}{2}\lambda T$. Then $\lambda T = 0.1$, but MTTF = $1/\lambda$. Therefore, $T = 0.1 \times$ MTTF.

Maintainability

We may now proceed to the quantitative description of repair processes and the definition of maintainability. Suppose that we let $\mathbf{t}$ be the time required to repair a system, measured from the time of failure. If all repairs take the same length of time, $\mathbf{t}$ is just a number, say $\mathbf{t} = \tau$. In reality, repairs require different lengths of time, and even the time to perform a given repair is uncertain because circumstances, skill level, and a host of other factors vary. Therefore $\mathbf{t}$ is normally not a constant but rather a random variable. This variable can be considered in terms of distribution functions as follows.

Suppose that we define the PDF for repair as

$$m(t)\ \Delta t = P\{t \le \mathbf{t} \le t + \Delta t\}. \tag{10.36}$$

That is, $m(t) \, \Delta t$ is the probability that repair will require a time between t and $t + \Delta t$. The CDF corresponding to Eq. 10.36 is defined as the maintainability

$$M(t) = \int_0^t m(t') \, dt', \tag{10.37}$$

and the mean time to repair or MTTR is then

$$\text{MTTR} = \int_0^\infty tm(t) \, dt. \tag{10.38}$$

Analogous to the derivations of the failure rate given in Chapter 6, we may define the instantaneous repair rate as

$$\nu(t) \, \Delta t = \frac{P\{t \le \mathbf{t} \le t + \Delta t\}}{P\{\mathbf{t} > t\}}; \tag{10.39}$$

$\nu(t) \, \Delta t$ is the conditional probability that the system will be repaired between t and $t + \Delta t$, given that it is failed at t. Noting that

$$M(t) = P\{\mathbf{t} \le t\} = 1 - P\{\mathbf{t} \ge t\}, \tag{10.40}$$

we then have

$$\nu(t) = \frac{m(t)}{1 - M(t)}. \tag{10.41}$$

Equations 10.37 and 10.41 may be used to express the maintainability and the PDF in terms of the repair rate. To do this, we differentiate Eq. 10.37 to obtain

$$m(t) = \frac{d}{dt} M(t), \tag{10.42}$$

and combine this result with Eq. 10.41 to yield

$$\nu(t) = [1 - M(t)]^{-1} \frac{d}{dt} M(t). \tag{10.43}$$

Moving dt to the left and integrating between 0 and t, we obtain

$$\int_0^t \nu(t') \, dt' = \int_0^{M(t)} \frac{dM}{1 - M}. \tag{10.44}$$

Evaluating the integral on the right-hand side and solving for the maintainability, we have

$$M(t) = 1 - \exp\left[-\int_0^t \nu(t') \, dt' \right]. \tag{10.45}$$

Finally, we may use Eq. 10.42 to express the PDF for repair times as

$$m(t) = \nu(t) \exp\left[-\int_0^t \nu(t') \, dt' \right]. \tag{10.46}$$

A great many factors go into determining both the mean time to repair and the PDF, $m(t)$, by which the uncertainties in repair time are characterized. These factors range from the ability to diagnose the cause of failure, on the one hand, to the availability of equipment and skilled personnel to carry out the repair procedures on the other. The determining factors in estimating repair time vary greatly with the type of system that is under consideration. This may be illustrated with the following comparison.

In many mechanical systems the causes of the failure are likely to be quite obvious. If a pipe ruptures, a valve fails to open, or a pump stops running, the diagnoses of the component in which the mechanical failure has occurred may be straightforward. The primary time entailed in the repair is then determined by how much time is required to extract the component from the system and install the new component, for each of these processes may involve a good deal of metal cutting, welding, or other time-consuming procedures.

In contrast, if a computer fails, maintenance personnel may spend most of the repair procedure time in diagnosing the problem, for it may take considerable effort to understand the nature of the failure well enough to be able to locate the circuit board, chip, or other component that is the cause. Conversely, it may be a rather straightforward procedure to replace the faulty component once it has been located.

In both of these examples we have assumed that the necessary repair parts are available at the time they are needed and that it is obvious how much of the system should be replaced to eliminate the fault. In fact, both the availability of parts and the level of repair involve subtle economic trade-offs between the cost of inventory, personnel, and system downtime.

For example, suppose that the pump fails because bearings have burned out. We must decide whether it is faster to remove the pump from the line and replace it with a new unit or to tear it down and replace only the bearings. If the entire pump is to be replaced, on-site inventories of spare pumps will probably be necessary, but the level of skill needed by repair personnel to install the new unit may not be great. Conversely, if most of the pump failures are caused by bearing failures, it may make sense to stock only bearings on site and to repack the bearings. In such a case repair personnel will need different and perhaps greater training and skill. Such trade-offs are typical of the many factors that must be considered in maintainability engineering, the discipline that optimizes $M(t)$ at a high level with as low a cost as possible.

10.4 REPAIR: REVEALED FAILURES

In this section we examine systems for which the failures are revealed, so that repairs can be immediately initiated. In these situations two quantities are of primary interest, the number of failures over a given span of time and the system availability. The number of failures is needed in order to calculate a variety of quantities including the cost of repair, the necessary repair parts inventory, and so on. Provided that the MTTR is much smaller than the MTTF, reasonable estimates for the number of failures can be obtained using the

Poisson distribution as in Chapter 6, and neglecting the system downtime for repair. For availability calculations, repair time must be considered or else we would obtain simply $A(t) = 1$. Ordinarily, this is not an acceptable approximation, for even small values of the unavailability $\tilde{A}(t)$ are frequently important, whether they be due to the risk incurred through the unavailability of a critical safety system or to the production loss during the downtimes of an assembly line.

In what follows, two models for repair are developed to estimate the availability of a system, constant repair rate, and constant repair time. It will be clear from comparing these that most of the more important results depend primarily on the MTTR, not on the details of the repair distribution.

Constant Repair Rates

To calculate availability, we must take the repair rate into account, even though it may be large compared to the failure rate. We assume that the distribution of times to repair can be characterized by a constant repair rate

$$\nu(t) = \nu. \tag{10.47}$$

The PDF of times to repair is then exponential,

$$m(t) = \nu e^{-\nu t}, \tag{10.48}$$

and the mean time to repair is simply

$$\text{MTTR} = 1/\nu. \tag{10.49}$$

Although the exponential distribution may not reflect the details of the distribution very accurately, it provides a reasonable approximation for predicting availabilities, for these tend to depend more on the MTTR than on the details of the distribution. As we shall illustrate, even when the PDF of the repair is bunched about the MTTR rather than being exponentially distributed, the constant repair rate model correctly predicts the asymptotic availability.

Suppose that we consider a two-state system; it is either operational, state 1, or it is failed, state 2. Then $A(t)$ and $\tilde{A}(t)$, the availability and unavailability, are the probabilities that the state is operational or failed, respectively, at time t, where t is measured from the time at which the system operation commences. We therefore have the initial conditions $A(0) = 1$ and $\tilde{A}(0) = 0$, and of course,

$$A(t) + \tilde{A}(t) = 1. \tag{10.50}$$

A differential equation for the availability may be derived in a manner similar to that used for the Poisson distribution in Chapter 6. We consider the change in $A(t)$ between t and $t + \Delta t$. There are two contributions. Since $\lambda \, \Delta t$ is the conditional probability of failure during Δt, given that the system is available at t, the loss of availability during Δt is $\lambda \, \Delta t \, A(t)$. Similarly, the gain in availability is equal to $\nu \, \Delta t \, \tilde{A}(t)$, where $\nu \, \Delta t$ is the conditional probability that the system is repaired during Δt, given that it is unavailable at t. Hence it follows that

$$A(t + \Delta t) = A(t) - \lambda \, \Delta t \, A(t) + \nu \, \Delta t \, \tilde{A}(t). \tag{10.51}$$

Rearranging terms and eliminating $\tilde{A}(t)$ with Eq. 10.50, we obtain

$$\frac{A(t + \Delta t) - A(t)}{\Delta t} = -(\lambda + \nu) A(t) + \nu. \tag{10.52}$$

Since the expression on the left-hand side is just the derivative with respect to time, Eq. 10.52 may be written as the differential equation,

$$\frac{d}{dt} A(t) = -(\lambda + \nu) A(t) + \nu. \tag{10.53}$$

We now may use an integrating factor of $e^{\lambda + \nu}$, along with the initial condition $A(0) = 1$ to obtain

$$A(t) = \frac{\nu}{\lambda + \nu} + \frac{\lambda}{\lambda + \nu} e^{-(\lambda + \nu)t}. \tag{10.54}$$

Note that the availability begins at $A(0) = 1$ and decreases monotonically to an asymptotic value $1/(1 + \lambda/\nu)$, which depends only on the ratio of failure to repair rate. The interval availability may be obtained by inserting Eq. 10.54 into Eq 10.31 to yield

$$A^*(T) = \frac{\nu}{\lambda + \nu} + \frac{\lambda}{(\lambda + \nu)^2 T} [1 - e^{-(\lambda + \nu)T}], \tag{10.55}$$

and the asymptotic availability is obtained by letting T go to infinity. Thus

$$A^*(\infty) = \frac{\nu}{\lambda + \nu}. \tag{10.56}$$

Finally, note from Eqs. 10.54 and 10.56 that for constant repair rates

$$A^*(\infty) = A(\infty). \tag{10.57}$$

Since, in most instances, repair rates are much larger than failure rates, a frequently used approximation comes from expanding Eq. 10.56 and deleting higher terms in λ/ν. We obtain after some algebra

$$A^*(\infty) \simeq 1 - \lambda/\nu. \tag{10.58}$$

The ratio in Eq. 10.56 may be expressed in terms of the mean time between failures and the mean time to repair. Since MTTF $= 1/\lambda$ and MTTR $= 1/\nu$, we have

$$A(\infty) = \frac{\text{MTTF}}{\text{MTTF} + \text{MTTR}}. \tag{10.59}$$

This expression is sometimes used for the availability even though neither failure or repair is characterized well by the exponential distribution. This is often quite adequate, for, in general, when availability is averaged over a reasonable period T of time, it is insensitive to the details of the failure or repair distributions. This is indicated for constant repair times in the following section.

EXAMPLE 10.4

In the following table are times (in days) over a 6-month period at which failure of a production line occurred (t_f) and times (t_r) at which the plant was brought back on line following repair.

i	t_{fi}	t_{ri}	i	t_{fi}	t_{ri}
1	12.8	13.0	6	56.4	57.3
2	14.2	14.8	7	62.7	62.8
3	25.4	25.8	8	131.2	134.9
4	31.4	33.3	9	146.7	150.0
5	35.3	35.6	10	177.0	177.1

(a) Calculate the 6-month-interval availability from the plant data.

(b) Estimate MTTF and MTTR from the data.

(c) Estimate the interval availability using the results of b and Eq. 10.59, and compare this result to that of a.

Solution During the 6 months (182.5 days) there are 10 failures and repairs.

(a) From the data we find that $\bar{A}(T)$ is just the fraction of that time for which the system is inoperable. Thus we find that

$$\bar{A}(T) = \frac{1}{T}\sum_{i=1}^{10}(t_{ri} - t_{fi})$$

$$= \frac{1}{182.5}(0.2 + 0.6 + 0.4 + 1.9 + 0.3 + 0.9 + 0.1 + 3.7 + 3.3 + 0.1)$$

$$\bar{A}(T) = 0.0630$$

$$A(T) = 1 - 0.063 = 0.937.$$

(b) Taking $t_{r0} = 0$, we first estimate the MTTF and MTTR from the data:

$$\text{MTTF} = \frac{1}{N}\sum_{i=1}^{10}(t_{fi} - t_{ri-1})$$

$$= \tfrac{1}{10}(12.8 + 1.2 + 10.6 + 5.6 + 2.0 + 20.8 + 5.4$$

$$+ 68.4 \div 11.8 + 27.0)$$

$$\text{MTTF} = \tfrac{1}{10}\,165.6 = 16.56.$$

$$\text{MTTR} = \frac{1}{N}\sum_{i=1}^{10}(t_{ri} - t_{fi}) = \frac{T}{10}\frac{1}{T}\sum_{i=1}^{10}(t_{fi} - t_{ri}) = \frac{182.5}{10}\bar{A}(T)$$

$$= 1.15 \text{ days.}$$

(c) $A(T) = \dfrac{\nu}{\nu + \lambda} = \dfrac{1}{1 + \dfrac{\text{MTTR}}{\text{MTTF}}} = \dfrac{1}{1 + \dfrac{0.85}{16.5}} = 0.935.$

Constant Repair Times

In the foregoing availability model we have used a constant repair rate, as we shall also do throughout much of the remainder of this chapter. Before proceeding, however, we repeat the calculation of the system availability using a repair model that is quite different; all the repairs are assumed to require exactly the same time, τ. Thus the PDF for time to repair has the form

$$m(t) = \delta(t - \tau), \tag{10.60}$$

where δ is the Dirac delta function discussed in Chapter 3. Although the availability is more difficult to calculate with this model, the result is instructive. It will be seen that whereas the details of the time dependence of $A(t)$ differ, the general trends are the same, and the asymptotic value is still given by Eq. 10.59.

A differential equation may be obtained for the availability, with the initial condition $A(0) = 1$. Since all repairs require a time τ, there are no repairs for $t < \tau$. Thus instead of Eq. 10.51, we have only the failure term on the right-hand side,

$$A(t + \Delta t) = A(t) - \lambda \, \Delta t \, A(t), \qquad 0 \le t \le \tau, \tag{10.61}$$

which corresponds to the differential equation

$$\frac{d}{dt} A(t) = -\lambda A(t), \qquad 0 \le t \le \tau. \tag{10.62}$$

For times greater than τ, repairs are also made; the number of repairs made during Δt is just equal to the number of failures during Δt at a time τ earlier: $\lambda \, \Delta t \, A(t - \tau)$. Thus the change in availability during Δt is

$$A(t + \Delta t) = A(t) - \lambda \, \Delta t \, A(t) + \lambda \, \Delta t \, A(t - \tau), \qquad t > \tau, \tag{10.63}$$

which corresponds to the differential equation

$$\frac{d}{dt} A(t) = -\lambda A(t) + \lambda A(t - \tau), \qquad t > \tau. \tag{10.64}$$

Equations 10.63 and 10.64 are more difficult to solve than those for the constant repair rate. During the first interval, $0 \le t \le \tau$, we have simply

$$A(t) = e^{-\lambda t}, \qquad 0 \le t \le \tau. \tag{10.65}$$

For $t > \tau$, the solution in successive intervals depends on that of the preceding interval. To illustrate, consider the interval $N\tau \le t \le (N + 1)\tau$. Applying an integrating factor $e^{\lambda t}$ to Eq. 10.64, we may solve for $A(t)$ in terms of $A(t - \tau)$:

$$A(t) = A(N\tau) e^{-\lambda(t-N\tau)} + \int_{N\tau}^{t} dt' \, \lambda e^{-\lambda(t-t')} A(t' - \tau), \qquad N\tau \le t \le (N+1)\tau.$$
$$(10.66)$$

For $N = 1$, we may insert Eq. 10.65 on the right-hand side to obtain

$$A(t) = e^{-\lambda t} + \lambda(t - \tau) e^{-\lambda(t-\tau)}, \qquad \tau \le t \le 2\tau. \qquad (10.67)$$

For $N = 2$ there will be three terms on the right-hand side, and so on. The general solution for arbitrary N appears quite similar to the Poisson distribution:

$$A(t) = \sum_{n=0}^{N} \frac{[\lambda(t - n\tau)]^n}{n!} e^{-\lambda(t-n\tau)}, \qquad N\tau \le t \le (N+1)\tau. \qquad (10.68)$$

The solutions for the constant repair rate and the constant repair time models are plotted for the point availability $A(t)$ in Fig. 10.5 for $\tau = 1/\nu$. Note that the discrete repair time leads to breaks in the slope of the availability curve, whereas this is not the case with the constant failure rate model. However, both curves follow the same general trend downward and converge to the same asymptotic value. Thus, if we are interested only in the general characteristics of availability curves, which ordinarily is the case, the constant repair rate model is quite adequate, even though some of the structure carried by a more precise evaluation of the repair time PDF may be lost. Moreover, to an even greater extent than with failure rates, not enough data are available in most cases to say much about the spread of repair times about the MTTR. Therefore, the single-parameter exponential distribution may be all that can be justified, and Eq. 10.59 provides a reasonable estimate of the availability.

10.5   TESTING AND REPAIR: UNREVEALED FAILURES

As long as system failures are revealed immediately, the time to repair is the primary factor in determining the system availability. When a system is not in continuous operation, however, failures may occur but remain undiscovered. This problem is most pronounced in backup or other emergency equipment that is operated only rarely, or in stockpiles of repair parts or other materials that may deteriorate with time. The primary loss of availability then may be

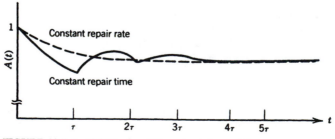

FIGURE 10.5 Availability for different repair models.

due to failures in the standby mode that are not detected until an attempt is made to use the system.

A primary weapon against these classes of failures is periodic testing. As we shall see, the more frequently testing is carried out, the more failures will be detected and repaired soon after they occur. However, this must be weighed against the expense of frequent testing, the loss of availability through down-time for testing, and the possibility of excessive component wear from too-frequent testing.

Idealized Periodic Tests

Suppose that we first consider the effect of a simple periodic test on a system whose reliability can be characterized by a constant failure rate:

$$R(t) = e^{-\lambda t}. \tag{10.69}$$

The first thing that should be clear is that system testing has no positive effect on reliability. For unlike preventive maintenance the test will only catch failures after they occur.

Testing, however, has a very definite positive effect on availability. To see this in the simplest case, suppose that we perform a system test at time interval T_0. In addition, we make the following three assumptions: (1) The time required to perform the test is negligible, (2) the time to perform repairs is negligible, and (3) the repairs are carried out perfectly and restore the system to an as-good-as-new condition. Later, we shall examine the effects of relaxing these assumptions.

Suppose that we test a system with reliability given by Eq. 10.69 at time interval T_0. As indicated, if there is no repair, the availability is equal to the reliability. Thus, before the first test,

$$A(t) = R(t), \qquad 0 \le t < T_0. \tag{10.70}$$

Since the system is repaired perfectly and restored to an as-good-as-new state at $t = T_0$, we will have $R(T_0) = 1$. Then since there is no repair between T_0 and $2T_0$, the availability will again be equal to the reliability, but now the reliability is evaluated at $t - T_0$:

$$A(t) = R(t - T_0), \qquad T_0 \le t < 2T_0. \tag{10.71}$$

This pattern repeats itself as indicated in Fig. 10.6. The general expression is

$$A(t) = R(t - NT_0), \qquad NT_0 \le t < (N+1)T_0. \tag{10.72}$$

For the situation indicated in Fig. 10.6, the interval and the asymptotic availability have the same value, provided that the integral in Eq. 10.31 is taken over a multiple of T_0, say mT_0. We have

$$A^*(mT_0) = \frac{1}{mT_0} \int_0^{mT_0} A(t)\, dt = \frac{1}{T_0} \int_0^{T_0} A(t)\, dt. \tag{10.73}$$

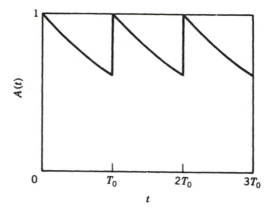

FIGURE 10.6 Availability with idealized periodic testing for unrevealed failures.

Since the interval availability is independent of the number of intervals over which $A^*(T)$ is calculated, so will the asymptotic availability $A^*(\infty)$:

$$A^*(\infty) = \lim_{m \to \infty} \frac{1}{mT_0} \int_0^{mT_0} A(t) \, dt = \frac{1}{T_0} \int_0^{T_0} A(t) \, dt. \tag{10.74}$$

The effect of the testing interval on availability may be seen by combining Eqs. 10.69 and 10.74. We obtain

$$A^*(\infty) = \frac{1}{\lambda T_0} (1 - e^{-\lambda T_0}). \tag{10.75}$$

Ordinarily, the test interval would be small compared to the MTTF: $\lambda T_0 \ll 1$. Therefore, the exponential may be expanded, and only the leading terms are retained to make the approximation

$$A^*(\infty) \simeq 1 - \tfrac{1}{2}\lambda T_0. \tag{10.76}$$

EXAMPLE 10.5

Annual inspection and repair are carried out on a large group of smoke detectors of the same design in public buildings. It is found that 15% of the smoke detectors are not functional. If it is assumed that the failure rate is constant,

(a) In what fraction of fires will the detectors offer protection?

(b) If the smoke detectors are required to offer protection for at least 99% of fires, how frequently must inspection and repair be carried out?

Solution With inspection and repair at interval T_0, the fraction of detectors that are operational at the time of inspection will be

$$R = e^{-\lambda T_0} = 0.85.$$

Then $\lambda T_0 = -\ln(0.85) = 0.162$. Since $T_0 = 1$ year, $\lambda = 0.162$/year.

(*a*) If we assume that the fires are uniformly distributed in time, the fractional protection is just equal to the interval availability; from Eq. 10.75

$$A^*(\infty) = \frac{1}{\lambda T_0}(1 - e^{-\lambda T_0}) = \frac{1}{0.162}(1 - 0.85) = 0.926.$$

(*b*) For this high availability the rare-event approximation, Eq. 10.76, may be used:

$$0.99 = A^*(\infty) \approx 1 - \tfrac{1}{2}\lambda T_0.$$

Thus from Eq. 10.76,

$$T_0 = \frac{2[1 - A^*(\infty)]}{\lambda} = \frac{2(1 - 0.99)}{0.162} = 0.123 \text{ year}$$

$$= 0.123 \times 12 \text{ months} \approx 1\tfrac{1}{2} \text{ months}.$$

Real Periodic Tests

Equation 10.76 indicates that we may achieve availabilities as close to one as desired merely by decreasing the test interval T_0. This is not the case, however, for as the test interval becomes smaller, a number of other factors—test time, repair time, and imperfect repairs—become more important in estimating availability.

When we examine these effects, it is useful to visualize them as modifications in the curve shown in Fig. 10.6. The interval or asymptotic availability may be pictured as proportional to the area under the curve within one test interval, divided by T. Thus we may view each of the factors listed earlier in terms of the increase or decrease that it causes in the area under the curve. In particular, with reasonable assumptions about the ratios of the various parameters involved, we may derive approximate expressions similar to Eq. 10.76 that are quite simple, but at the same time are not greatly in error.

Consider first the effect of a nonnegligible test time, t_t. During the test we assume that the system must be taken off line, and the system has an availability of zero during the test. The point availability will then appear as the solid line in Fig. 10.7. Provided that we again assume that $\lambda T_0 \ll 1$, so that Eq. 10.76 holds, and that $t_t \ll T_0$, the test time, is small compared to the test interval, we may approximate the contribution of the test to system downtime as t_t/T_0. The availability indicated in Eq. 10.76 is therefore decreased to

$$A^*(\infty) \simeq 1 - \tfrac{1}{2}\lambda T_0 - \frac{t_t}{T_0}. \tag{10.77}$$

We next consider the effect of a nonzero time to repair on the availability. The probability of finding a failed system at the time of testing is just one minus the point availability at the time the test is carried out. For small T_0 this probability may be shown to be approximately λT_0. Since $1/\nu$ is the mean time to repair, the contribution to be unavailability over the period T_0 is $\lambda T_0/\nu$, or dividing by the interval T_0, we find, as in Eq. 10.58, the loss of

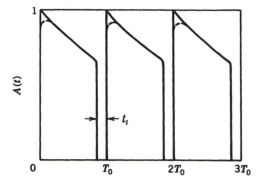

FIGURE 10.7 Availability with realistic periodic testing for unrevealed failures.

availability to be approximately λ/ν. We may therefore modify our availability by subtracting this term to yield

$$A^*(\infty) \simeq 1 - \tfrac{1}{2}\lambda T_0 - \frac{t_t}{T_0} - \frac{\lambda}{\nu}. \tag{10.78}$$

The effect of this contribution to the system unavailability is indicated by the dotted line in Fig. 10.7.

Examination of Eq. 10.78 is instructive. Clearly, decreases in failure rate and in test time t_t increase the availability, as do increases in the repair rate ν. It may also be shown that the more perfect the repair, the higher the availability. Decreasing the test interval, however, may either increase or decrease the availability, depending on the value of the other parameters. For, as indicated in Eq. 10.78, it appears in both the numerator and the denominator of terms.

Suppose that we differentiate Eq. 10.78 with respect to T_0 and set the result equal to zero in order to determine the maximum availability:

$$\frac{d}{dT_0} A^*(\infty) = -\tfrac{1}{2}\lambda + \frac{t_t}{T_0^2} = 0. \tag{10.79}$$

The optimal test interval is then

$$T_0 = \left(\frac{2t_t}{\lambda}\right)^{1/2}. \tag{10.80}$$

Substitution of this expression back into Eq. 10.78 yields a maximum availability of

$$A^*(\infty) = 1 - (2\lambda t_t)^{1/2} - \frac{\lambda}{\nu}. \tag{10.81}$$

If the test interval is longer than Eq. 10.80, undetected failures will lower availability. However, if a shorter test interval is employed, the loss of availability during testing will not be fully compensated for by earlier detection of failures.

The test interval should increase as the failure rate decreases, and decrease as the testing time can be decreased. Other trade-offs may need to be considered as well. For example, will hurrying to decrease the test time increase the probability that failures will be missed?

EXAMPLE 10.6

A sulfur dioxide scrubber is known to have a MTBF of 137 days. Testing the scrubber requires half a day, and the mean time to repair is 4 days. (*a*) Choose the test period to maximize the availability. (*b*) What is the maximum availability?

Solution (*a*) From Eq. 10.80, with MTBF $= 1/\lambda$,

$$T_0 = (2t_t \, \text{MTBF})^{1/2} = (2 \times 0.5 \times 137)^{1/2} = 11.7 \text{ days.}$$

(*b*) From Eq. 10.81,

$$A^*(\infty) = 1 - \left(\frac{2t_t}{\text{MTBF}}\right)^{1/2} - \frac{\text{MTTR}}{\text{MTBF}},$$

$$A^*(\infty) = 1 - \left(\frac{2 \times 0.5}{137}\right)^{1/2} - \frac{4}{137} = 0.885.$$

10.6 SYSTEM AVAILABILITY

Thus far we have examined only the effects on availability of the failure and repair of a system as a whole. But just as for reliability, it is often instructive to examine the availability of a system in terms of the component availabilities. Not only are data more likely to be available at the component level, but the analysis can provide insight into the gains made through redundant configurations, and through different testing and repair strategies.

Since availability, like reliability, is a probability, system availabilities can be determined from parallel and series combinations of component availabilities. In fact, the techniques developed in Chapter 9 for combining reliabilities are also applicable to point availabilities, but only provided that both the failure and repair rates for the components are independent of one another. If this is not the case, either the β-factor method described in Chapter 9 or the Markov methods discussed in the following chapter may be required to model the component dependencies. In this chapter we consider situations in which the component properties are independent of one another, deferring analysis of component dependencies to the following chapter.

In what follows we estimate point availabilities of systems in terms of components. The appropriate integral is then taken to obtain interval and asymptotic availabilities. When the component availabilities become time-independent after a long period of operation, steady-state availabilities may be calculated simply by letting $t \to \infty$ in the point availabilities. In testing or other situations in which there is a periodicity in the point availability, the point availability must be averaged over a test period, even though the system has been in operation for a substantial length of time. Very often when repair rates are much higher than failure rates, simplifying approximations, in which λ/ν is assumed to be very small, are of sufficient accuracy and lead to additional physical insight in comparing systems.

For systems without redundancy the availability obeys the product law introduced in Chapter 9. Suppose that we let $\tilde{X}$ represent the failed state of the system, and X the unfailed or operational state of the system. Similarly, let $\tilde{X}_i$ represent the failed state of component i, and X_i the unfailed state of the same component. In a nonredundant system, all the components must be available for the system to be available:

$$X = X_1 \cap X_2 \cap \ldots \cap X_M. \tag{10.82}$$

Since the availability is defined as just the probability that the system is available, we have

$$A(t) = \prod_i A_i(t). \tag{10.83}$$

where the $A_i(t)$ are the independent component availabilities.

For redundant (i.e., parallel) systems, all the components must be unavailable if the system is to be unavailable. Thus, if $\tilde{X}$ signifies a failed system and $\tilde{X}_i$ the failed state of component i, we have

$$\tilde{X} = \tilde{X}_1 \cap \tilde{X}_2 \cap \tilde{X}_3 \cap \ldots \cap \tilde{X}_M. \tag{10.84}$$

Since the unavailability is one minus the availability, we have

$$1 - A(t) = [1 - A_1(t)][1 - A_2(t)] \ldots [1 - A_M(t)], \tag{10.85}$$

or more compactly,

$$A(t) = 1 - \prod_i [1 - A_i(t)]. \tag{10.86}$$

Comparing Eqs. 10.83 and 10.86 with Eqs. 9.1 and 9.38 indicates that the same relationships hold for point availabilities as for reliabilities. The other relationships derived in Chapter 9 also hold when the assumption that the components are mutually independent is made throughout.

Revealed Failures

Suppose that we now apply the constant repair rate model to each component. According to Eq. 10.54, the component availabilities are then

$$A_i(t) = \frac{\nu_i}{\nu_i + \lambda_i} + \frac{\lambda_i}{\nu_i + \lambda_i} e^{-(\lambda_i + \nu_i)t}. \tag{10.87}$$

This relationship may be applied in the foregoing equations to estimate system availability.

If we are interested only in asymptotic availability, we may delete the second term of Eq. 10.87 to obtain

$$A_i(\infty) = \frac{\nu_i}{\nu_i + \lambda_i}. \tag{10.88}$$

Combining this expression with Eq. 10.83, we have for a nonredundant system

$$A(\infty) = \prod_i \frac{\nu_i}{\nu_i + \lambda_i}. \tag{10.89}$$

If we further make the reasonable assumption that repair rates are large compared to failure rates, $\nu_i \gg \lambda_i$, then

$$A_i(\infty) \simeq 1 - \frac{\lambda_i}{\nu_i}. \tag{10.90}$$

With this expression substituted into Eq. 10.83 to estimate the availability of a nonredundant system, we obtain

$$A(\infty) \simeq \prod_i \left(1 - \frac{\lambda_i}{\nu_i}\right). \tag{10.91}$$

But since we have already deleted higher-order terms in the ratios λ_i/ν_i, for consistency we also should eliminate them from this equation. This yields

$$A(\infty) \simeq 1 - \sum_i \frac{\lambda_i}{\nu_i}. \tag{10.92}$$

Thus the rapid deterioration of the availability with an increased number of components is seen. If we further assume that all the repair rates can be replaced by an average value $\nu_i = \nu$, Eq. 10.92 becomes

$$A(\infty) \simeq 1 - \lambda/\nu, \tag{10.93}$$

where

$$\lambda = \sum_i \lambda_i. \tag{10.94}$$

Therefore, we obtain the same result as given for the system as a whole, provided that we sum the component failure rates as in Chapter 6.

The effect of redundancy may be seen by inserting Eq. 10.88 into Eq. 10.86, the availability of a parallel system. For N identical units with $\lambda_i = \lambda$ and $\nu_i = \nu$, we have

$$A(\infty) = 1 - \left(\frac{\lambda}{\lambda + \nu}\right)^N. \tag{10.95}$$

If we consider the case where $\nu \gg \lambda$, then

$$A(\infty) \simeq 1 - \left(\frac{\lambda}{\nu}\right)^N, \tag{10.96}$$

or correspondingly for the unavailability,

$$\tilde{A}(\infty) \simeq \left(\frac{\lambda}{\nu}\right)^N. \tag{10.97}$$

The analogy to the reliability of parallel systems is clear; both unreliability and unavailability are proportional to the Nth power of the failure rate. The foregoing relationships assume that there are no common-mode failures. If there are, the β-factor method of Chapter 9 may be adapted, putting a fictitious component in series with a failure and a repair rate for the common-mode failure. Once again the presence of common-mode failure limits the gains that can be made through the use of parallel configurations, although not as severely as for systems that cannot be repaired. Suppose we consider as an example N units in parallel, each having a failure rate λ divided into independent and common-mode failures as in Eqs. 9.24 through 9.30. We have

$$A(\infty) = \{1 - [1 - A_I(\infty)]^N\}A_c(\infty), \tag{10.98}$$

where A_I are the availabilities with only the independent failure rate λ_I taken into account, and A_c is the common-mode availability with failure rate λ_c. We assume that both common and independent failure modes have the same repair rate. Thus

$$A(\infty) = \left[1 - \left(\frac{\lambda_I}{\lambda_I + \nu}\right)^N\right]\frac{\nu}{\lambda_c + \nu}. \tag{10.99}$$

This may also be written in terms of β factors by recalling that $\lambda_I \equiv (1 - \beta)\lambda$ and $\lambda_c \equiv \beta\lambda$.

EXAMPLE 10.7

A system has a ratio of $\nu/\lambda = 100$. What will the asymptotic availability be (*a*) for the system, (*b*) for two of the systems in parallel with no common-mode failures, and (*c*) for two systems in parallel with $\beta = 0.2$?

Solution   (*a*) $A(\infty) = \dfrac{100}{1 + 100} = 0.990.$

(*b*) $A(\infty) = 1 - \left(\dfrac{1}{1 + 100}\right)^2 = 0.99990.$

(*c*) $\dfrac{\lambda_I}{\nu} = (1 - \beta)\dfrac{\lambda}{\nu} = (1 - 0.2)\dfrac{1}{100} = 0.8 \times 10^{-2}$

$\dfrac{\lambda_c}{\nu} = \beta\dfrac{\lambda}{\nu} = 2 \times 10^{-3}.$

Therefore, from Eq. 10.99,

$$A(\infty) = \left[1 - \left(\frac{0.8 \times 10^{-2}}{1 + 0.8 \times 10^{-2}}\right)^2\right]\frac{1}{2 \times 10^{-3} + 1} = 0.9979.$$

Unrevealed Failures

In the derivations just given it is assumed that component failures are detected immediately and that repair is initiated at once. Situations are also encountered in which the component failures go undetected until periodic testing takes place. The evaluation of availability then becomes more complex, for several testing strategies may be considered. Not only is the test interval T_0 subject to change, but the testing may be carried out on all the components simultaneously or in a staggered sequence. In either event the calculation of the system availability is now more subtle, for the point availabilities will have periodic structures, and they must be averaged over a test period in order to estimate the asymptotic availability.

To illustrate, consider the effects of simultaneous and staggered testing patterns on two simple component configurations: the nonredundant configuration consisting of two identical components in series, and the completely redundant configuration consisting of two identical components in parallel. For clarity we consider the idealized situation in which the testing time and the time to repair can be ignored. The failure rates are assumed to be constant.

We begin by letting $A_1(t)$ and $A_2(t)$ be the component point availabilities. Since the testing is carried out at intervals of T_0, we need only determine the system point availability $A(t)$ between $t = 0$ and $t = T_0$, for the asymptotic mission availability is then obtained by averaging $A(t)$ over the test period:

$$A^*(\infty) = A^*(T_0) = \frac{1}{T_0} \int_0^{T_0} A(t) \, dt. \tag{10.100}$$

Simultaneous Testing When both components are tested at the same time, $t = 0, T_0, 2T_0, \ldots$, the point availabilities are given by

$$A_1(t) = e^{-\lambda t}, \qquad 0 \le t < T_0, \tag{10.101}$$

and

$$A_2(t) = e^{-\lambda t}, \qquad 0 \le t < T_0. \tag{10.102}$$

For the series system we have

$$A(t) = A_1(t) A_2(t), \tag{10.103}$$

or

$$A(t) = e^{-2\lambda t}, \qquad 0 \le t < T_0. \tag{10.104}$$

For the parallel system we obtain

$$A(t) = A_1(t) + A_2(t) - A_1(t) A_2(t), \tag{10.105}$$

or

$$A(t) = 2e^{-\lambda t} - e^{-2\lambda t}, \qquad 0 \le t < T_0. \tag{10.106}$$

The availabilities are plotted as solid lines in Fig. 10.8a and b, respectively. The asymptotic availability obtained from Eq. 10.100 for the series system is

$$A_s^*(T_0) = \frac{1}{2\lambda T_0}(1 - e^{-2\lambda T_0}) \tag{10.107}$$

whereas that of the parallel system is

$$A_p^*(T_0) = \frac{1}{2\lambda T_0}(3 - 4e^{-\lambda T_0} + e^{-2\lambda T_0}). \tag{10.108}$$

Staggered Testing We now consider the testing of components at staggered intervals of $T_0/2$. We assume that component 1 is tested at $0, T_0, 2T_0, \ldots,$ whereas component 2 is tested at the half-intervals $T_0/2, 3T_0/2, \ldots$. The point availabilities within any interval after the first one are given by

$$A_1(t) = e^{-\lambda t}, \quad 0 \le t < T_0, \tag{10.109}$$

and

$$A_2(t) = \begin{cases} \exp\left[-\lambda\left(t + \dfrac{T_0}{2}\right)\right] & 0 \le t < \dfrac{T_0}{2}, \\[3mm] \exp\left[-\lambda\left(t - \dfrac{T_0}{2}\right)\right] & \dfrac{T_0}{2} \le t < T_0. \end{cases} \tag{10.110}$$

To determine the point system availability, we combine these two equations with Eqs. 10.103 and 10.105, respectively, for the series and parallel configurations. The results are plotted as dotted lines in Figs. 10.8a and 10.8b.

To calculate the asymptotic availabilities for staggered testing, we first note from Fig. 10.8 that the system point availabilities for both series and parallel situations have a periodicity over the half-intervals $T_0/2$. Therefore, instead of averaging $A(t)$ over an entire interval as in Eq. 10.100, we need to

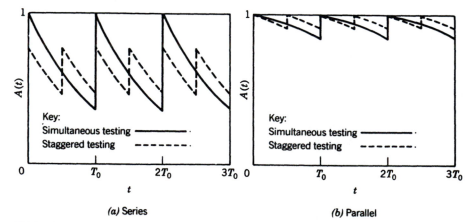

(a) Series   *(b)* Parallel

FIGURE 10.8 Availability for a two-component system with unrevealed failures.

TABLE 10.1 Availability $A^*(T_0)$ for Unrevealed Failures

Testing	Series system	Parallel system
Simultaneous	$1 - \lambda T_0 + \frac{2}{3}(\lambda T_0)^2$	$1 - \frac{1}{3}(\lambda T_0)^2$
Staggered	$1 - \lambda T_0 + \frac{13}{24}(\lambda T_0)^2$	$1 - \frac{5}{24}(\lambda T_0)^2$

average it over only the half-interval. Hence

$$A^*(T_0) = \frac{2}{T_0} \int_0^{T_0/2} A(t) \, dt. \tag{10.111}$$

For the series configuration we calculate $A_1(t)A_2(t)$ from Eqs. 10.109 and 10.110, substitute the result into Eq. 10.111, and carry out the integral to obtain

$$A_s^*(T_0) = \frac{1}{2\lambda T_0} (e^{-\lambda T_0/2} - e^{-3\lambda T_0/2}). \tag{10.112}$$

Similarly, for the parallel configuration we form $A(t)$ by substituting Eqs. 10.109 and 10.110 into Eq. 10.105, combine the result with Eq. 10.111, and perform the integral to obtain

$$A_p^*(T_0) = \frac{1}{\lambda T_0} (2 - 2e^{-\lambda T_0} - e^{-\lambda T_0/2} + e^{-3\lambda T_0/2}). \tag{10.113}$$

Although the point availabilities plotted as dotted lines in Fig. 10.8 are interesting in understanding the effects of staggering on the availability, the asymptotic values are often more useful, for they allow us to compare the strategies with a single number. Evaluation of the appropriate expressions indicates that in the nonredundant (series) configuration higher availability is obtained from simultaneous testing, whereas staggered testing yields the higher availability for redundant (parallel) configurations.

This behavior can be understood explicitly if the expressions for the asymptotic availability are expanded in powers of λT_0, since for small failure rates the lowest-order terms in λT_0 will dominate the expressions. The results of such expansions are presented in Table 10.1.

The effects of staggered testing become more pronounced when repair time, testing time, or both are not negligible. We can see, for example, that even for a zero failure rate, the testing time t_t will decrease the availability of the series system by t_t/T_0 if the systems are tested simultaneously. If the tests are staggered in the series system, the availability will decrease by $2t_t/T_0$. Conversely, in the parallel system simultaneous testing with no failures will decrease the availability by t_t/T_0, but if the tests are staggered so that they do not take both components out at the same time, the availability does not decrease.

EXAMPLE 10.8

A voltage monitor achieves an average availability of 0.84 when it is tested monthly; the repair time is negligible. Since the 0.84 availability is unacceptably low, two monitors are placed in parallel. What will the availability of this twin system be (*a*) if the monitors are tested monthly at the same time, (*b*) if they are tested monthly at staggered intervals?

Solution   First we must find λT_0. Try Eq. 10.76, the rare-event approximation:

$$0.84 = 1 - \tfrac{1}{2}\lambda T_0; \qquad \lambda T_0 \approx 0.32.$$

This is too large for the exponential expansion to be used. Therefore, we use Eq. 10.75 instead. We obtain a transcendental equation

$$0.84 = \frac{1}{\lambda T_0}(1 - e^{-\lambda T_0}).$$

Solving iteratively, we find that

λT_0	0.320	0.340	.360	.380
$(1/0.84)(1 - e^{-\lambda T_0})$	0.326	0.343	.3599	.376

Therefore,

$$\lambda T_0 \approx .36;$$

(*a*) From Eq. 10.108 we find for simultaneous testing

$$A_p^*(T_0) = \frac{1}{2 \times 0.36}(3 - 4e^{-0.36} + e^{-2\times0.36}) = 0.967.$$

(*b*) From Eq. 10.113 we find for staggered testing

$$A_p^*(T_0) = \frac{1}{0.36}(2 - 2e^{-0.36} - e^{-0.36/2} + e^{-3\times0.36/2}) = 0.978.$$

These results can be generalized to combinations of series and parallel configurations. However, the evaluation of the integral in Eq. 10.100 over the test period may become tedious. Moreover, the evaluation of maintenance, testing, and repair policies become more complex in real systems that contain combinations of revealed and unrevealed failures, large numbers of components, and dependencies between components. Some of the more common types of dependencies are included in the following chapter.

Bibliography

Ascher, H., and H. Feingold, "Repairable Systems Reliability: Modeling, Inference, Misconceptions, and Their Causes," *Lecture Notes in Statistics Series*, Vol 7, Marble Decker, NY, 1984.

Barlow, R.E., and F. Proschan, *Mathematical Theory of Reliability*, Wiley, NY, 1965.

Gertsbakh, I. B., *Models for Preventive Maintenance*, North-Holland Publishing Co., Amsterdam, 1977.

Jardine, A. K. S., *Maintenance, Replacement, and Reliability*, Wiley, NY, 1973.

Sandler, G. H., *System Reliability Engineering*, Prentice-Hall, Englewood Cliffs, NJ, 1963.

Smith, D. J., *Reliability, Maintainability and Risk*, 46h ed., Butterworth–Heinemann, Oxford, 1993

Exercises

10.1 Without preventive maintenance the reliability of a condensate demineralizer is characterized by

$$\int_0^t \lambda(t')\, dt' = 1.2 \times 10^{-2}t + 1.1 \times 10^{-9}t^2$$

where t is in hours. The design life is 10,000 hr.

(a) What is the design-life reliability?

(b) Suppose that by overhaul the demineralizer is returned to as-good-as-new condition. How frequently should such overhauls be performed to achieve a design-life reliability of at least 0.95?

(c) Repeat b for a target reliability of at least 0.975.

10.2 Discuss under what conditions preventative maintenance can increase the reliability of a simple active parallel system, even though the component failure rates are time-independent. Justify your results.

10.3 Repeat b of Exercise 10.1 assuming that there is a 1% probability that faulty overhaul will cause the demineralizer to fail destructively immediately following start-up. Is it possible to achieve the 0.95 reliability? If so, how many overhauls are required?

10.4 Derive an equation analogous to Eqs. 10.27 and 10.28 that includes a probability p_I of independent maintenance failure and a probability p_c of common-mode maintenance failure.

10.5 Suppose that a device has a failure rate of

$$\lambda(t) = (0.015 + 0.02t)/\text{year},$$

where t is in years.

(a) Calculate the reliability for a 5-year design life assuming that no maintenance is performed.

(b) Calculate the reliability for a 5-year design life assuming that annual preventive maintenance restores the system to an as-good-as-new condition.

(c) Repeat b assuming that there is a 5% chance that the preventive maintenance will cause immediate failure.

10.6 A machine has a failure rate given by $\lambda(t) = at$. Without maintenance the reliability at the end of one year is $R(1) = 0.86$.

(a) Determine the value of "a".

(b) If as-good-as-new preventive maintenance is performed at two-month intervals, what will the one-year reliability be?

(c) If in *b* there is a 2% probability that each maintenance will cause system failure, what will be the value of the reliability at the end of one year?

10.7 Suppose that the times to failure of an unmaintained component may be given by a Weibull distribution with $m = 2$. Perfect preventive maintenance is performed at intervals $T = 0.25\theta$.

(a) Find the MTTF of the maintained system in terms of θ.

(b) Determine the percentage increase in the MTTF over that of the unmaintained system.

10.8 Solve Exercise 10.7 approximately for the situation in which $T \ll \theta$.

10.9 The reliability of a device is given by the Rayleigh distribution

$$R(t) = e^{-(t/\theta)^2}.$$

The MTTF is considered to be unacceptably short. The design engineer has two alternatives: a second identical system may be set in parallel or (perfect) preventive maintenance may be performed at some interval T. At what interval T must the preventive maintenance be performed to obtain an increase in the MTTF equal to what would result from the parallel configuration without preventive maintenance? (*Note:* See the solution for Exercise 9.19.)

10.10 Show that preventive maintenance has no effect on the MTTF for a system with a constant failure rate.

10.11 The following table gives a series of times to repair (man-hours) obtained for a diesel engine.

11.6	7.9	27.7	17.8	8.9	22.5
3.3	33.3	75.3	9.4	28.5	5.4
10.3	1.1	7.8	41.9	13.3	5.3

(a) Estimate the MTTR.

(b) Estimate the repair rate and its 90% confidence interval assuming that the data is exponentially distributed.

10.12 Find the asymptotic availability for the systems shown in Exercise 9.38, assuming that all the components are subject only to revealed failures and that the repair rate is ν. Then approximate your result for the case $\nu/\lambda \gg 1$.

10.13 A computer has an MTTF = 34 hr and an MTTR = 2.5 hr.

(a) What is the availability?

(b) If the MTTR is reduced to 1.5 hr, what MTTF can be tolerated without decreasing the availability of the computer?

10.14 A generator has a long-term availability of 72%. Through a management reorganization the MTTR (mean time to *repair*) is reduced to one half of its former value. What is the generator availability following the reorganization?

10.15 A system consists of two subsystems in series, each with $\nu/\lambda = 10^2$ as its ratio of repair rate to failure rate. Assuming revealed failures, what is the availability of the system after an extended period of operation?

10.16 A robot has a failure rate of $0.05\ \mathrm{hr}^{-1}$. What repair rate must be achieved if an asymptotic availability of 95% is to be maintained?

10.17 Reliability testing has indicated that without repair a voltage inverter has a 6-month reliability of 0.87; make a rough estimate of the MTTR that must be achieved if the inverter is to operate with an availability of 0.95. (Assume revealed failures and a constant failure rate.)

10.18 The control unit on a fire sprinkler system has an MTTF for unrevealed *failures* of 30 months. How frequently must the unit be tested/repaired if an average *availability* of 99% is to be maintained.

10.19 A device has a constant failure rate, and the failures are unrevealed. It is found that with a test interval of 6 months the interval availability is 0.98. Use the "rare-event" approximation to estimate the failure rate. (Neglect test and repair times.)

10.20 Starting with Eqs. 10.107 and 10.112, derive the results for series systems with simultaneous and staggered testing given in Table 10.1.

10.21 The following table gives the times at which a system failed (t_f) and the times at which the subsequent repairs were completed (t_r) over a 2000-hr period.

t_f	t_r	t_f	t_r
51	52	1127	1134
90	92	1236	1265
405	412	1297	1303
507	529	1372	1375
535	539	1424	1439
615	616	1531	1552
751	752	1639	1667
760	766	1789	1795
835	839	1796	1808
881	884	1859	1860
933	941	1975	1976
1072	1091		

    (a) Calculate the average availability over the time interval $0 \le t \le t_{max}$ directly from the data.

(b) Assuming constant failure and repair rates, estimate λ and μ from the data.

(c) Use the values of λ and μ obtained in *b* to estimate $A(t)$ and the time-averaged availability for the interval $0 \leq t \leq t_{max}$. Compare your results to *a*.

10.22 Starting with Eqs. 10.108 and 10.113, derive the results for parallel systems with simultaneous and staggered testing given in Table 10.1.

10.23 An auxiliary feedwater pump has an availability of 0.960 under the following conditions: The failures are unrevealed; periodic testing is carried out on a monthly (30-day) basis; and testing and repair require that the system be shut down for 8 hr.

(a) What will the availability be if the shutdown time can be reduced to 2 hr?

(b) What will the availability be if the tests are performed once per week, with the 8-hr shutdown time?

(c) Given the 8-hr shutdown time, what is the optimal test interval?

10.24 A pressure relief system consists of two valves in parallel. The system achieves an availability of 0.995 when the valves are tested on a staggered basis, each valve being tested once every 3 months.

(a) Estimate the failure rate of the valves.

(b) If the test procedure were relaxed so that each valve is tested once in 6 months, what would the availability be?

10.25 In annual test and replacement procedures 8% of the emergency respirators at a chemical plant are found to be inoperable.

(a) What is the availability of the respirators?

(b) How frequently must the test and replacement be carried out if an availability of 0.99 is to be reached? (Assume constant failure rates.)

10.26 Consider three units in parallel, each tested at equally staggered intervals of T_0. Assume constant failure rates.

(a) What is $A(t)$?

(b) Plot $A(t)$.

(c) What is $A^*(T_0)$?

(d) Find the rare-event approximate for $A^*(T_0)$.

10.27 Unrevealed bearing failures follow a Weibull distribution with $m = 2$ and $\theta = 5000$ operating hours. How frequently must testing and repair take place if bearing availability is to be maintained at least 95%?

10.28 The reliability of a system is represented by the Rayleigh distribution

$$R(t) = e^{-(t/\theta)^2}.$$

Suppose that all failures are unrevealed. The system is tested and re-paired to an as-good-as-new condition at intervals of T_0. Neglecting the times required for test and repair, and assuming perfect maintenance:

(a) Derive an expression for the asymptotic availability $A^*(\infty)$.

(b) Find an approximation for $A^*(\infty)$ when $T_0 \ll \theta$.

(c) Evaluate $A^*(\infty)$ for $T_0/\theta = 0.1, 0.5, 1.0$, and 2.0.

CHAPTER 11

Failure Interactions

"If anything can go wrong, it will."

Murphy

11.1 INTRODUCTION

In reliability analysis perhaps the most pervasive technique is that of estimating the reliability of a system in terms of the reliability of its components. In such analysis it is frequently assumed that the component failure and repair properties are mutually independent. In reality, this is often not the case. Therefore, it is necessary to replace the simple products of probabilities with more sophisticated models that take into account the interactions of component failures and repairs.

Many component failure interactions—as well as systems with independent failures—may be modeled effectively as Markov processes, provided that the failure and repair rates can be approximated as time-independent. Indeed, we have already examined a particular example of a Markov process; the derivation of the Poisson process contained in Chapter 6. In this chapter we first formulate the modeling of failures as Markov processes and then apply them to simple systems in which the failures are independent. This allows us both to verify that the same results are obtained as in Chapter 9 and to familiarize ourselves with Markov processes. We then use Markov methods to examine failure interactions of two particular types, shared-load systems and standby systems, and follow with demonstrations of how to incorporate such failure dependencies into the analysis of larger systems. Finally, the analysis is generalized to take into account operational dependencies such as those created by shared repair crews.

11.2 MARKOV ANALYSIS

We begin with the Markov formulation by designating all the possible states of a system. A state is defined to be a particular combination of operating

TABLE 11.1 Markov States of Three-Component Systems

	State #							
Component	1	2	3	4	5	6	7	8
a	*O*	*X*	*O*	*O*	*X*	*X*	*O*	*X*
b	*O*	*O*	*X*	*O*	*X*	*O*	*X*	*X*
c	*O*	*O*	*O*	*X*	*O*	*X*	*X*	*X*

Note: *O* = operating; *X* = failed.

and failed components. Thus, for example, if we have a system consisting of three components, we may easily show that there are eight different combinations of operating and failed components and therefore eight states. These are enumerated in Table 11.1, where *O* indicates an operational component and *X* a failed component. In general, a system with *N* components will have 2^N states so that the number of states increases much faster than the number of components.

For the analysis that follows we must know which of the states correspond to system failure. This, in turn, depends on the configuration in which the components are used. For example, three components might be arranged in any of the three configurations shown in Fig. 11.1. If all the components are in series, as in Fig. 11.1*a*, any combination of one or more component failures will cause system failure. Thus states 2 through 8 in Table 11.1 are failed system states. Conversely, if the three components are in parallel as in Fig. 11.1*b*, all three components must fail for the system to fail. Thus only state 8 is a system failure state. Finally, for the configuration shown in Fig. 11.1*c* both components 1 and 2 or component 3 must fail for the system to fail. Thus states 4 through 8 correspond to system failure.

The object of Markov analysis is to calculate $P_i(t)$, the probability that the system is in state *i* at time *t*. Once this is known, the system reliability can be calculated as a function of time from

$$R(t) = \sum_{i \in O} P_i(t), \qquad (11.1)$$

where the sum is taken over all the operating states (i.e., over those states for which the system is not failed). Alternately, the reliability may be calculated

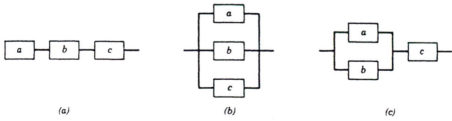

(a)　　　　　　　(b)　　　　　　　(c)

FIGURE 11.1 Reliability block diagrams for three-component systems.

from

$$R(t) = 1 - \sum_{i \in X} P_i(t), \tag{11.2}$$

where the sum is over the states for which the system is failed.

In what follows, we designate state 1 as the state for which all the components are operating, and we assume that at $t = 0$ the system is in state 1. Therefore,

$$P_1(0) = 1, \tag{11.3}$$

and

$$P_i(0) = 0, \quad i \neq 1. \tag{11.4}$$

Since at any time the system can only be in one state, we have

$$\sum_i P_i(t) = 1, \tag{11.5}$$

where the sum is over all possible states.

To determine the $P_i(t)$, we derive a set of differential equations, one for each state of the system. These are sometimes referred to as state transition equations because they allow the $P_i(t)$ to be determined in terms of the rates at which transitions are made from one state to another. The transition rates consist of superpositions of component failure rates, repair rates, or both. We illustrate these concepts first with a very simple system, one consisting of only two independent components, a and b.

Two Independent Components

A two-component system has only four possible states, those enumerated in Table 11.2. The logic of the changes of states is best illustrated by a state transition diagram shown in Fig. 11.2. The failure rates λ_a and λ_b for components a and b indicate the rates at which the transitions are made between states. Since $\lambda_a \Delta t$ is the probability that a component will fail between times t and $t + \Delta t$, given that it is operating at t (and similarly for λ_b), we may write the net change in the probability that the system will be in state 1 as

$$P_1(t + \Delta t) - P_1(t) = -\lambda_a \Delta t P_1(t) - \lambda_b \Delta t P_1(t), \tag{11.6}$$

TABLE 11.2 Markov States of Three-Component Systems

	State #			
Component	1	2	3	4
a	O	X	O	X
b	O	O	X	X

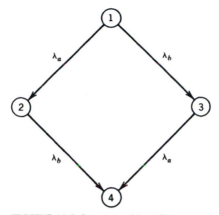

FIGURE 11.2 State transition diagram
with independent failures.

or in differential form

$$\frac{d}{dt}P_1(t) = -\lambda_a P_1(t) - \lambda_b P_1(t).$$ (11.7)

To derive equations for state 2, we first observe that for every transition out of state 1 by failure of component a, there must be an arrival in state 2. Thus the number of arrivals during Δt is $\lambda_a \Delta t\, P_1(t)$. Transitions can also be made out of state 2 during Δt; these will be due to failures of component b, and they will make a contribution of $-\lambda_b \Delta t\, P_2(t)$. Thus the net increase in the probability that the system will be in state 2 is given by

$$P_2(t + \Delta t) - P_2(t) = \lambda_a \Delta t\, P_1(t) - \lambda_b \Delta t\, P_2(t),$$ (11.8)

or dividing by Δt and taking the derivative, we have

$$\frac{d}{dt}P_2(t) = \lambda_a P_1(t) - \lambda_b P_2(t).$$ (11.9)

Identical arguments can be used to derive the equation for $P_3(t)$. The result is

$$\frac{d}{dt}P_3(t) = \lambda_b P_1(t) - \lambda_a P_3(t).$$ (11.10)

We may derive one more differential equation, which is for state 4. We note from the diagram that the transitions into state 4 may come either as a failure of component b from state 2 or as a failure of component a from state 3; the transitions during Δt are $\lambda_b \Delta t\, P_2(t)$ and $\lambda_a \Delta t\, P_3(t)$, respectively. Consequently, we have

$$P_4(t + \Delta t) - P_4(t) = \lambda_b \Delta t\, P_2(t) + \lambda_a \Delta t\, P_3(t)$$ (11.11)

or, correspondingly,

$$\frac{d}{dt}P_4(t) = \lambda_b P_2(t) + \lambda_a P_3(t).$$ (11.12)

State 4 is called an absorbing state, since there is no way to get out of it. The other states are referred to as nonabsorbing states.

From the foregoing derivation we see that we must solve four coupled ordinary differential equations in time in order to determine the $P_i(t)$. We begin with Eq. 11.7 for $P_1(t)$, since it does not depend on the other $P_i(t)$. By substitution, it is clear that the solution to Eq. 11.7 that meets the initial condition, Eq. 11.3, is

$$P_1(t) = e^{-(\lambda_a + \lambda_b)t}. \tag{11.13}$$

To find $P_2(t)$, we first insert Eq. 11.13 into Eq. 11.9,

$$\frac{d}{dt} P_2(t) = \lambda_a e^{-(\lambda_a + \lambda_b)t} - \lambda_b P_2(t), \tag{11.14}$$

yielding an equation in which only $P_2(t)$ appears. Moving the last term to the left-hand side, and multiplying by an integrating factor $e^{\lambda_b t}$, we obtain

$$\frac{d}{dt} [e^{\lambda_b t} P_2(t)] = \lambda_a e^{-\lambda_a t}. \tag{11.15}$$

Multiplying by dt, and integrating the resulting equation from time equals zero to t, we have

$$[e^{\lambda_b t} P_2(t)]_0^t = \lambda_a \int_0^t e^{-\lambda_a t'} \, dt'. \tag{11.16}$$

Carrying out the integral on the right-hand side, utilizing Eq. 11.4 on the left-hand side, and solving for $P_2(t)$, we obtain

$$P_2(t) = e^{-\lambda_b t} - e^{-(\lambda_a + \lambda_b)t}. \tag{11.17}$$

Completely analogous arguments can be applied to the solution of Eq. 11.10. The result is

$$P_3(t) = e^{-\lambda_a t} - e^{-(\lambda_a + \lambda_b)t}. \tag{11.18}$$

We may now solve Eq. 11.11 for $P_4(t)$. However, it is more expedient to note that it follows from Eq. 11.5 that

$$P_4(t) = 1 - \sum_{i=1}^{3} P_i(t), \tag{11.19}$$

Therefore, inserting Eqs. 11.13, 11.17, and 11.18 into this expression yields the desired solution

$$P_4(t) = 1 - e^{-\lambda_a t} - e^{-\lambda_b t} + e^{-(\lambda_a + \lambda_b)t}. \tag{11.20}$$

With the $P_i(t)$ known, we may now calculate the reliability. This, of course, depends on the configuration of the two components, and there are only two possibilities, series and parallel. In the series configuration any failure causes system failure. Hence

$$R_s(t) = P_1(t) \tag{11.21}$$

or

$$R_s(t) = e^{-(\lambda_a + \lambda_b)t}. \tag{11.22}$$

Since, for the active parallel configuration both components a and b must fail to have system failure,

$$R_p(t) = P_1(t) + P_2(t) + P_3(t), \tag{11.23}$$

or, using Eq. 11.19, we have

$$R_p(t) = 1 - P_4(t). \tag{11.24}$$

Therefore,

$$R_p(t) = e^{-\lambda_a t} + e^{-\lambda_b t} - e^{-(\lambda_a + \lambda_b)t}. \tag{11.25}$$

This analysis assumes that the failure rate of each component is independent of the state of the other component. As can be seen from Fig. 11.2, the transitions $1 \to 2$ and $3 \to 4$, which involve the failure of component a, have the same failure rate, even though one takes place with component b in operating order and the other with failed component b. The same argument applies in comparing the transitions $1 \to 3$ and $2 \to 4$. Since the failure rates—and therefore the failure probabilities—are independent of the system state, they are mutually independent. Therefore, the expressions derived in Chapter 9 should still be valid. That this is the case may be seen from the following. For constant failure rates the component reliabilities derived in Chapter 9 are

$$R_l(t) = e^{-\lambda_l t}, \qquad l = a, b. \tag{11.26}$$

Thus the series expression, Eq. 11.22, reduces to

$$R_s(t) = R_a(t) R_b(t), \tag{11.27}$$

and the parallel expression, Eq. 11.25, is

$$R_p(t) = R_a(t) + R_b(t) - R_a(t) R_b(t). \tag{11.28}$$

These are just the expressions derived earlier for independent components, without the use of Markov methods.

Load-Sharing Systems

The primary value of Markov methods appears in situations in which component failure rates can no longer be assumed to be independent of the system state. One of the common cases of dependence is in load-sharing components, whether they be structural members, electric generators, or mechanical pumps or valves. Suppose, for example, that two electric generators share an electric load that either generator has enough capacity to meet. It is nevertheless true that if one generator fails, the additional load on the second generator is likely to increase its failure rate.

To model load-sharing failures, consider once again two components, *a* and *b*, in parallel. We again have a four-state system, but now the transition diagram appears as in Fig. 11.3. Here λ_a^* and λ_b^* denote the increased failure rates brought about by the higher loading after one failure has taken place.

The Markov equations can be derived as for independent failures if the changes in failure rates are included. Comparing Fig. 11.2 with 11.3, we see that the resulting generalizations of Eqs. 11.7, 11.9, 11.10, and 11.12 are

$$\frac{d}{dt}P_1(t) = -(\lambda_a + \lambda_b)P_1(t), \tag{11.29}$$

$$\frac{d}{dt}P_2(t) = \lambda_a P_1(t) - \lambda_b^* P_2(t), \tag{11.30}$$

$$\frac{d}{dt}P_3(t) = \lambda_b P_1(t) - \lambda_a^* P_3(t) \tag{11.31}$$

and

$$\frac{d}{dt}P_4(t) = \lambda_b^* P_2(t) + \lambda_a^* P_3(t). \tag{11.32}$$

The solution procedure is also completely analogous. The results are

$$P_1(t) = e^{-(\lambda_a + \lambda_b)t}, \tag{11.33}$$

$$P_2(t) = e^{-\lambda_b^* t} - e^{-(\lambda_a + \lambda_b^*)t}, \tag{11.34}$$

$$P_3(t) = e^{-\lambda_a^* t} - e^{-(\lambda_a^* + \lambda_b)t} \tag{11.35}$$

and

$$P_4(t) = 1 - e^{-\lambda_a^* t} - e^{-\lambda_b^* t} - e^{-(\lambda_a + \lambda_b^*)t} + e^{-(\lambda_a + \lambda_b^*)t} + e^{-(\lambda_a^* + \lambda_b)t}. \tag{11.36}$$

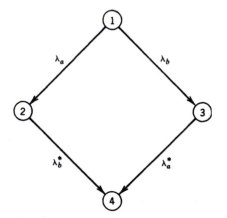

FIGURE 11.3 State transition diagram with load sharing.

Finally, since both components must fail for the system to fail, the reliability is equal to $1 - P_4(t)$, yielding

$$R_p(t) = e^{-\lambda_a^* t} + e^{-\lambda_b^* t} + e^{-(\lambda_a + \lambda_b)t} - e^{-(\lambda_a + \lambda_b^*)t} - e^{-(\lambda_a^* + \lambda_b)t}. \tag{11.37}$$

It is easily seen that if $\lambda_a^* = \lambda_a$ and $\lambda_b^* = \lambda_b$, there is no dependence between failure rates, and Eq. 11.37 reduces to Eq. 11.25. The effects of increased loading on a load-sharing redundant system can be seen graphically by considering the situation in which the two components are identical: $\lambda_a = \lambda_b = \lambda$ and $\lambda_a^* = \lambda_b^* = \lambda^*$. Equation 11.37 then reduces to

$$R(t) = 2e^{-\lambda^* t} + e^{-2\lambda t} - 2e^{-(\lambda + \lambda^*)t}. \tag{11.38}$$

In Fig. 11.4 we have plotted $R(t)$ for the two-component parallel system, while varying the increase in failure rate caused by increased loading (i.e., the ratio λ^*/λ). The two extremes are the system in which the two components are independent, $\lambda^* = \lambda$, and the totally dependent system in which the failure of one component brings on the immediate failure of the other, $\lambda^* = \infty$. Notice that these two extremes correspond to Eqs. 11.25 and 11.22, for independent failures of parallel and series configurations, respectively.

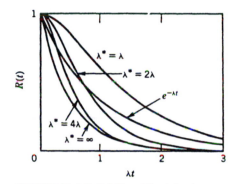

FIGURE 11.4 Reliability of load-sharing systems.

EXAMPLE 11.1

Two diesel generators of known MTTF are hooked in parallel. Because the failure of one of the generators will cause a large additional load on the other, the design engineer estimates that the failure rate will double for the remaining generator. For how many MTTF can the generator system be run without the reliability dropping below 0.95?

Solution Take $\lambda^* = 2\lambda$. Then Eq. 11.38 is

$$R = 0.95 = 2e^{-2\lambda t} + e^{-2\lambda t} - 2e^{-3\lambda t},$$

where t is the time at which the reliability drops below 0.95. Let $x = e^{-\lambda t}$. Then

$$2x^3 - 3x^2 + 0.95 = 0.$$

The solution must lie in the interval $0 < x < 1$. By plotting the left-hand side of the equation, we may show that the equation is satisfied at only one place, at

$$x = 0.8647.$$

Therefore, $\lambda t = \ln(1/x) = 0.1454$. Since $\lambda = 1/\text{MTTF}$ for the diesel generators, the maximum time of operation is $t = 0.1454/\lambda = 0.1454$ MTTF. Note that if only a single generator had been used, it could have operated for only $t = \ln(1/R)/\lambda = 0.0513$ MTTF without violating the criterion.

11.3   RELIABILITY WITH STANDBY SYSTEMS

Standby or backup systems are a widely applied type of redundancy in fault tolerant systems, whether they be in the form of extra logic chips, navigation components, or emergency power generators. They differ, however, from active parallel systems in that one of the units is held in reserve and only brought into operation in the event that the first unit fails. For this reason they are often referred to as passive parallel systems. By their nature standby systems involve dependency between components; they are nicely analyzed by Markov methods.

Idealized System

We first consider an idealized standby system consisting of a primary unit a and a backup unit b. If the states are numbered according to Table 11.2, the system operation is described by the transition diagram, Fig. 11.5. When the primary unit fails, there is a transition $1 \rightarrow 2$, and then when the backup unit fails, there is a transition $2 \rightarrow 4$, with state 4 corresponding to system failure. Note that there is no possibility of the system's being in state 3, since we have

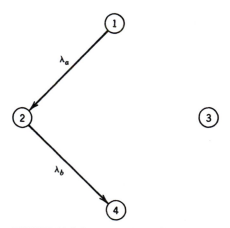

FIGURE 11.5 State transition diagram for a standby configuration.

assumed that the backup unit does not fail while in the standby state. Hence $P_3(t) = 0$. Later we consider the possibility of failure in this standby state as well as the possibility of failures during the switching from primary to backup unit.

From the transition diagram we may construct the Markov equations for the three states quite easily. For state 1 there is only a loss term from the transition $1 \rightarrow 2$. Thus

$$\frac{d}{dt} P_1(t) = -\lambda_a P_1(t). \tag{11.39}$$

For state 2 we have one source term, from the $1 \rightarrow 2$ transition, and one loss term from the $2 \rightarrow 4$ transition. Thus

$$\frac{d}{dt} P_2(t) = \lambda_a P_1(t) - \lambda_b P_2(t). \tag{11.40}$$

Since state 4 results only from the transition $2 \rightarrow 4$, we have

$$\frac{d}{dt} P_4(t) = \lambda_b P_2(t). \tag{11.41}$$

The foregoing equations may be solved sequentially in the same manner as those of the preceding sections. We obtain

$$P_1(t) = e^{-\lambda_a t}, \tag{11.42}$$

$$P_2(t) = \frac{\lambda_a}{\lambda_b - \lambda_a} (e^{-\lambda_a t} - e^{-\lambda_b t}), \tag{11.43}$$

$$P_3(t) = 0 \tag{11.44}$$

and

$$P_4(t) = 1 - \frac{1}{\lambda_b - \lambda_a} (\lambda_b e^{-\lambda_a t} - \lambda_a e^{-\lambda_b t}), \tag{11.45}$$

where we have again used the initial conditions, Eqs. 11.3 and 11.4. Since state 4 is the only state corresponding to system failure, the reliability is just

$$R(t) = P_1(t) + P_2(t), \tag{11.46}$$

or

$$R(t) = e^{-\lambda_a t} + \frac{\lambda_a}{\lambda_b - \lambda_a} (e^{-\lambda_a t} - e^{-\lambda_b t}). \tag{11.47}$$

This, in turn, may be simplified to

$$R(t) = \frac{1}{\lambda_b - \lambda_a} (\lambda_b e^{-\lambda_a t} - \lambda_a e^{-\lambda_b t}). \tag{11.48}$$

The properties of standby systems are nicely illustrated by comparing their reliability versus time with that of an active parallel system. For brevity we consider the situation $\lambda_a = \lambda_b = \lambda$. In this situation we must be careful in evaluating the reliability, for both Eqs. 11.47 and 11.48 contain $\lambda_b - \lambda_a$ in the denominator. We begin with Eq. 11.47 and rewrite the last term as

$$R(t) = e^{-\lambda_a t} + \frac{\lambda_a}{(\lambda_b - \lambda_a)} e^{-\lambda_a t}[1 - e^{-(\lambda_b - \lambda_a)t}]. \tag{11.49}$$

Then, going to the limit as λ_b approaches λ_a, we have $(\lambda_b - \lambda_a)t \ll 1$, and we can expand

$$e^{-(\lambda_b - \lambda_a)t} = 1 - (\lambda_b - \lambda_a)t + \tfrac{1}{2}(\lambda_b - \lambda_a)^2 t^2 - \cdots \tag{11.50}$$

Combining Eqs. 11.49 and 11.50, we have

$$R(t) = e^{-\lambda_a t} + \lambda_a e^{-\lambda_a t}[t - \tfrac{1}{2}(\lambda_a - \lambda_b)t^2 + \cdots]. \tag{11.51}$$

Thus as λ_b and λ_a become equal, only the first two terms remain, and we have for $\lambda_b = \lambda_a = \lambda$:

$$R(t) = (1 + \lambda t)e^{-\lambda t}. \tag{11.52}$$

In Fig. 11.6 are compared the reliabilities of active and standby parallel systems whose two components have identical failure rates. Note that the standby parallel system is more reliable than the active parallel system because the backup unit cannot fail before the primary unit, even though the reliability of the primary unit is not affected by the presence of the backup unit.

The gain in reliability is further indicated by the increase in the system MTTF for the standby configuration, relative to that for the active configuration. Substituting Eq. 11.52 into Eq. 6.22, we have for the standby parallel system

$$\text{MTTF} = 2/\lambda \tag{11.53}$$

compared to a value of

$$\text{MTTF} = 3/2\lambda \tag{11.54}$$

for the active parallel system.

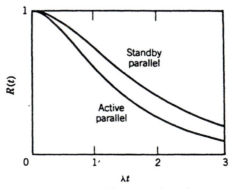

FIGURE 11.6 Reliability comparison for standby and active parallel systems.

Failures in the Standby State

We next model the possibility that the backup unit fails before it is required. We generalize the state transition diagram as shown in Fig. 11.7. The failure rate λ_b^+ represents failure of the backup unit while it is inactive; state 3 represents the situation in which the primary unit is operating, but there is an undetected failure in the backup unit.

There are now two paths for transition out of state 1. Thus for $P_1(t)$ we have

$$\frac{d}{dt}P_1(t) = -\lambda_a P_1(t) - \lambda_b^+ P_1(t). \tag{11.55}$$

The equation for state 2 is unaffected by the additional failure path; as in Eq. 11.40, we have

$$\frac{d}{dt}P_2(t) = \lambda_a P_1(t) - \lambda_b P_2(t). \tag{11.56}$$

We must now set up an equation to determine $P_3(t)$. This state is entered through the $1 \rightarrow 3$ transition with rate λ_b^+ and is exited through the $3 \rightarrow 4$ transition with rate λ_a. Thus

$$\frac{d}{dt}P_3(t) = \lambda_b^+ P_1(t) - \lambda_a P_3(t). \tag{11.57}$$

Finally, state 4 is entered from either states 2 or 3;

$$\frac{d}{dt}P_4(t) = \lambda_b P_2(t) + \lambda_a P_3(t). \tag{11.58}$$

The Markov equations may be solved in the same manner as before. We obtain, with the initial conditions Eqs. 11.3 and 11.4,

$$P_1(t) = e^{-(\lambda_a + \lambda_b^+)t}, \tag{11.59}$$

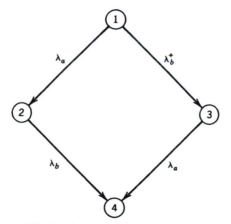

FIGURE 11.7 State transition diagram with failure in the backup mode.

$$P_2(t) = \frac{\lambda_a}{\lambda^a + \lambda_b^+ - \lambda_b} [e^{-\lambda_b t} - e^{-(\lambda_a + \lambda_b^+)t}] \tag{11.60}$$

and

$$P_3(t) = e^{-\lambda_a t} - e^{-(\lambda_a + \lambda_b^+)t}. \tag{11.61}$$

There is no need to solve for $P_4(t)$, since once again it is the only state for which there is system failure, and therefore,

$$R(t) = P_1(t) + P_2(t) + P_3(t), \tag{11.62}$$

yielding

$$R(t) = e^{-\lambda_a t} + \frac{\lambda_a}{\lambda_a + \lambda_b^+ - \lambda_b} [e^{-\lambda_b t} - e^{-(\lambda_a + \lambda_b^+)t}]. \tag{11.63}$$

Once again it is instructive to examine the case $\lambda_a = \lambda_b = \lambda$ and $\lambda_b^+ = \lambda^+$, in which Eq. 11.63 reduces to

$$R(t) = \left(1 + \frac{\lambda}{\lambda^+}\right) e^{-\lambda t} - \frac{\lambda}{\lambda^+} e^{-(\lambda + \lambda^+)t}. \tag{11.64}$$

In Fig. 11.8 the results are shown, having values of λ^+ ranging from zero to λ. The deterioration of the reliability is seen with increasing λ^+. The system MTTF may be found easily by inserting Eq. 11.64 into Eq. 6.22. We have

$$\text{MTTF} = \frac{1}{\lambda} + \frac{1}{\lambda^+} - \frac{\lambda}{\lambda^+} \frac{1}{\lambda + \lambda^+}. \tag{11.65}$$

When $\lambda^+ = \lambda$, the foregoing results reduce to those of an active parallel system. This is sometimes referred to as a "hot-standby system," since both units are then running and only a switch from one to the other is necessary. Fault-tolerant control systems, which can use only the output of one device at a time but which cannot tolerate the time required to start up the backup

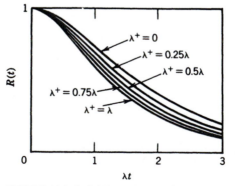

FIGURE 11.8 Reliability of a standby system with different rates of failure in the backup mode.

unit, operate in this manner. Unlike active parallel systems, however, they must switch from primary unit to backup unit. We consider switching failures next.

EXAMPLE 11.2

A fuel pump with an MTTF of 3000 hr is to operate continuously on a 500-hr mission.

(*a*) What is the mission reliability?

(*b*) Two such pumps are put in a standby parallel configuration. If there are no failures of the backup pump while in the standby mode, what is the system MTTF and the mission reliability?

(*c*) If the standby failure rate is 15% of the operational failure rate, what is the system MTTF and the mission reliability?

Solution

(*a*) The component failure rate is $\lambda = 1/3000 = 0.333 \times 10^{-3}$/hr. Therefore, the mission reliability is

$$R(T) = \exp\left(-\frac{1}{3000} \times 500\right) = 0.846.$$

(*b*) In the absence of standby failures, the system MTTF is found from Eq. 11.53 to be

$$\text{MTTF} = \frac{2}{\lambda} = 2 \times 3000 = 6000 \text{ hr.}$$

The system reliability is found from Eq. 11.52 to be

$$R(500) = \left(1 + \frac{1}{3000} \times 500\right) \times \exp\left(-\frac{1}{3000} \times 500\right) = 0.988.$$

(*c*) We find the system MTTF from Eq. 11.65 with $\lambda^+ = 0.15/3000 = 0.5 \times 10^{-4}$/hr:

$$\text{MTTTF} = \frac{1}{0.333 \times 10^{-3}} + \frac{1}{0.5 \times 10^{-4}}$$

$$-\frac{0.333 \times 10^{-3}}{0.5 \times 10^{-4}} \frac{1}{0.333 \times 10^{-3} + 0.5 \times 10^{-4}}$$

$$\text{MTTT} = 5609 \text{ hr.}$$

From Eq. 11.64 the system reliability for the mission is $R(500) = 0.986$.

Switching Failures

A second difficulty in using standby systems stems from the switch from the primary unit to the backup. This switch may take action by electric relays, hydraulic valves, electronic control circuits, or other devices. There is always the possibility that the switching device will have a demand failure probability p large enough that switching failures must be considered. For brevity we do not consider backup unit failure while it is in the standby mode.

The state transition diagram with these assumptions is shown in Fig. 11.9. Note that the transition out of state 1 in Fig. 11.5 has been divided into two paths. The primary failure rate is multiplied by $1 - p$ to get the successful transition into state 2, in which the backup system is operating. The second path with rate $p\lambda_a$ indicates a transition directly to the failed-system state that results when there is a demand failure on the switching mechanism.

For the situation depicted in Fig. 11.9, state 1 is still described by Eq. 11.39. Now, however, the $1 \rightarrow 2$ transition is decreased by a factor $1 - p$ and so, instead of Eq. 11.40, state 2 is described by

$$\frac{d}{dt}P_2(t) = (1 - p)\lambda_a P_1(t) - \lambda_b P_2(t) \tag{11.66}$$

and state 4 is described by

$$\frac{d}{dt}P_4(t) = \lambda_b P_2(t) + p\lambda_a P_1(t). \tag{11.67}$$

Since $P_1(t)$ is again given by Eq. 11.42, we need solve only Eq. 11.66 to obtain

$$P_2(t) = (1 - p)\frac{\lambda_a}{\lambda_b - \lambda_a}(e^{-\lambda_a t} - e^{-\lambda_b t}). \tag{11.68}$$

Accordingly, since state 4 is the only failed state and $P_3(t) = 0$, we may write

$$R(t) = P_1(t) + P_2(t), \tag{11.69}$$

or inserting Eqs. 11.42 and 11.68, we obtain for the reliability

$$R(t) = e^{-\lambda_a t} + \frac{(1 - p)\lambda_a}{\lambda_b - \lambda_a}(e^{-\lambda_a t} - e^{-\lambda_b t}). \tag{11.70}$$

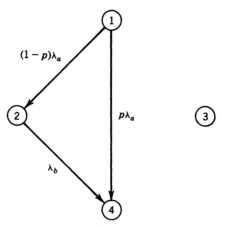

FIGURE 11.9 State transition diagram with standby switching failures.

Once again it is instructive to consider the case $\lambda_a = \lambda_b = \lambda$, for which we obtain

$$R(t) = [1 + (1 - p)\lambda t]e^{-\lambda t}. \tag{11.71}$$

Clearly, as p increases, the value of the backup system becomes less and less, until finally if p is one (i.e., certain failure of the switching system), the backup system has no effect on the system reliability.

EXAMPLE 11.3

An annunciator system has a mission reliability of 0.9. Because reliability is considered too low, a redundant annunciator of the same design is to be installed. The design engineer must decide between an active parallel and a standby parallel configuration. The engineer knows that failures in standby have a negligible effect, but there is a significant probability of a switching failure.

(a) How small must the probability of a switching failure be if the standby configuration is to be more reliable than the active configuration?

(b) Discuss the switching failure requirement of a for very short mission times.

Solution

(a) Assuming a constant failure rate, we know that for the mission time T,

$$\lambda T = \ln\left[\frac{1}{R(T)}\right] = \ln\left(\frac{1}{0.9}\right) = 0.1054.$$

To find the failure probability, we equate Eq. 11.71 with Eq. 9.11 for the active parallel system:

$$[1 + (1 - p)\lambda T]e^{-\lambda T} = 2e^{-\lambda T} - e^{-2\lambda T}.$$

Thus

$$p = 1 - \frac{1}{\lambda T}(1 - e^{-\lambda T})$$

$$= 1 - \frac{1}{0.1054}(1 - e^{-0.1054}) = 0.05.$$

(b) For active parallel Eq. 9.19 gives the short mission time approximation:

$$R_a \approx 1 - (\lambda t)^2.$$

For standby parallel we expand 11.71 for small λt:

$$R_{sb} = [1 + (1 - p)\lambda t]e^{-\lambda t} = [1 + (1 - p)\lambda t][1 - \lambda t + \tfrac{1}{2}(\lambda t)^2 \cdots]$$
$$\approx 1 - p\lambda t - (\tfrac{1}{2} - p)(\lambda t)^2.$$

Then we calculate p for $R_a - R_{sb} = 0$:

$$1 - (\lambda t)^2 - 1 + p\lambda t + (\tfrac{1}{2} - p)(\lambda t)^2 = 0$$

or

$$p = \frac{\tfrac{1}{2}\lambda t}{1 - \lambda t} \approx \frac{1}{2}\lambda t.$$

The shorter the mission, the smaller p must be, or else switching failures will be more probable than the failures of the second annunciator in the active parallel configuration.

The combined effects of failures in the standby mode and switching failures may be included in the foregoing analysis. For two identical units the reliability may be shown to be

$$R(t) = \left[1 + (1 - p) \frac{\lambda}{\lambda^+} \right] e^{-\lambda t} - (1 - p) \frac{\lambda}{\lambda^+} e^{-(\lambda + \lambda^+)t}, \qquad (11.72)$$

which reduces to Eq. 11.71 as $\lambda^+ \to 0$. For a hot-standby system in which identical primary and backup systems are both running so that $\lambda^+ = \lambda$, we obtain from Eq. 11.72

$$R(t) = (2 - p) e^{-\lambda t} - (1 - p) e^{-2\lambda t}. \qquad (11.73)$$

Thus the reliability is less than that of an active parallel system because there is a probability of switching failure. As stated earlier, in hot-standby systems, such as for control devices, the output of only one unit can be used at a time. If the probability of switching failure is too great, an alternative is to add a third unit and use a 2/3 voting system, as discussed in Chapter 9.

Primary System Repair

Two considerable benefits are to be gained by using redundant system components. The first is that more than one failure must occur in order for the system to fail. A second is that components can be repaired while the system is on line. Much higher reliabilities are possible if the failed component has a high probability of being repaired before a second one fails.

Component repair increases the reliability of either active parallel or standby parallel systems. Moreover, either system may be analyzed using Markov methods. In what follows we derive the reliability for a system consisting of a primary and a backup unit. We assume that the primary unit can be repaired on line. For clarity, we assume that failure of the backup unit in standby mode and switching failures can be neglected.

The state transition diagram shown in Fig. 11.10 differs from Fig. 11.5 only in that the repair transition has been added. This creates an additional source term of $\nu P_2(t)$ in Eq. 11.39,

$$\frac{d}{dt} P_1(t) = -\lambda_a P_1(t) + \nu P_2(t), \qquad (11.74)$$

and the corresponding loss term is substracted from Eq. 11.40,

$$\frac{d}{dt} P_2(t) = \lambda_a P_1(t) - (\lambda_b + \nu) P_2(t). \qquad (11.75)$$

The reliability, once again, is calculated from Eq. 11.46.

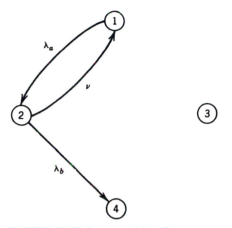

FIGURE 11.10 State transition diagram with primary system repair.

The equations can no longer be solved one at a time, sequentially, as in the previous examples, for now $P_1(t)$ depends on $P_2(t)$. Laplace transforms may be used to solve Eqs. 11.74 and 11.75, but to avoid introducing additional nomenclature we use the following technique instead. Suppose that we look for solutions of the form

$$P_1(t) = Ce^{-\alpha t}; \qquad P_2(t) = C'e^{-\alpha t}, \tag{11.76}$$

where C, C', and α are constants. Substituting these expressions into Eqs. 11.74 and 11.75, we obtain

$$-\alpha C = -\lambda_a C + \nu C'; \qquad -\alpha C' = \lambda_a C - (\lambda_b + \nu)C'. \tag{11.77}$$

The constants C and C' may be eliminated between these expressions to yield the form

$$\alpha^2 - (\lambda_a + \lambda_b + \nu)\alpha + \lambda_a\lambda_b = 0 \tag{11.78}$$

Solving this quadratic equation, we find that there are two solutions for α:

$$\alpha_{\pm} = \frac{(\nu + \lambda_a + \lambda_b)}{2} \pm \tfrac{1}{2}[(\nu + \lambda_a + \lambda_b)^2 - 4\lambda_a\lambda_b]^{1/2}. \tag{11.79}$$

Thus our solutions have the form

$$P_1(t) = C_+ e^{-\alpha_+ t} + C_- e^{-\alpha_- t}, \tag{11.80}$$

$$P_2(t) = C'_+ e^{-\alpha_+ t} + C'_- e^{-\alpha_- t}. \tag{11.81}$$

We must use the initial conditions along with Eq. 11.79 to evaluate $C_{\pm}$ and $C'_{\pm}$. Combining Eqs. 11.80 and 11.81 with the initial conditions $P_1(0) = 1$ and $P_2(0) = 0$, we have

$$C_+ + C_- = 1; \qquad C'_+ + C'_- = 0. \tag{11.82}$$

Furthermore, adding Eqs. 11.77, we may write, for α_+ and α_-,

$$\alpha_{\pm} C_{\pm} = (\lambda_b - \alpha_{\pm}) C'_{\pm}. \tag{11.83}$$

These four equations can be solved for $C_{\pm}$ and $C'_{\pm}$. Then, after some algebra, we may add Eqs. 11.80 and 11.81 to obtain from Eq. 11.46

$$R(t) = \frac{\alpha_+}{\alpha_+ - \alpha_-} e^{-\alpha_- t} - \frac{\alpha_-}{\alpha_+ - \alpha_-} e^{-\alpha_+ t}. \tag{11.84}$$

The improvement in reliability with standby systems is indicated in Fig. 11.11, where the two units are assumed to be identical, $\lambda_a = \lambda_b = \lambda$, and plots are shown for different ratios of ν/λ. In the usual case, where $\nu \gg \lambda$, it is easily shown that $\alpha_+ \gg \alpha_-$, so that the second term in Eq. 11.84 can be neglected, and that $\alpha_- \approx -\lambda_a \lambda_b / \nu$. Hence we may write, approximately,

$$R(t) \approx \exp\left(-\frac{\lambda_a \lambda_b}{\nu} t\right). \tag{11.85}$$

In the situation in which $\nu \gg \lambda_a, \lambda_b$, the deterioration of reliability is likely to be governed not by the possibility that the backup system will fail before the primary system is repaired, but rather by one of the two other possibilities: (1) that switching to the backup system will fail, or (b) that the backup system has failed. These failures are dealt with either by improving the switching and standby mode reliabilities or by utilizing an active parallel system with repairable components. Then the switching is obviated, and the configuration is more likely to be designed so that failures in either component are revealed immediately.

11.4 MULTICOMPONENT SYSTEMS

The models described in the two preceding sections concern the dependencies between only two components. In order to make use of Markov methods in

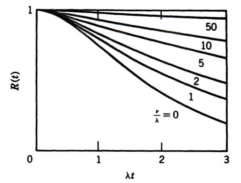

FIGURE 11.11 The effect of primary system repair rate on the reliability of a standby system.

realistic situations, however, it is often necessary to consider dependencies between more than two components or to build the dependency models into many-component systems. In this section we first undertake to generalize Markov methods for the consideration of dependencies between more than two components. We then examine how to build dependency models into larger systems in which some of the component failures are independent of the others.

Multicomponent Markov Formulations

The treatment of larger sets of components by Markov methods is streamlined by expressing the coupled set of state transition equations in matrix form. Moreover, the resulting coefficient matrix can be used to check on the formulation's consistency and to gain some insight into the physical processes at play. To illustrate, we first put one of the two-component, four-state systems discussed earlier into matrix form. The generalization to larger systems is then obvious.

Consider the backup configuration shown in Fig. 11.7, in which we allow for failure of the unit in the standby mode. The four equations for the $P_i(t)$ are given by Eqs. 11.55 through 11.58. If we define a vector $\mathbf{P}(t)$, whose components are $P_1(t)$ through $P_4(t)$, we may write the set of simultaneous differential equations as

$$\frac{d}{dt}\begin{bmatrix} P_1(t) \\ P_2(t) \\ P_3(t) \\ P_4(t) \end{bmatrix} = \begin{bmatrix} -\lambda_a - \lambda_b^+ & 0 & 0 & 0 \\ \lambda_a & -\lambda_b & 0 & 0 \\ \lambda_b^+ & 0 & -\lambda_a & 0 \\ 0 & \lambda_b & \lambda_a & 0 \end{bmatrix}\begin{bmatrix} P_1(t) \\ P_2(t) \\ P_3(t) \\ P_4(t) \end{bmatrix}. \tag{11.86}$$

Consider next a system with three components in parallel, as shown in Fig. 11.1b. Suppose that this is a load-sharing system in which the component failure rate increases with each component failure:

λ_1 = component failure rate with no component failures,

λ_2 = component failure rate with one component failure,

λ_3 = component failure rate with two component failures.

If we again enumerate the possible system states in Table 11.1, the state transition diagram will appear as in Fig. 11.12. From this diagram we may construct the equations for the $P_i(t)$. In matrix form they are

$$\frac{d}{dt}\begin{bmatrix} P_1(t) \\ P_2(t) \\ P_3(t) \\ P_4(t) \\ P_5(t) \\ P_6(t) \\ P_7(t) \\ P_8(t) \end{bmatrix} = \begin{bmatrix} -3\lambda_1 & 0 & 0 & 0 & 0 & 0 & 0 & 0 \\ \lambda_1 & -2\lambda_2 & 0 & 0 & 0 & 0 & 0 & 0 \\ \lambda_1 & 0 & -2\lambda_2 & 0 & 0 & 0 & 0 & 0 \\ \lambda_1 & 0 & 0 & -2\lambda_2 & 0 & 0 & 0 & 0 \\ 0 & \lambda_2 & \lambda_2 & 0 & -\lambda_3 & 0 & 0 & 0 \\ 0 & \lambda_2 & 0 & \lambda_2 & 0 & -\lambda_3 & 0 & 0 \\ 0 & 0 & \lambda_2 & \lambda_2 & 0 & 0 & -\lambda_3 & 0 \\ 0 & 0 & 0 & 0 & \lambda_3 & \lambda_3 & \lambda_3 & 0 \end{bmatrix}\begin{bmatrix} P_1(t) \\ P_2(t) \\ P_3(t) \\ P_4(t) \\ P_5(t) \\ P_6(t) \\ P_7(t) \\ P_8(t) \end{bmatrix},$$

$$\tag{11.87}$$

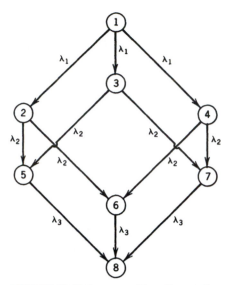

FIGURE 11.12 State transition diagram for a three-component parallel system.

where there are now $2^3 = 8$ states in all. The generalization to more components is straightforward, provided that the logical structure of the dependencies is understood.

Equations 11.86 and 11.87 may be used to illustrate an important property of the coefficient matrix, one which serves as an aid in constructing the set of equations from the state transition diagram. Each transition out of a state must terminate in another state. Thus, for each negative entry in the coefficient matrix, there must be a positive entry in the same column, and the sum of the elements in each column must be zero. Thus the matrix may be constructed systematically by considering the transitions one at a time. If the transition originates from the ith state, the failure rate is subtracted from the ith diagonal element. If the transition is to the jth state, the failure rate is then added to the jth row of the same column.

A second feature of the coefficient matrix involves the distinction between operational and failed states. In reliability calculations we do not allow a system to be repaired once it fails. Hence there can be no way to leave a failed state. In the coefficient matrix this is indicated by the zero in the diagonal element of each failed state. This is not the case, however, when availability rather than reliability is being calculated. Availability calculations are discussed in the following section.

For larger systems of equations it is often more convenient to write Markov equations in the matrix form

$$\frac{d}{dt}\mathsf{P}(t) = \mathsf{MP}(t),\qquad(11.88)$$

where **P** is a column vector with components $P_1(t), P_2(t), \ldots$, and **M** is referred to as the Markov transition matrix. Instead of repeating the entire set of equations, as in Eqs. 11.86 and 11.87, we need write out only the matrix. Thus, for example, the matrix for Eq. 11.86 is

$$\mathbf{M} = \begin{bmatrix} -\lambda_a - \lambda_b^+ & 0 & 0 & 0 \\ \lambda_a & -\lambda_b & 0 & 0 \\ \lambda_b^+ & 0 & -\lambda_a & 0 \\ 0 & \lambda_b & \lambda_a & 0 \end{bmatrix}. \tag{11.89}$$

The dimension of the matrix increases as 2^N, where N is the number of components. For larger systems, particularly those whose components are repaired, the simple solution algorithms discussed earlier become intractable. Instead, more general Laplace transform techniques may be required. If there are added complications, such as time-dependent failure rates, the equations may require solution by numerical integration or by Monte Carlo simulation.

EXAMPLE 11.4

A 2/3 system is constructed as follows. After the failure of either component a or c, whichever comes first, component b is switched on. The system fails after any two of the components fail. The components are identical with failure rate λ.

(*a*) Draw a state transition diagram for the system.

(*b*) Write the corresponding Markov transition matrix.

(*c*) Find the system reliability $R(t)$.

(*d*) Determine the reliability when time is set equal to the MTTF one component.

Solution   For this three-component system, there are eight states. We define these according to Table 11.1.

(*a*) The state transition diagram is shown in Fig. 11.13. Note that states 3 and 8 are not reachable.

(*b*) The Markov transition matrix is

$$\mathbf{M} = \begin{bmatrix} -2\lambda & 0 & 0 & 0 & 0 & 0 & 0 & 0 \\ \lambda & -2\lambda & 0 & 0 & 0 & 0 & 0 & 0 \\ 0 & 0 & 0 & 0 & 0 & 0 & 0 & 0 \\ \lambda & 0 & 0 & -2\lambda & 0 & 0 & 0 & 0 \\ 0 & \lambda & 0 & 0 & 0 & 0 & 0 & 0 \\ 0 & \lambda & 0 & \lambda & 0 & 0 & 0 & 0 \\ 0 & 0 & 0 & \lambda & 0 & 0 & 0 & 0 \\ 0 & 0 & 0 & 0 & 0 & 0 & 0 & 0 \end{bmatrix} \blacktriangleleft$$

(*c*) The reliability is given by $R(t) = P_1(t) + P_2(t) + P_4(t)$; thus only three of the eight equations need be solved. First, $dP_1/dt = -2\lambda P_1$, with $P_1(0) = 1$ yields $P_1(t) = e^{-2\lambda t}$. The equations for $P_2 + P_4$ are the same:

$$\frac{dP_n}{dt} = \lambda P_1 - 2\lambda P_n, \quad P_n(0) = 0; \qquad n = 2, 4.$$

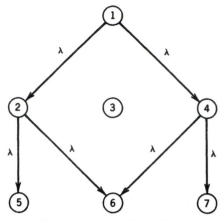

FIGURE 11.13 State transition diagram for Example 11.4.

Therefore,

$$\frac{dP_n}{dt} = \lambda e^{-2\lambda t} - 2\lambda P_n.$$

We use the integrating factor $e^{2\lambda t}$ to obtain

$$\frac{d}{dt}(P_n e^{-2\lambda t}) = \lambda.$$

Then integrating between 0 and t, we obtain

$$P_n(t) e^{2\lambda t} - P_n(0) = \lambda t.$$

Thus

$$P_n(t) = \lambda t e^{-2\lambda t}, \qquad n = 2, 4.$$

Substituting into $R(t) = P_1 + P_2 + P_4$ yields

$$R(t) = (1 + 2\lambda t) e^{-2\lambda t}.$$

(d) $t = \text{MTTF} \equiv 1/\lambda$. Then

$$R(\text{MTTF}) = (1 + 2 \times 1) e^{-2 \times 1} = 0.406.$$

Combinations of Subsystems

In principle, we can treat systems of many components using Markov methods. However, with 2^N equations the solutions soon become unmanageable. A more efficient approach is to define one or more subsystems containing the components with dependencies between them. These subsystems can then

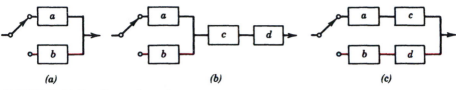

FIGURE 11.14 Standby configurations.

be treated as single blocks in a reliability block diagram, and the system reliability can be calculated using the techniques of Chapter 9, since the failures in the subsystem defined in this way are independent of one another.

To understand this procedure, consider the system configurations shown in Fig. 11.14. In Fig. 11.14a is shown the convention for drawing a two-component standby system of the type discussed in the preceding section as a reliability block diagram. In Fig. 11.14b the standby parallel subsystem, consisting of components a and b, is in series with two other components. The reliability of the standby subsystem (with no switching errors) is given by Eq. 11.63. Therefore, we define the reliability of the standby subsystem as

$$R_{sb}(t) = e^{-\lambda_a t} + \frac{\lambda_a}{\lambda_a + \lambda_b^+ - \lambda_b} [e^{-\lambda_b t} - e^{-(\lambda_a + \lambda_b^+)t}]. \qquad (11.90)$$

Then, if the failures in components c and d are independent of those in the standby subsystem, the system reliability can be calculated using the product rule

$$R(t) = R_{sb}(t) R_c(t) R_d(t). \qquad (11.91)$$

Generalization of this technique to more complex configurations is straightforward.

The configuration in Fig. 11.14c illustrates a somewhat different situation. Here the primary and standby subsystems themselves each consist of two components, a and c, and b and d, respectively. Here we may simplify the Markov analysis by first combining the four components into two subsystems, each having a composite failure rate. Thus we define

$$\lambda_{ac} = \lambda_a + \lambda_c, \qquad (11.92)$$

$$\lambda_{bd} = \lambda_b + \lambda_d, \qquad (11.93)$$

and

$$\lambda_{bd}^+ = \lambda_b^+ + \lambda_d^+. \qquad (11.94)$$

We may again apply Eq. 11.90 to calculate the system reliability if we replace λ_a, λ_b, and λ_b^+ with λ_{ac}, λ_{bd}, and λ_{bd}^+, respectively.

11.5 AVAILABILITY

In availability, as well as in reliability, there are situations in which the component failures cannot be considered independent of one another. These include shared-load and backup systems in which all the components are repairable. They may also include a variety of other situations in which the dependency is introduced by the limited number of repair personnel or by replacement parts that may be called on to put components into working order. Thus, for example, the repair of two redundant components cannot be considered independent if only one crew is on station to carry out the repairs.

The dependencies between component failure and repair rates may be approached once more with Markov methods, provided that the failures are revealed, and that the failure and repair rates are time-independent. Although we have already treated the repair of components in reliability calculations, there is a fundamental difference in the analysis that follows. In reliability calculations components can be repaired only as long as the system has not failed; the analysis terminates with the first system failure. In availability calculations we continue to repair components after a system failure in order to bring the system back on line, that is, to make it available once again.

The differences between Markov reliability and availability calculations for systems with repairable components can be illustrated best in terms of the matrix notion developed in the preceding section. For this reason we first illustrate an availability calculation with a system for which the reliability was calculated in the preceding section, standby redundance. We then illustrate the limitation placed on the availability of an active parallel configuration by the availability of only one repair crew.

Standby Redundancy

Suppose that we consider the reliability of a two-component system, consisting of a primary and a backup unit. We assume that switching failures and failure in the standby mode can be neglected. In the preceding section the analysis of such a system is carried out assuming that the primary unit can be repaired with a rate ν. Since there are only three states with nonzero probabilities the state transition diagram may be drawn as in Fig. 11.15a, where state 3 is the

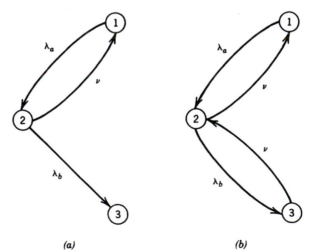

(a)   *(b)*

FIGURE 11.15 State transition diagrams for a standby system: (*a*) for reliability, (*b*) for availability.

failed state. The transition matrix for Eq. 11.88 is then given by

$$
\mathbf{M} = \begin{bmatrix} -\lambda_a & \nu & 0 \\ \lambda_a & -\lambda_b - \nu & 0 \\ 0 & \lambda_b & 0 \end{bmatrix}. \tag{11.95}
$$

The estimate of the availability of this system involves one additional state transition. In order for the system to go back into operation after both units have failed, we must be able to repair the backup unit. This requires an added repair transition from state 3 to state 2, as indicated in Fig. 11.15*b*. This repair transition is represented by two additional terms in the Markov transition matrix. We have

$$
\mathbf{M} = \begin{bmatrix} -\lambda_a & \nu & 0 \\ \lambda_a & -\lambda_b - \nu & \nu \\ 0 & \lambda_b & -\nu \end{bmatrix}. \tag{11.96}
$$

Here we assume that when both units have failed, the backup unit will be repaired first; we also assume that the repair rates are equal. More general cases may also be considered.

An important difference can be seen in the structures of Eqs. 11.95 and 11.96. In Eq. 11.96 all the diagonal elements are nonzero. This is a fundamental difference from reliability calculations. In availability calculations the system must always be able to recover from any failed state. Thus there can be no zero diagonal elements, for these would represent an absorbing or inescapable failed state; transitions can always be made out of operating states through the failure of additional components.

The availability of the system is given by

$$
A(t) = \sum_{i \in 0} P_i(t), \tag{11.97}
$$

where the sum is over the operational states. The Markov equations, Eq. 11.88, may be solved using Laplace transforms or other methods to determine the $P(t)$, and Eq. 11.97 may be evaluated for the detailed time dependence of the point availability.

We are usually interested in the asymptotic or steady-state availability, $A(\infty)$, rather than in the time dependence. This quantity may be calculated more simply. We note that as $t \to \infty$, the derivative on the right-hand side of Eq. 11.88 vanishes and we have the time-independent relationship

$$
\mathbf{MP}(\infty) = 0. \tag{11.98}
$$

In our problem this represents the three simultaneous equations

$$
-\lambda_a P_1(\infty) + \nu P_2(\infty) = 0, \tag{11.99}
$$

$$
\lambda_a P_1(\infty) - (\lambda_b + \nu) P_2(\infty) + \nu P_3(\infty) = 0, \tag{11.100}
$$

and

$$
\lambda_b P_2(\infty) - \nu P_3(\infty) = 0. \tag{11.101}
$$

This set of three equations is not sufficient to solve for the $P_i(\infty)$. For all Markov transition matrices are singular; that is, the equations are linearly dependent, yielding only $N - 1$ (in our case two) independent relationships. This is easily seen, since adding Eqs. 11.99 and 11.101 yields Eq. 11.100. The needed piece of additional information is the condition that all of the probabilities must sum to one:

$$\sum_i P_i(\infty) = 1. \tag{11.102}$$

In the situation in which we take $\lambda_a = \lambda_b = \lambda$, our problem is easily solved. Combining Eqs. 11.99, 11.101, and 11.102, we obtain

$$P_1(\infty) = \left[1 + \frac{\lambda}{\nu} + \left(\frac{\lambda}{\nu}\right)^2\right]^{-1}, \tag{11.103}$$

$$P_2(\infty) = \left[1 + \frac{\lambda}{\nu} + \left(\frac{\lambda}{\nu}\right)^2\right]^{-1} \frac{\lambda}{\nu}, \tag{11.104}$$

and

$$P_3(\infty) = \left[1 + \frac{\lambda}{\nu} + \left(\frac{\lambda}{\nu}\right)^2\right]^{-1} \left(\frac{\lambda}{\nu}\right)^2. \tag{11.105}$$

The steady-state availability may be found by setting $t = \infty$ Eq. 11.97:

$$A(\infty) = 1 - \left[1 + \frac{\lambda}{\nu} + \left(\frac{\lambda}{\nu}\right)^2\right]^{-1} \left(\frac{\lambda}{\nu}\right)^2. \tag{11.106}$$

If we further assume that $\lambda/\nu \ll 1$, we may write

$$A(\infty) \approx 1 - \left(\frac{\lambda}{\nu}\right)^2. \tag{11.107}$$

EXAMPLE 11.5

Suppose that the system availability for standby systems must be 0.9. What is the maximum acceptable value of the failure to repair rate ratio λ/ν?

Solution   Let $x = \lambda/\nu$ in Eq. 11.106. Then

$$A(\infty) = 1 - (1 + x + x^2)^{-1}(x^2).$$

Converting to a quadratic equation, we have $x^2 - \gamma x - \gamma = 0$, where

$$\gamma = \frac{1 - A}{A} = \frac{1 - 0.9}{0.9} = \frac{1}{9}$$

and

$$\frac{\lambda}{\nu} \equiv x = \frac{+\gamma + \gamma\sqrt{1 + 4/\gamma}}{2} = 0.393.$$

If instead the rare-event approximation is used,

$$\frac{\lambda}{\nu} \approx \sqrt{1 - A(\infty)} = \sqrt{1 - 0.9} = 0.316.$$

Other configurations are also possible. If two repair crews are available, repairs may be carried out on the primary and backup units simultaneously; the result is the four-state system of Table 11.2. As indicated in Fig. 11.16a, it is possible to get the primary unit running before the backup unit is repaired. In this situation states 1, 2, and 3 are operating states and must be included in the sum in Eq. 11.97. The Markov matrix now becomes

$$\mathbf{M} = \begin{bmatrix} -\lambda_a & \nu & \nu & 0 \\ \lambda_a & -\nu - \lambda_b & 0 & \nu \\ 0 & 0 & -\nu - \lambda_a & \nu \\ 0 & \lambda_b & \lambda_a & -2\nu \end{bmatrix}. \tag{11.108}$$

Other possibilities may also be added. For example, if switching failures and failures of the backup unit while in standby are not negligible, the state transition diagram is modified as shown in Fig. 11.16b, where p represents the probability of failure in switching from the primary to the backup, and λ_b^+ the standby failure rate of the backup unit. The Markov transition matrix corresponding to Fig. 11.16b is

$$\mathbf{M} = \begin{bmatrix} -\lambda_a - \lambda_b^+ & \nu & \nu & 0 \\ (1 - p)\lambda_a & -\lambda_b - \nu & 0 & \nu \\ \lambda_b^+ & 0 & -\lambda_a - \nu & \nu \\ p\lambda_a & \lambda_b & \lambda_a & -2\nu \end{bmatrix}. \tag{11.109}$$

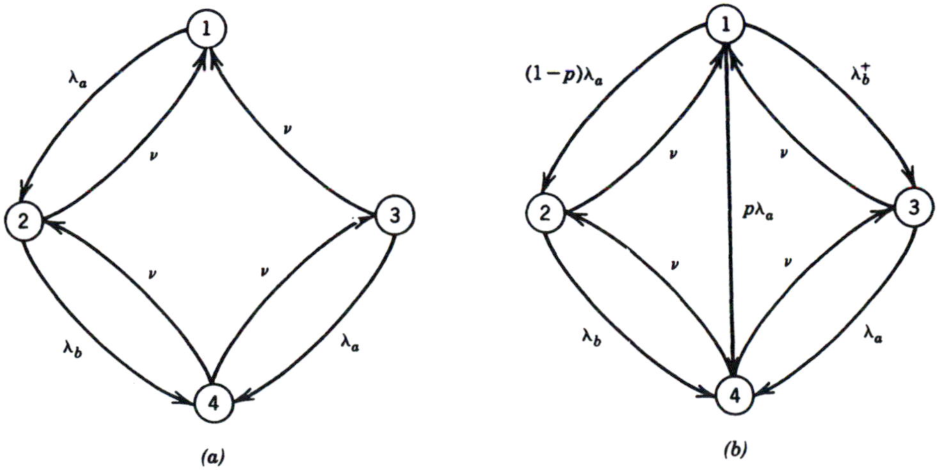

FIGURE 11.16 State transition diagrams for repairable standby systems.

To recapitulate, steady-state availability problems are solved by the same procedure. Any $N - 1$ of the N equations represented by Eq. 11.98 are combined with the condition, Eq. 11.102, that the probabilities must add to one, to solve for the components of $P(\infty)$. These are then substituted into Eq. 11.97 with the sum taken over all operating states to obtain the availability.

Shared Repair Crews

We conclude with the analysis of an active parallel system consisting of two identical units. We assume that the failure rates are identical and that they are independent of the state of the other unit. We also assume that the repair rates for the two units are the same. In this situation the failures and repairs of the two units are independent, provided that each unit has its own repair crew. The availability is then given by Eq. 10.95. The dependency is introduced not by a hardware failure, as in the case of standby redundancy, but by an operational decision to provide a single repair crew that can handle only one unit at a time.

The state transition diagram for the system using two repair crews is shown in Fig. 11.17a. Since the availability can be calculated from the component availabilities, as in Eq. 10.95, we shall not pursue the Markov solution further. Our attention is directed to the system using one repair crew, indicated by the state transition diagram given in Fig. 11.17b.

The transition matrix corresponding to Fig. 11.17b is

$$M = \begin{bmatrix} -2\lambda & \nu & \nu & 0 \\ \lambda & -\lambda-\nu & 0 & \nu \\ \lambda & 0 & -\lambda-\nu & 0 \\ 0 & \lambda & \lambda & -\nu \end{bmatrix}. \qquad (11.110)$$

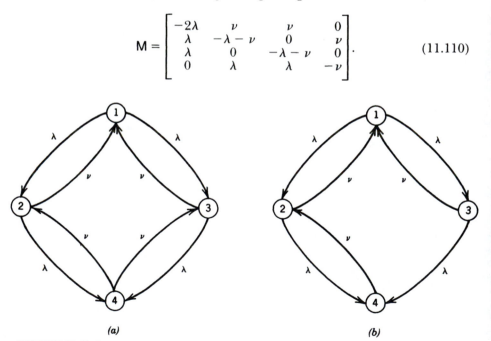

(a)   **(b)**

FIGURE 11.17 State transition diagrams for an active parallel system: (a) two repair crews, (b) one repair crew.

We solve the equations obtained from this matrix along with Eq. 11.102 to yield, after some algebra,

$$P_1(\infty) = \left[1 + 2\frac{\lambda}{\nu} + 2\left(\frac{\lambda}{\nu}\right)^2\right]^{-1}, \tag{11.111}$$

$$P_2(\infty) + P_3(\infty) = \left[1 + 2\frac{\lambda}{\nu} + 2\left(\frac{\lambda}{\nu}\right)^2\right]^{-1}\frac{2\lambda}{\nu}, \tag{11.112}$$

and

$$P_4(\infty) = \left[1 + 2\frac{\lambda}{\nu} + 2\left(\frac{\lambda}{\nu}\right)^2\right]^{-1}\frac{2\lambda^2}{\nu^2}. \tag{11.113}$$

Substitution of the results into Eq. 11.97 then yields for the steady-state availability

$$A(\infty) = 1 - \left[1 + 2\frac{\lambda}{\nu} + 2\left(\frac{\lambda}{\nu}\right)^2\right]^{-1}\frac{2\lambda^2}{\nu^2}. \tag{11.114}$$

For the usual case where $\lambda/\nu \ll 1$, this may be approximated by

$$A(\infty) \simeq 1 - 2\left(\frac{\lambda}{\nu}\right)^2. \tag{11.115}$$

The loss in availability because a second repair crew is not on hand can be determined by comparing these expressions to those obtained for system availability when there are two repair crews. From Eq. 10.95, with $N = 2$, we have

$$A(\infty) = 1 - \left[1 + 2\frac{\lambda}{\nu} + \left(\frac{\lambda}{\nu}\right)^2\right]^{-1}\left(\frac{\lambda}{\nu}\right)^2, \tag{11.116}$$

or for the case where $\lambda/\nu \ll 1$,

$$A(\infty) \simeq 1 - \left(\frac{\lambda}{\nu}\right)^2. \tag{11.117}$$

Thus the unavailability is roughly doubled if only one repair crew is present.

EXAMPLE 11.6

A system has an availability of 0.90. Two such systems, each with its own repair crew, are placed in parallel. What is the availability

(a) for a standby parallel configuration with perfect switching and no failure of the unit in standby;

(b) for an active parallel configuration?

(c) What is the availability if only one repair crew is assigned to the active parallel configuration?

Solution   The system availability is given by $A(\infty) = \nu/(\nu + \lambda)$. Therefore $\nu/\lambda = A(\infty)/[1 - A(\infty)] = 0.9/(1 - 0.9) = 9; \lambda/\nu = 0.1111$.

(*a*) From Eq. 11.106,

$$A(\infty) = 1 - \frac{(0.1111)^2}{1 + 0.1111 + (0.1111)^2} = 0.989.$$

(*b*) From Eq. 11.116,

$$A(\infty) = 1 - \frac{(0.1111)^2}{1 + 2 \times 0.1111 + (0.1111)^2} = 0.990.$$

(*c*) From Eq. 11.114,

$$A(\infty) = 1 - \frac{2 \times (0.1111)^2}{1 + 2 \times 0.1111 + 2 \times (0.1111)^2} = 0.980.$$

Bibliography

Barlow, R. E., and F. Proschan, *Mathematical Theory of Reliability*, Wiley, New York, 1965.

Green, A. E., and A. J. Bourne, *Reliability Technology*, Wiley, New York, 1972.

Henley, E. J., and H. Kumamoto, *Reliability Engineering and Risk Assessment*, Prentice-Hall, Englewood Cliffs, NJ, 1981.

McCormick, N. J., *Reliability and Risk Analysis*, Academic Press, NY, 1981.

Sandler, G. H., *System Reliability Engineering*, Prentice-Hall, Englewood Cliffs, NJ, 1963.

Exercises

11.1 Two stamping machines operate in parallel positions on an assembly line, each with the same MTTF at the rated speed. If one fails, the other takes up the load by doubling its operating speed. When this happens, however, the failure rate also doubles. Assuming no repair, how many MTTF for a machine at the rated speed will elapse before the system reliability drops below (*a*) 0.99, (*b*) 0.95, (*c*) 0.90?

11.2 Enumerate the 16 possible states of a four-component system by writing a table similar to Table 11.1. For the following configurations which are the failed states?

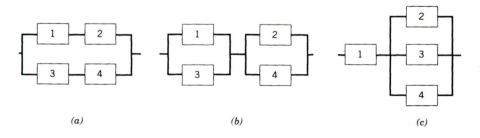

(*a*)   (*b*)   (*c*)

11.3 Consider a system consisting of two identical units in an active parallel configuration. The units cannot be repaired. Moreover, because they share loads, the failure rate λ^* of the remaining unit is substantially larger than the unit failure rates when both are operating.

(a) Find an approximation for the system reliability for a short period of time (i.e., $\lambda t \ll 1$ and $\lambda^* t \ll 1$).

(b) How large must the ratio of λ^*/λ become before the MTTF of the system is no greater than that for a single unit with failure rate λ?

11.4 Repeat Exercise 11.1 for the standby configurations shown in Fig. 11.14.

11.5 For the idealized standby system for which the reliability is given by Eq. 11.52,

(a) Calculate the MTTF in terms of λ.

(b) Plot the time-dependent failure rate $\lambda(t)$ and compare your results to the active parallel system depicted in Fig. 9.2*b*.

11.6 Verify Eqs. 11.42 through 11.45.

11.7 Calculate the variance for the time-to-failure for two identical units, each with a failure rate λ, placed in standby parallel configuration, and compare your results to the variance of the same two units placed in active parallel configuration. (Ignore switching failures and failures in the standby mode.)

11.8 Derive Eq. 11.52 assuming that $\lambda_b = \lambda_a$ from the beginning.

11.9 Under a specified load the failure rate of a turbogenerator is decreased by 30% if the load is shared by two such generators. A designer must decide whether to put two such generators in active or standby parallel configuration. Assuming that there are no switching failures or failures in the standby mode,

(a) Which system will yield the larger MTTF?

(b) What is the ratio of MTTF for the two systems?

11.10 Show that Eq. 11.64 reduces to Eq. 11.52 as $\lambda^+ \to 0$.

11.11 Consider the following configuration consisting of four identical units with failure rate λ and with negligible switching and standby failure rates. There is no repair.

(a) Show that the reliability can be expressed in terms of the Poisson distribution discussed in Chapter 6.

(b) Evaluate the reliability in the rare-event approximation for small λt.

(c) Compare the result from *b* to the rare-event approximation for four identical units in active parallel configuration, as developed in Chapter 9, and evaluate the reliabilities for $\lambda t = 0.1$.

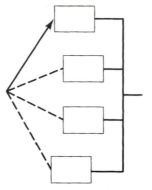

11.12 Verify Eq. 11.68.

11.13 For the following system, assume unit failure rates λ, no repair, and no switching or standby failures.

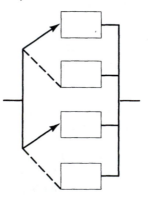

(a) Calculate the reliability.

(b) Approximate the result by the rare-event approximation for small λt, and compare your result to that for four units in an active parallel configuration.

11.14 Consider a standby system in which there is a switching failure probability p and a failure rate in the standby mode of λ_b^+.

(a) Draw the transition diagram.

(b) Write the Markov equations.

(c) Solve for the system reliability.

(d) Reduce the reliability to the situation in which the units are identical, $\lambda_a = \lambda_b = \lambda$, $\lambda_b^+ = \lambda$.

11.15 A design team is attempting to optimize the reliability of a navigation device. The choices for the rate gyroscopes are (*a*) a hot standby system consisting of two gyroscopes, and (*b*) a 2/3 voting system consisting of three gyroscopes. The mission time is 20 hr, and the gyroscope failure rate is 3×10^{-5}/hr. What is the greatest probability of switching failure in the hot standby system for which mission reliability is greater than that of the $\frac{2}{3}$ system? Assume that failures in logic on the 2/3 system can be neglected. (*Hint:* Assume rare-event approximations for the gyroscope failures.)

11.16 Derive Eq. 11.72.

11.17 (a) Find the asymptotic availability for a standby system with two repair crews; the Markov matrix is given by Eq. 11.108. Assume that $\lambda_a = \lambda_b = 0.01$/hr and $\nu = 0.5$/hr.

   (b) Evaluate the asymptotic availability for a standby system for the same data, except that there is only one repair crew. The Markov matrix is given by Eq. 11.96.

11.18 Derive Eqs. 11.82 and 11.83.

11.19 A system has an asymptotic availability of 0.93. A second redundant system is added, but only the original repair crew is retained. Assuming that all failures are revealed, estimate the asymptotic availability.

11.20 Derive Eqs. 11.103 through 11.105.

11.21 Assume that the units in Exercise 11.11 all have failure and repair rates λ and ν. A single crew repairs the most recently failed unit first.

   (a) Determine the asymptotic availability in terms of ν and λ.
   (b) Approximate your result for the case $\lambda/\nu \ll 1$.
   (c) Compare your result to that for the same units in active parallel configuration when $\lambda/\nu = 0.02$.

11.22 Consider the 2/3 standby configuration shown on the following page. It consists of three identical units; two units are required for operation. If either unit *a* or *c* fails, unit *b* is switched on. Ignore switching failures and repair, but assume failure rate λ and λ^* in the operating and standby modes.

   (a) Enumerate the possible system states and draw a transition diagram.
   (b) Write the Markov equations for the system.

11.23 Two ventilation units are in active parallel configuration. Each has an MTTF of 120 hr. Each is attended by a repair crew, and the MTTR is known to be 8 hr.

   (a) Calculate the availability, assuming that either unit can provide adequate ventilation.

(b) The units are replaced by new models with an MTTF of 200 hr. Can the staff be reduced to one repair crew without a net loss of availability? (Assume that the MTTR remains the same.)

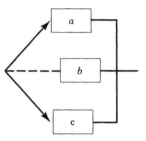

11.24 Assume that the units in Exercise 11.22 have identical repair rates ν.

(a) Enumerate the system states and draw a transition diagram.
(b) Write the transition matrix, **M**, for the Markov equations.
(c) Determine the asymptotic value of the system availability.

CHAPTER 12

System Safety Analysis

"Human error, lack of imagination, and blind ignorance. The practice of engineering is in large measure a continuing struggle to avoid making mistakes for these reasons."

Samuel C. Florman,

The Existential Pleasures of Engineering,

1976

12.1 INTRODUCTION

The discussion of system safety analysis in this chapter presents a different emphasis from the more general reliability considerations considered thus far. Whereas all failures are included in the determination of reliability, our attention now is turned specifically to those that may create safety hazards. The analysis of such hazards is often difficult, for with proper precautions taken in design, manufacture, and operation, failures causing safety problems should occur infrequently. Thus, the small probabilities encountered complicates the collection of data needed for analysis and making improvements. As a result, increased importance is assumed by more qualitative methods as well as by the engineer's understanding of the hazards that may arise. These difficulties notwithstanding, the potentially life-threatening nature of the hazards under consideration make safety analysis an indispensable component of reliability engineering.

Safety systems analysis has derived much of its importance from its association with industrial activities that may engender accidents of grave consequences. If we examine, in detail, historic accidents such as the disastrous chemical leak at Bhopal, India in 1984, or the 1986 destruction of the nuclear reactor at Chernobyl, some of the difficulties in the safety assessment of such systems begins to become apparent. First, the system is likely to have very small probabilities of a catastrophic failure, because it has redundant configurations of critical components. It then follows that the events to be avoided have either never occurred, or if they have, only rarely. There are few if any statistics on the probabilities of failures of the system as a whole, and reliability testing on the system level is likely to be impossible. Secondly, whatever accidents have occurred have rarely been the result of component failures of a type that would be easy to predict through reliability testing. Rather, the web of events leading to the accident is usually a complex of equipment failures, faulty maintenance, instrumentation and control problems, and human errors.

Safety analysis is essential for the full range of products and systems, from the large technological systems just discussed to small consumer items. For even though the later may not pose the threat of single catastrophic accidents, their production in large quantities leads to the possibility of many individual incidents, each capable of causing injury or death. Here again, the limitations of standard reliability testing and evaluation procedures are apparent. The primary challenge to the product development personnel is to understand the wide variety of environments and circumstances under which the product will be used, and to try to anticipate and protect against faulty installation or maintenance, misuse, inappropriate environments, and other hazards that may not be revealed through standard reliability tests. An additional imperative is to examine not only how the product may fail in a hazardous manner, but also how the user may be harmed during normal operation. Adequate protection must be afforded from the rotating blades, electrical filaments, flammable liquids, heated surfaces, and other potential hazardous features that are necessary constituents of many industrial and consumer products.

Even though hazard creation most often involves the intertwined effects of equipment failure and human behavior, analysis is expedited by examining them separately. Thus in the following section we build on the discussion in the preceding chapters to focus on those particular aspects of equipment failure most closely related to safety hazards. In Section 12.3 the importance of the human element is emphasized. In that discussion the primary focus is on the operations of industrial facilities where efforts may be much more effective in reducing human error than they are likely to be in modifying consumer psychology. With the background gained in examining the hazardous aspects of equipment and of human causes, we are prepared in Section 12.4 for an overview of those analytical methods that have been developed to rationalize the discussion of safety analysis. Sections 12.5 through 12.7 then focus on the construction and evaluation of fault trees.

12.2   PRODUCT AND EQUIPMENT HAZARDS

In examining equipment with safety repercussions, it is useful once again to frame the analysis in terms of the bathtub curve, and consider infant mortality, random events, and aging as hazard causes. Most of the materials discussed in earlier chapters regarding these causes remains relevant. Now, however, we must extend the level of analysis to even less probable and therefore possibly more bizarre sets of causes. We also must consider not only product or equipment failures but also potential hazards created in the course of product usage.

Design shortcomings or variability in the production process are the most likely causes of early or infant mortality failures. Changes in details late in the design process to facilitate manufacture or construction, which are not thoroughly checked to ensure that a new hazard hasn't been introduced, may be particularly dangerous. Such a change was implicated, for example, in the 1981 collapse of the Kansas City Hyatt Regency walkways that resulted in 114 fatalities. Failure to meet materials specification, improvisation in construction procedures or unsafe economic choices made in manufacturing processes may all defeat the integrity of the original design and result in weakened systems that are then prone to infant mortality hazards. Faulty installations of hot water heaters, stoves or other consumer products are also prone to create infant mortality hazards.

Random failures or hazards are characterized by chance occurrences that are independent of product age. In general they are caused by an environment that is unanticipated or for which the product does not have the strength to withstand. They tend to be brought about because the product is used—or misused—under conditions that were not contemplated in the design, or were thought to be so improbable that they were lost in the cost-performance trade-offs. The largest danger in creating a new product is arguably not that there is an inadequate safety margin against a known hazard, but that a potential hazard completely escapes the attention of the design team. Even if a thorough study reveals all significant hazards, however, many decisions must be faced with safety implications.

Governmental bodies, professional organizations and insurance under-writers' codes of standards provide a basis for assessing the level of potential hazards for many products. Often such standards must be promulgated by specialized bodies cognizant of unique hazard combinations of particular industries. The safety of food processing equipment, for example, is compli-cated by the conflicting requirements that machinery be readily accessible for cleaning to prevent unsanitary conditions from arising, and the need for extensive guard equipment to protect workers from hot surfaces, cutting blades, and other mechanical hazards. While standards and codes of good practice provide a point of departure for the analysis of hazards, new designs and novel applications may be expected to present potentially hazardous conditions that have not been contemplated in the standards. Thus to make informed safety decisions it is incumbent upon the product development personnel to gain a thorough understanding of the product and its re-quired use.

To understand the difficult trade-offs that must be faced, consider a television monitor. Ventilation slits are required to prevent overheating and to allow the electronics to operate at a reasonable temperature. More and larger ventilation paths will likely improve reliability and prolong the life of the set. However, the designer must also consider unusual locations where ventilation is curtailed, where debris is piled on top or stacked against the monitor or where other cooling impediments are encountered. Safety analysis then requires not only the determination of the effects of these situations on set life, but also whether there is an unacceptable risk of fire. Conversely, if the ventilation slits are made larger to add an extra margin of cooling capacity, then the increased danger that a child will succeed in inserting a kitchen knife or other object through a slit and come into contact with high voltage must be addressed. Thirdly, the magnitude of the hazard created if fluid is spilled or the monitor immersed must be considered to determine whether fluid entering through the ventilation slits will result in a benign failure or an unacceptable risk of electrical shock.

The engineering for safety must go beyond the contemplation of unusual accidents and inadvertent misuse to consider situations where the user behavior compounds potential hazards. From the nineteenth-century captains of Mississippi river boats, who blocked safety valves in order to get more pressure and more performance from their boilers, to present day motorists, who negate the effects of antilock breaks by driving more aggressively on wet pavements, product users frequently overcome safety features in order to enhance performance at the cost of increased risk. Operational limits exceeded to increase performance, safety guards removed to facilitate maintenance, and warnings ignored as a result of past false alarms are among the plethora of causes of increased risk induced by unintended usage. Such behavior further complicates the already difficult legal and ethical issues raised in determining the extent to which users must be protected from their deliberate unsafe practices.

Product modifications or modernizations likewise may introduce new and unanticipated hazards. Motors modified for racing, aircraft converted from civilian to military or from passenger to cargo use, robots or machinery devoted to new and novel manufacturing tasks all require careful scrutiny to ensure that the safety integrity of the original design is not compromised. But often modifications take place years into the product life, when knowledge of the original design calculations has faded, components suppliers have changed, and technology has evolved. An example of particularly ill-conceived design modifications were those made to the steamship *Birkenhead*. In converting this warship to a troop carrier large passageways were cut through the water-tight bulkheads to provide more light, air and spaciousness for the troops. But the penetrations not only destroyed the water-tight compartmentalization of the ship but also greatly weakened the bulkheads. Thus when the ship struck a rock in 1852, it both flooded very rapidly and broke in two, resulting in over 400 fatalities. While engineering safety practices have matured a great deal since that time, it, like other historical disasters, serves as a reminder of the potential consequences of ignorance in making ad-hoc modifications to existing systems.

Even after provisions have been made to minimize the dangers of infant mortality or random hazards, there remains the problem of dealing with the aging failures that may be expected to become increasingly pronounced as the product approaches the end of its useful life. Normally, a target life is stipulated as a part of the design process. Assuming adequate maintenance is provided to replace those components with shorter lives—such as spark plugs, brake linings, and tires on automobiles, for example—failures attributable to aging should not create significant risk within the design life. In relatively few situations, however, can it be guaranteed that a product or system will not continue to be used well beyond its design life. To be sure, in some areas of rapid technological development, such as in microprocessor development, products may become obsolescent and be replaced long before aging effects become important. Likewise, safety-critical systems may be licensed or controlled for removal from service after the number of operating hours for which previous analysis and/or life tests have verified their capability. Military aircraft and nuclear reactor pressure vessels, for example may fall into this category. More often than not however, the increasing cost of maintenance and recovery from breakdown is weighed against replacement cost in determining at what point a product is retired.

Even where there are strong safety implications, a system can be allowed to operate well beyond its target design life provided dependable inspection and repair protocols are employed. The knowledge of the aging process that has been gained through the years of operation, however, must provide inspection methods capable of detecting the aging phenomena early enough to repair or take the system out of service before the deterioration reaches a hazardous threshold. Many commercial aircraft, for example, have been allowed to operate under such scrutiny beyond the design life originally targeted.

With consumer products the situation is likely to be quite different. For unless there is a clear and obvious danger, the user is prone to run the product until it fails and then decide whether to replace or repair it. The critical design consideration here is to ensure that the wearout modes are benign. The challenge is simply illustrated with a hot plate, coffee maker, or other appliance with a heating element. Suppose the design includes a fuse to prevent fire in the event that the heater fails in a dangerous mode. Then, the heater failure had better occur before the fuse deterioration becomes a problem. One complicated situation, in fact, was recently in the courts, where a consumer product design was "improved" by incorporating a heater with a longer design life. However, after the new design resulted in a number of fires it was discovered that the melting temperature of the fuse gradually increased with time to the point where by the time the heater finally failed, the fuss was no longer operable.

The foregoing discussion provides only the beginnings for the level of sophistication needed to ferret out the potential hazards that may be brought about by infant mortality, random and aging phenomena, and their interactions. The analytical methods introduced in Section 12.4 provide techniques for more structured analysis. Use of these should reduce the possibility of potentially significant hazards that escape consideration altogether. In addition, the reading of case histories in newspapers and the professional literature over a period of years is invaluable in enhancing one's ability to identify and eliminate potential hazards before they become safety problems.

12.3 HUMAN ERROR

All engineering is a human endeavor, and in the broadest sense most failures are due to human causes, whether they be ignorance, negligence, or limitations of vigilance, strength, and manual dexterity. Designers may fail to fully understand system characteristics or to anticipate properly the nature and magnitudes of the loading to which a system may be subjected or the environmental conditions under which it must operate. Indeed, much of engineering education is devoted to understanding these and related phenomena. Similarly, errors committed during manufacture or construction are attributable either to the personnel involved or to the engineers responsible for the setup of the manufacturing process. Quality assurance programs have a central role in detecting and eliminating such errors in manufacture and construction.

We shall consider here only human errors that are committed after design and manufacture; those that are committed in the operation and maintenance of a system. This is a convenient separation, since design and manufacturing errors, whether they are considered human or not, appear in the as-built system as shortcomings in the reliability of the hardware.

Even with our attention confined to human errors appearing in the operation and maintenance of a system, we find that the uncertainties involved are generally much greater than in the analysis of hardware reliability. There are three categories of uncertainty. First, the natural variability of human performance is considerable. Not only do the capabilities of people differ, but the day-to-day and hour-to-hour performance of any one individual also varies. Second, there is a great deal of uncertainty about how to model probabilistically the variability of human performance, since the interactions with the environment, with stress, and with fellow workers are extremely complex and to a large extent psychological. Third, even when tractable models for limited aspects of human performance can be formulated, the numerical probabilities or model parameters that must be estimated in order to apply them are usually only very approximate, and the range of situations to which they apply is relatively narrow.

It is, nevertheless, necessary to include the effects of human error in the safety analysis of any complex system. For as the consequences of accidents become more serious and more emphasis is put on reliable hardware and highly redundant configurations, an increasing proportion of the risk is likely to come from human error, or more accurately from complex interactions of human shortcomings and equipment problems. Even though accurate predictions of failure probabilities are problemmatical, a great deal may be gained from studying the characteristics of human reliability and contrasting them with those of hardware. From such study comes an insight into how systems may be designed and operated in order to minimize and mitigate

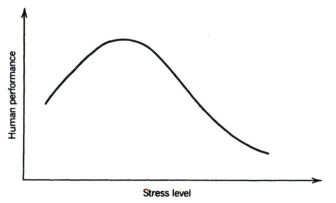

FIGURE 12.1 The effect of stress level on human performance.

accidents in which the operating and maintenance staff may play an important role.

It has been pointed out* that increasingly there is a centralization of systems, whether they be larger-capacity power and chemical plants, aircraft carrying greater number of passengers, or structures with larger capacities. Since human error in the operation of many such centralized systems may lead to accidents of major consequence to life and property, there has been an increased emphasis on plant automation. There are certainly limitations on such automation, particularly when the uncertainty of how an operator may react to a situation is overriden by the need for human adaptability in dealing with conditions that have not or could not be incorporated into the automated control system. Moreover, automated operation does not tend to eliminate humans from consideration, but rather to remove them to tasks of two quite dissimilar varieties; routine tasks of maintaining, testing, and calibrating equipment; and protective tasks of watching for plant malfunctions and preventing their accident propagation. These two classes of tasks tend to enter system safety considerations in different ways. When humans err in routine testing, maintenance, and repair work, they may introduce latently risky conditions into the plant. Any errors that they make in taking protective actions under emergency conditions may increase the severity of an accident.

The problems inherent in maximizing human reliability for the two classes of tasks may be viewed graphically in Fig. 12.1. Generally, there is an optimum level of psychological stress for human performance. When the level is too low, humans are bored and make careless errors; too high a level may cause them to make a number of inappropriate, near-panic responses to a situation. To illustrate, consider the example of flying a commercial airliner. The pilot's monitoring of controls during level, uneventful flight in a highly automated

* J. Rasmussen, "Human Factors in High Risk Technology," in *High Risk Technology*, A. E. Green (ed.), Wiley, NY, 1982.

aircraft would fall on the low level of the curve. The principal danger here is carelessness or lack of attention. Normal take-offs and landings are likely to be closer to the optimum stress level for attentive behavior. At the other extreme pilot reaction to major inflight emergencies, such as onboard fires or power failures, is likely to be degraded by the high stress level present. Because of the quite different factors that come into play, we shall now consider human reliability and its degradation under the two limiting situations of very routine tasks and tasks performed in emergency situations.

Routine Operations

For purposes of analysis it is useful to classify human errors as random, systematic, or sporadic. These classes may be illustrated by considering the simple example, shown in Fig. 12.2, of the ability to hit a target.* Random errors are dispersed about the desired value without bias; that is, they have the true mean value (in x and y), but the variance may be too large. These errors may be corrected if they are attributable to an inappropriate tool or man—machine interface. For example, if it is not possible to read instruments finely enough or to adjust setting precisely enough, such improvements are in order. Similarly, training in the particular task may reduce the dispersion of random errors. Figure 12.2b illustrates systematic errors whose dispersion is sufficiently small, but with a bias departing from the mean value. Such bias may be caused by tools or instruments that are out of calibration, or it may come from incorrect performance of a procedure. In either case corrective measures may be taken. More subtle psychological factors—such as the desire of an inspector not to miss any faulty parts, and thus declaring a good many faulty even though they are not—may also cause bias errors.

Perhaps sporadic errors, pictured in Fig. 12.2c, are the most difficult to deal with, for they rarely show observable patterns. They are committed when the person acts in an extreme or careless way: forgetting to do something altogether, performing an action that was not called for, or reversing the order in which things are done. For example, a meter reader might, in taking a series of meter readings, read a wrong meter. Again, careful design of the man—machine interface can minimize the number of sporadic errors. Color, shape, and other means can be used to differentiate instruments and control and to minimize confusion. Sporadic errors, in particular, are amplified by the carelessness inherent in low-stress situations, as well as by the confusion of high-stress situations.

Let us first examine sporadic errors made in routine situations. Certainly, under any circumstances, errors are minimized by a well-designed work environment. Such design would take into account all the standard considerations or human factors engineering: comfortable seating, adequate light, temperature and humidity control, and well-designed control and instrument panels to minimize the possibilities for confusion. The attention span that can be

---

* H. R. Guttmann, unpublished lecture notes, Northwestern University, 1982.

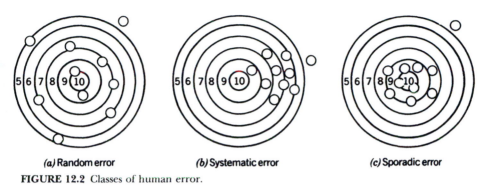

(a) Random error   *(b)* Systematic error   *(c)* Sporadic error

FIGURE 12.2 Classes of human error.

expected for routine tasks is still limited. As indicated in Fig. 12.3, attention spans for detailed monitoring tend to deteriorate rapidly after about half an hour, indicating the need for frequent rotation of such duties for optimal performance. The same deterioration may be expected for very repetitive tasks, unless there is careful checking or other intervention to insure that such deterioration does not take place.

Probably one of the most important ways in which system reliability is degraded is through the dependencies introduced between redundant components during the course of routine maintenance, testing, and repair. An example is the turning off of both of the redundant auxiliary feedwater systems at the Three Mile Island reactor. The point is that if technicians perform a task incorrectly on one piece of equipment, they are likely to do it incorrectly on all like pieces of equipment. This problem may be countered, at least in part, by a variety of techniques. Diversity of equipment is one, for just as the hardware will not be subjected to the same failure modes, the maintenance procedures will also be different. Staggering the times or the personnel doing

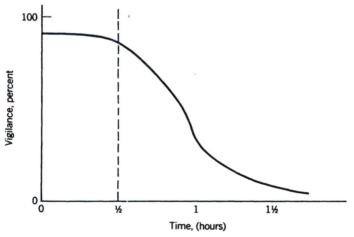

FIGURE 12.3 Vigilance versus time.

maintenance on redundant equipment also tends to reduce dependencies, although some smaller degree of dependency may remain through the use of common tools or incorrect training procedures.

Independent checking of procedures also decreases both the probability of failure and the degree of dependency. Even here, however, psychological factors limit effectiveness. When the inspector and the person performing the maintenance have worked with each other for an extended period of time, the inspector may tend to become less careful as he or she grows more confident of the colleague's abilities. Similarly, if two independent checks are to be performed, they are unlikely to be truly independent, for often the very knowledge that a procedure is being checked twice will tend to decrease the care with which it is done.

Reliability is also degraded when operating and maintenance personnel inappropriately modify or make shortcuts in operating and maintenance procedures. Often operating and maintenance personnel gain an understanding of the system that was not available at the time of design and modify procedures to make them more efficient and safer. The danger is that, without a thorough design review, new loadings and environment degradation may be introduced, and component dependencies may increase inadvertently. For example, in the 1979 crash of the DC-10 in Chicago, it is thought that a modified procedure for removing the engines for inspection and preventive maintenance led to excessive fatigue stresses on the engine support pylon, causing the engine to break off during takeoff.

Although the methodology is not straightforward, data are available on the errors committed in the course of routine tasks. Extensive efforts have been made to develop task analysis and simulation methods.* Failure probabilities are first estimated for rudimentary functions. Then, by combining these factors, we can estimate probabilities that more extensive procedures will engender errors.

Emergency Operations

At the high-stress end of the spectrum shown in Fig. 12.1 are the protective tasks that must be performed by operations personnel under emergency conditions to prevent potentially dangerous situations from getting completely out of hand. Here a well-designed, man–machine interface, clear-cut procedures, and thorough training are critical, for in such situations actions that are not familiar from routine use must be taken quickly, with the knowledge that mistakes may be disastrous. Moreover, since such situations are likely to be caused by subtle combinations of malfunctions, they may be confusing and call for diagnostic and problem-solving ability, not just the skill and rule-based actions exercised for routine tasks.

* A. D. Swain, and H. R. Guttmann, *Handbook of Human Reliability Analysis with Emphasis on Nuclear Power Plant Applications,* U.S. Nuclear Regulatory Commission, NUREG/CR-1287, 1980.

Under emergency conditions conflicting information may well confuse operators who then act in ways that further propagate the accident. With proper training and the ability to function under psychological stress, however, they may be able to solve the problem and save the day. For example, the confusion of the operators at the Three Mile Island reactor caused them to turn off the emergency core-cooling system, thus worsening the accident. In contrast, the pilot of a Boeing 767 managed to make use of his earlier experience as an amateur glider pilot and safely land his aircraft after a series of equipment failures and maintenance errors had caused the plane to run out of fuel while in flight over Canada.

There are a number of common responses to emergency situations that must be taken into consideration when designing systems and establishing operating procedures. Perhaps the most important is the incredulity response. In the rare event of a major accident, it is common for an operator not to believe that an accident is taking place. The operator is more likely to think that there is a problem with the instruments or alarms, causing them to produce spurious signals. At installations that have been subjected to substantial numbers of false alarms, a real one may very well be disbelieved. Systems should be carefully designed to keep spurious alarms to a minimum, and straightforward checks to distinguish accidents from faulty instrument performance should be provided. In some situations it is desirable to mandate that safety actions be taken, even though the operator may feel that faulty instruments are the cause of the problem.

A second common reaction to emergencies is reverting to stereotype. The operator reverts to the stereotypical response of the population of which he or she is a part, even though more recent training has been to the contrary. For example, in the United States turning a light or other switch "up" means that it is "on." In Europe, however, "down" is "on." Thus, although Americans may be trained to put a particular switch down to turn it on, under the time pressure of an emergency they are likely to revert to the population stereotype and try to put the switch up. The obvious solution to this problem is to take great care in human factors engineering not to violate population stereotypes in the design of instrumentation and control systems. This problem may be aggravated if operators from one culture are transferred to another, or if care is not taken in the use of imported equipment.

Finally, once a mistake is made, such as placing a switch in the wrong position, in a panic an operator is likely to repeat the mistake rather than think through the problem. This reaction, as well as other inappropriate emergency responses, must be considered when deciding the extent to which emergency actions should or can be automated. On the one hand, when there is extreme time pressure, automated protection systems may eliminate the errors discussed. At the same time, such systems do not have the flexibility and problem-solving ability of human operators, and these advantages may be of overwhelming importance, assuming that there is time for the situation to be properly assessed.

In summary, to ensure a high degree of human reliability in emergency situations, control rooms, whether they be aircraft cockpits or chemical plant control installations, must be carefully designed according to good human factors practice. It is also important that the procedures for all anticipated situations are readily understandable, and finally, that operators are drilled at frequent intervals on emergency procedures, preferably with simulators that model the real conditions.

Even though we may characterize human behavior under emergency conditions and suggest actions that will improve human reliability, it is difficult indeed to obtain quantitative data on failure probabilites. As we have indicated, such situations happen only infrequently and often they are not well documented. Moreover, it is difficult to obtain a realistic response from simulator experiments when the subjects know that they are in an experiment and not a life-threatening situation.

12.4   METHODS OF ANALYSIS

Probably the most important task in eliminating or reducing the probability of accidents is to identify the mechanisms by which they may take place. The ability to make such identifications in turn requires that the analyst have a comprehensive understanding of the system under consideration, both in how it operates and in the limitations of its components. Even the most knowledgeable analysts are in danger of missing critical failure modes, however, unless the analysis is carried out in a very systematic manner. For this reason a substantial number of formal approaches have been developed for safety analysis. In this section we introduce three of the most widely used: failure modes and effects analysis, event trees, and fault trees. In later sections the use of fault trees is developed in more detail.

Failure Modes and Effects Analysis

Failure modes and effects analysis, usually referred to by the acronym FMEA, is one of the most widely employed techniques for enumerating the possible modes by which components may fail and for tracing through the characteristics and consequences of each mode of failure on the system as a whole. The method is primarily qualitative in nature, although some estimates of failure probabilities are often included.

Although there are many variants of FMEA, its general characteristics can be illustrated with the analysis of a rocket shown in Fig. 12.4. In the left-hand column the major components or subsystems are listed; then, in the next column the physical modes by which each of the components may fail are given. This is followed, in the third column, by the possible causes of each of the failure modes. The fourth column lists the effects of the failure. The method becomes more quantitative if an estimate of the probability of each failure mode is made. Criticality or an alternative ranking of the failure's importance is usually included to separate failure modes that are catastrophic

FAILURE MODES AND EFFECTS ANALYSIS

1 SUBSYSTEM _____ 2. DWG. NR. _____ 3. PREPARED BY _____ 4. DATE _____

ITEM	FAILURE MODES	CAUSE OF FAILURE	POSSIBLE EFFECTS	PROBABILITY OF OCCURRENCE	CRITICALITY	POSSIBLE ACTION TO REDUCE FAILURE RATE OR EFFECTS
Motor case	Rupture	a Poor workmanship b. Defective materials c. Damage during transportation d. Damage during handling e. Overpressurization	Destruction of missile	0.0006	Critical	Close control of manufacturing processes to ensure that workmanship meets prescribed standards. Rigid quality control of basic materials to eliminate defectives. Inspection and pressure testing of completed cases. Provision of suitable packaging to protect motor during transportation.
Propellant grain	a. Cracking b. Voids c. Bond separation	a Abnormal stresses from cure b Excessively low temperatures c Aging effects	Excessive burning rate; overpressurization, motor case rupture during otherwise normal operation	0.0001	Critical	Carefully controlled production. Storage and operation only within prescribed temperature limits. Suitable formulation to resist effects of aging.
Liner	a. Separation from motor case b Separation from motor grain or insulation	a. Inadequate cleaning of motor case after fabrication b. Use of unsuitable bonding material c. Failure to control bonding process properly	Excessive burning rate Overpressurization Case rupture during operation	0.0001	Critical	Strict observance of proper cleaning procedures. Strict inspection after cleaning of motor case to ensure that all contaminants have been removed.

FIGURE 12.4 Failure modes and effects analysis. (From Willie Hammer, *Handbook of System and Product Safety*, © 1972, p. 153, with permission from Prentice-Hall, Englewood Cliffs, NJ.)

from those that merely cause inconvenience or moderate economic loss. The final column in most FMEA charts is a listing of possible remedies.

In a more extensive FMEA the information shown in Figure 12.4 may be expanded. For example, failures are not categorized as simply critical or not critical but by four levels denoting seriousness.

1. Negligible—loss of function that has no effect on the system.
2. Marginal—a fault that will degrade the system to some extent but will not cause the system to be unavailable, for example, the loss of one of two redundant pumps, either of which can perform a required function.
3. Critical—a fault that will completely degrade system performance, for example, the loss of a component that renders a safety system unavailable.
4. Catastrophic—a fault that will have severe consequences and perhaps cause injuries or fatalities, for example, catastrophic pressure vessel failure.

Additional columns also may be included in FMEA. A list of symptoms or methods of detection of each failure mode may be very important for safe operations. A list of compensating provisions for each failure mode may be provided to emphasize the relative seriousness of the modes. In order to concentrate improvement efforts on eliminating those having the widest effects, it is common also to rank the various causes of a particular mode according to the percentage of the mode's failures that they incur.

The emphasis in FMEA is usually on the basic physical phenomena that can cause a device or component to fail. Therefore, it often serves as a suitable starting point for enumerating and understanding the failure mechanisms before proceeding to one of the other techniques for safety analysis. To understand better the progression of accidents when they pass through several stages and to analyze the effects of component redundancies on system safety, engineers often supplement FMEA with the more graphic event-tree and fault-tree methods for quantifying system behavior during accidents.

Event Trees

In many accident scenarios the initiating event—say, the failure of a component—may have a wide spectrum of results, ranging from inconsequential to catastrophic. The consequences may be determined by how the accident progression is affected by subsequent failure or operation of other components or subsystems, particularly safety or protection devices, and by human errors made in responding to the initiating event. In such situations an inductive method may be very useful. We begin by asking "what if" the initiating event occurs and then follow each of the possible sequences of events that result from assuming failure or success of the components and humans affected as the accident propagates. After such sequences are defined, we may attempt to attach probabilities to them if such a quantitative estimate is needed.

The event tree is a quantitative technique for such inductive analysis. It begins with a specific initiating event, a particular cause of an accident, and then follows the possible progressions of the accident according to the success or failure of other components or pieces of equipment. Event trees are a particular adaptation of the more general decision-tree formalism that is widely employed for business and economic analysis. They are quite useful in analyzing the effects of the functioning or failure of safety systems in response to an accident, particularly when events follow with a particular time progression. The following is a very simple application of event-tree analysis.

Suppose that we want to examine the effects of the power failure in a hospital in order to determine the probability of a blackout, along with other likely consequences. For simplicity we assume that the situations may be analyzed in terms of just three components: (1) the off-site local utility power system that supplies electricity to the hospital; (2) a diesel generator that supplies emergency power, and (3) a voltage-monitoring system that monitors the off-site power supply and, in the event of a failure, transmits a signal that starts the diesel generator.

We are concerned with a sequence of three events. The initiating event is the loss of off-site power. The second event is detection of the loss and subsequent functioning of the voltage-monitoring system; and the third event is the start-up and operation of the diesel generator. This sequence is shown in the event tree in Fig. 12.5. Note that at each event there is a branch corresponding to whether a system operates or fails. By convention, the upward branches signify successful operation, and the lower branches failure.

Note that for a sequence of N events there will be 2^N branches of the tree. The number may be reduced, however, by eliminating impossible branches. For example, the generator cannot start unless the voltage monitor functions. Thus the path is impossible (has a zero probability) and can be pruned from the tree, as in Fig. 12.6.

We may follow an event tree from left to right to find the probabilities and consequences of differing sequences of events. The probabilities of the various outcomes are determined by attaching a probability to each event on the tree. In our tree the probabilities are P_i for the initial event, P_v for the failure of the voltage monitoring system, and P_g for the failure of the diesel generator. With the assumption that the failures are independent, the probability of a blackout is therefore $P_iP_v + P_i(1 - P_v)P_g$.

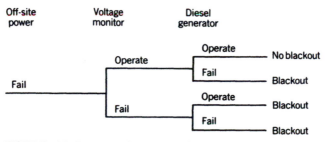

FIGURE 12.5 Event tree for power failure.

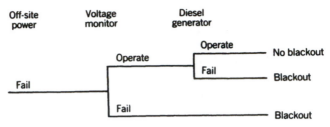

FIGURE 12.6 Reduced event tree for power failure.

Fault Trees

Fault-tree analysis is a deductive methodology for determining the potential causes of accidents, or for system failures more generally, and for estimating the failure probabilities. In its narrowest sense fault-tree analysis may be looked on as an alternative to the use of reliability block diagrams in determining system reliability in terms of the corresponding components. However, fault-tree analysis differs both in the approach to the problem and in the scope of the analysis.

Fault-tree analysis is centered about determining the causes of an undesired event, referred to as the top event, since fault trees are drawn with it at the top of the tree. We then work downward, dissecting the system in increasing detail to determine the root causes or combinations of causes of the top event. Top events are usually failures of major consequence, engendering serious safety hazards or the potential for significant economic loss.

The analysis yields both qualitative and quantitative information about the system at hand. The construction of the fault tree in itself provides the analyst with a better understanding of the potential sources of failure and thereby a means to rethink the design and operation of a system in order to eliminate many potential hazards. Once completed, the fault tree can be analyzed to determine what combinations of component failures, operational errors, or other faults may cause the top event. Finally, the fault tree may be used to calculate the demand failure probability, unreliability, or unavailability of the system in question. This task of quantitative evaluation is often of primary importance in determining whether a final design is considered to be acceptably safe.

The rudiments of fault-tree analysis may be illustrated with a very simple example. We use the same problem of a hospital power failure treated inductively by event-tree analysis earlier to demonstrate the deductive logic of fault-tree analysis. We begin with blackout as the top event and look for the causes, or combination of causes, that may lead to it. To do this, we construct a fault tree as shown in Fig. 12.7. In examining its causes, we see that both the off-site power system *and* the emergency power supply must fail. This is represented by a ∩ gate in the fault tree, as shown. Moving down to the second level, we see that the emergency power supply fails if the voltage monitor *or* the diesel generator fails. This is represented by a ∪ gate in the fault tree as shown.

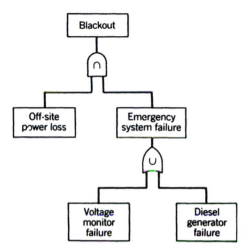

FIGURE 12.7 Fault tree for blackout.

We see that the fault tree consists of a structure of OR and AND gates, with boxes to describe intermediate events. Using the same probabilities as in the event tree, we can determine the probability of a blackout in terms of P_l, and P_v, and P_g, the failure probabilities for off-site power, voltage monitor, and diesel generator.

The most straightforward fault trees to draw are those, such as in the preceding example, in which all the significant primary failures are component failures. If a reliability block diagram can be drawn, a fault tree can also be drawn. This can be seen in an additional example.

Consider the system shown in Fig. 9.9. We may look at the system as consisting of an upper subsystem ($a1$, $a2$, and $b1$) and a lower subsystem ($a3$, $a4$, and $b2$), in addition to component c. For a system to fail, either component c must fail or the upper and lower subsystems must fail. Proceeding downward, for the upper subsystem to fail either component $b1$ must fail or both $a1$ and $a2$ must fail. Treating the lower subsystem analogously, we obtain the tree shown in Fig. 12.8.

EXAMPLE 12.1

Construct a reliability block diagram corresponding to the fault tree in Fig. 12.7.

Solution The reliability block diagram having the same logic and failure probability as the fault tree of Fig. 12.7 is depicted in Fig. 12.9.

12.5 FAULT-TREE CONSTRUCTION

Of the methods discussed in the preceding section, fault-tree analysis has been the most thoroughly developed and is finding increased use for system

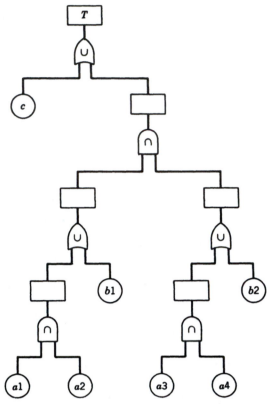

FIGURE 12.8 Fault tree.

safety analysis in a wide variety of applications. It is particularly well suited to situations in which tracing a failure to its root causes requires dissecting the system into subsystems, components, and parts to get at the level where failure data are available. For example, in the aforetreated hospital blackout we may not have the test data that is required to determine P_v for the voltage monitor or P_g for the diesel generator. We must then delve more deeply and examine the components of these devices; we may need to construct the probability that the voltage monitor will fail from the failure rates of its components.

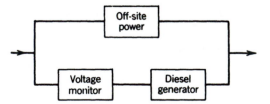

FIGURE 12.9 Reliability block diagram for electrical power.

It may be argued that such dissection can also be done by subdividing the blocks appearing in reliability block diagrams. Although this is true, there are some important differences. Reliability block diagrams are success-oriented; that is, all failures are lumped together to obtain the probability that a system will fail. In most reliability studies we are interested only in knowing the reliability (i.e., the probability that the system does not fail). Conversely, in fault-tree analysis we are often interested only in a particular undesirable event (i.e., a failure that leads to a safety hazard) and in calculating the probability that it will happen. Hence failures that do not cause the safety hazard defined by the top event are excluded from consideration.

The difference between reliability analysis and safety analysis may be illustrated by the example of a hot-water heater. In reliability analysis—carried out with a reliability block diagram—failure of any kind will cause failure of the system to supply hot water. Most of these failures have no safety implications: The heater unit fails to turn on, the tank develops a leak, and so on. In safety analysis—using a fault tree—we would be interested in a particular safety hazard such as the explosion of the tank. The other failures listed would not be included in the fault-tree construction.

Because of the increasing importance of fault-tree analysis, the remainder of this chapter is devoted to it. In this section we discuss the construction of fault trees by first giving the standardized nomenclature. Then following a brief discussion of fault classifications, we supply several illustrative examples. In Sections 12.6 and 12.7 fault trees are evaluated. In qualitative evaluation the fault tree is reduced to a logical expression, giving the top event in terms of combinations of primary-failure events. In quantitative evaluation the probability of the top event is expressed in terms of the probabilities of the primary-failure events.

Nomenclature

As we have seen, the fault tree is made up of events, expressed as boxes, and gates. Two types of gates appear, the OR and the AND gate. The OR gate as indicated in Fig. 12.10a is used to show that the output event occurs only if one or more of the input events occur. There may be any number of input events of an OR gate. The AND gate as indicated in Fig. 12.10b is used to

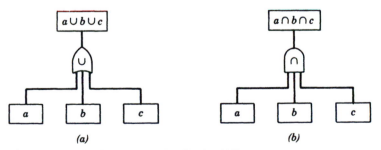

FIGURE 12.10 Fault-tree gates: (a) OR, (b) AND.

show that the output fault occurs only if all the input faults occur. There may be any number of input faults to an AND gate.

Generally, OR and AND gates are distinguished by their shape. In free-hand drawings, however, it may be desirable to put the ∪ and ∩ symbols on the gates. Or the so-called engineering notation, in which OR is represented by a "+" and AND by "·", may be used. Obviously, if these notations are included, the care with which the shape of the gate is drawn becomes of secondary importance.

In addition to the AND and OR gates, the INHIBIT gate shown in Fig. 12.11*a* is also widely used. It is a special case of the AND gate. The output is caused by a single input, but some qualifying condition must be satisfied before the input can produce the output. The condition that must exist is indicated conventionally by an ellipse, which is located to the right of the gate. In other words the output happens only if the input occurs under the conditions specified within the ellipse. The ellipse may also be used to indicate conditions on OR or AND gates. This is shown in Figs. 12.11*b* and *c*.

The rectangular boxes in the foregoing figures indicate top or intermediate events; they appear as outputs of gates. Shape also distinguishes different types of primary or input events appearing at the bottom of the fault tree. The primary events of a fault tree are events that, for one of a number of reasons, are not developed further. They are events for which probabilities must be provided if the fault tree is to be evaluated quantitatively (i.e., if the probability of the top event is to be calculated).

In general, four different types of primary events are distinguished. These make up part of the list of symbols in Table 12.1. The circle describes a basic event. This is a basic initiating fault event that requires no further development. The circle indicates that the appropriate resolution of the fault tree has been reached.

The undeveloped event is indicated by a diamond. It refers to a specific fault event, although it is not further developed, either because the event is of insufficient consequence or because information relevant to the event is unavailable. In contrast, the external event, signified by a house-shaped figure, indicates an event that is normally expected to occur. Thus house symbol displays are not of themselves faults.

The last symbols in Table 12.1 are the triangles indicating transfers into

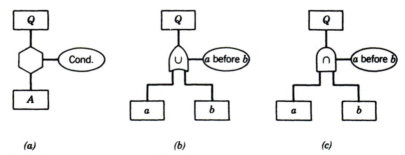

(a)   (b)   (c)

FIGURE 12.11 Fault-tree conditional gates.

TABLE 12.1 Fault-Tree Symbols Commonly Used

Symbol	Name	Description
	Rectangle	Fault event; it is usually the result of the logical combination of other events.
	Circle	Independent primary fault event.
	Diamond	Fault event not fully developed, for its causes are not known; it is only an assumed primary fault event.
	House	Normally occurring basic event; it is not a fault event.
	OR Gate	The union operation of events; i.e., the output event occurs if one or more of the inputs occur.
	AND Gate	The intersection operation of events; i.e., the output event occurs if and only if all the inputs occur.
	INHIBIT Gate	Output exists when X exists and condition A is present; this gate functions somewhat like an AND gate and is used for a secondary fault event X.
	Triangle-in	Triangle symbols provide a tool to avoid repeating sections of a fault tree or to transfer the tree construction from one sheet to the next. The triangle-in appears at the bottom of a tree and represents the branch of the tree (in this case A) shown someplace else. The triangle-out appears at the top of a tree and denotes that the tree A is a subtree to one shown someplace else.
	Triangle-out	

Source: Adapted from H. R. Roberts, W. E. Vesley, D. F. Haast, and F. F. Goldberg, *Fault Tree Handbook*, U.S. Nuclear Regulatory Commission, NUREG-0492, 1981.

and out of the fault tree. These are used when more than one page is required to draw a fault tree. A transfer-in triangle indicates that the input to a gate is developed on another page. A transfer-out triangle at the top of a tree indicates that it is the input to a gate appearing on another page.

In fault-tree construction a distinction is made between a fault and a failure. The word *failure* is reserved for basic events such as a burned-out bearing in a pump or a short circuit in an amplifier. The word *fault* is more all-encompassing. Thus, if a valve closes when it should not, this may be considered a valve fault. However, if the valve fault is due to a spurious signal from the shorted amplifier, it is not a valve failure. Thus all failures are faults, but not all faults are failures.

Fault Classification

The dissection of a system to determine what combinations of primary failures may lead to the top event is central to the construction of a fault tree. This dissection is likely to proceed most smoothly when the system can be divided into subsystems, components, or parts in order to associate the faults with discrete pieces of the system. Even then, a great deal of attention must be given to the component interactions, particularly common-mode failures. Beyond decomposing the system into components, however, we must also examine which components are more likely to fail and study with care the various modes by which component failure may occur.

In the material already covered, we have examined several ways of classifying failures that are very useful for fault-tree construction. Distinguishing between hardware faults and human error is essential, as is the classification of hardware failures into early, random, and aging, each with its own characteristics and causes. In what follows we discuss briefly two additional classifications. The division of failures into primary, secondary, and command faults is particularly useful in determining the logical structure of a fault tree. The classification of components as passive or active is important in determining which ones are likely to make larger contributions to system failure.

Primary, Secondary, and Command Faults   Failures may be usefully classified as primary, secondary, and command faults.* A primary fault by definition occurs in an environment and under a loading for which the component is qualified. Thus a pressure vessel's bursting at less than the design pressure is classified as a primary fault. Primary faults are most often caused by defective design, manufacture, or construction and are therefore most closely correlated to wear-in failures. Primary faults may also be caused by excessive or unanticipated wear, or they may occur when the system is not properly maintained and parts are not replaced on time.

Secondary faults occur in an environment or under loading for which the component is not qualified. For example, if a pressure vessel fails through excessive pressure for which it was not designed, it has a secondary fault. As indicated by the name, the basic failure is not of the vessel but in the excessive loading or adverse environment. Such failures often occur randomly and are characterized by constant failure rates.

Although a component fails when it has primary and secondary faults, it operates correctly when it has a command fault, but at the wrong time or place. Thus, our pressure vessel might lose pressure through the unwanted opening of a relief valve, even though there is no excessive pressure. If the valve opens through an erroneous signal, it has a command fault. For com-

---

* *Fault Tree Handbook,* op. cit.

mand failures we must look beyond the component failure to find the source of the erroneous command.

Passive and Active Faults   Components may be designated as either passive or active. Passive components include such things as pipes, cables, bearings, welds, and bolts. They function in a more or less static manner, often acting as transmitters of energy, such as a buss bar or cable, or of fluids such as piping. Transmitters of mechanical loads, such as structural members, beams, columns, and so on, and connectors, such as welds, bolts, or other fasteners, are also passive components. A passive component may usually be thought of as a mechanism for transmitting the output of one active component to the input of another. In the broadest sense, the quantity transmitted may be an electric signal, a fluid, mechanical loading, or any number of other quantities.

Active components contribute to the system function in a dynamic manner, altering in some way the system's behavior. For example, pumps and valves modify fluid flow; relays, switches, amplifiers, rectifiers, and computer chips modify electric signals; motors, clutches, and other machinery modify the transmission of mechanical loading.

Our primary reason for distinguishing between active and passive components is that failure rates are normally much higher for active components than for passive components, often by two or three orders of magnitude. The terms *active* and *passive* refer to the primary function of the component. Indeed, an active component may have many passive parts that are prone to failure. For example, a pump and its function are active, but the pump housing is considered passive, even though a housing rupture is one mode by which the pump may fail. In fact, one of the reasons that active components have higher failure rates than passive ones is that they tend to be made up of many nonredundant parts both active and passive.

Examples

We present here four examples of rather simple systems, and ones that are, moreover, readily understandable without specialized knowledge. This is consistent with the philosophy that one should not attempt to construct a fault tree until the design and function of the system is thoroughly understood. The first example is a demand failure, the failure of a motor to start; and the second is the failure of a continuously operating system. The third involves both start-up and operation; in the fourth the top event is a catastrophic failure, and its causes involve faulty procedures and operator actions as well as equipment failures.

EXAMPLE 12.2*

Draw a fault tree for the motor circuit shown in Fig. 12.12. The top event for the fault-tree analysis is simply failure of the motor to operate.

* Adapted from J. B. Fussel in *Generic Techniques in System Reliability Assessment,* E. J. Henley and J. W. Lynn (eds.), Nordhoff, Leyden, Holland, 1976.

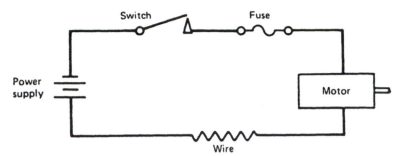

FIGURE 12.12 Electric motor circuit. (From J. B. Fussel, in *Generic Techniques in System Reliability Assessment*, pp. 133–162, E. J. Henley and J. W. Lynn (eds.), Martinus Nijhoff/Dr. Junk Publishers (was Sijthoff Noordhoff), Leyden, 1976, reprinted by permission.)

Solution The fault tree is shown in Figure 12.13. Note that failures are distinguished as primary and secondary. For primary failures we would expect data to be available to determine the failure probabilities. If not, further dissection of the component into its parts might be necessary. The secondary faults are either command faults, such as no current to the motor, or excessive loading, such as an overload in the circuits. For these we must delve deeper to locate the causes of the faults.

EXAMPLE 12.3*

Draw a fault tree for the coolant supply system pictured in Fig. 12.14. Here the top event is loss of minimum flow to a heat exchanger.

Solution The fault tree is shown in Fig. 12.15. Not all of the faults at the bottom of the tree are primary failures. Thus it may be desirable to develop some of the faults, such as loss of the pump inlet supply, further. Conversely, the faults may be considered too insignificant to be traced further, or data may be available even though they are not primary failures.

EXAMPLE 12.4†

Consider the sump pump system shown in Fig. 12.16. Redundance is provided by a battery-driven backup system that is activated when the utility power supply fails. Draw a fault tree for the flooding of a basement protected by this system.

---

* Adapted from J. A. Burgess, "Spotting Trouble Before It Happens," *Machine Design*, **42**, No. 23, 150 (1970).
† Adapted from A. H-S. Ang and W. H. Tang, *Probability Concepts in Engineering Planning and Design*, Vol. 2, Wiley, New York, 1984.

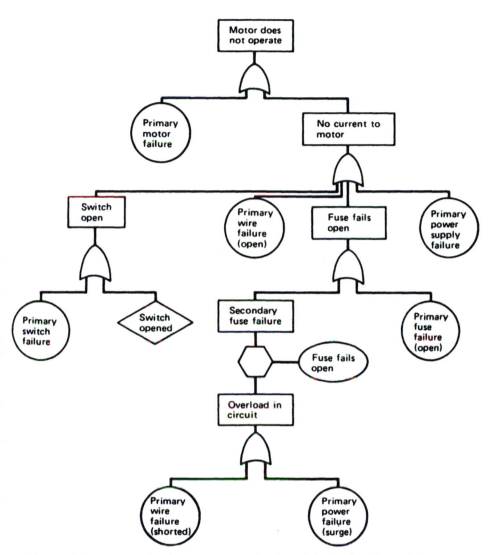

FIGURE 12.13 Fault tree for electric motor circuit. (From J. B. Fussel in *Generic Techniques in System Reliability Assessment*, pp. 133–162, E. J. Henley and J. W. Lynn (eds.), Martinus Nijhoff/ Dr. Junk Publishers (was Sijthoff Noordhoff), Leyden, 1976, reprinted by permission.)

Solution   The fault tree is shown in Fig. 12.17. The tree accounts for the fact that flooding can occur if the rate of inflow from the storm exceeds the pump capacity. Moreover, flooding can occur from storms within the system's capacity if there are malfunctions of both pumps and the inflow is large enough to fill the sump. Primary pump failures may be caused either by the failure of the pump itself or by loss of ac power. Similarly, the second pump may malfunction or it may be lost through failure of the battery. The battery fails only if all three events at the bottom of the tree take place.

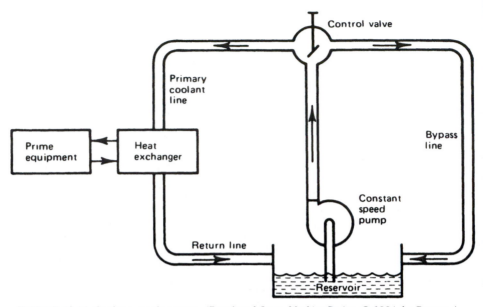

FIGURE 12.14 Coolant supply system. (Reprinted from *Machine Design,* © 1984, by Penton/ IPC, Cleveland, Ohio.)

EXAMPLE 12.5*

The final example that we consider is the pumping system shown in Fig. 12.18. The top event here is rupture of the pressure tank. This situation has the added complication that operator errors as well as equipment failures may lead to the top event. Before a fault tree can be drawn, the procedure by which the system is operated must be specified. The tank is filled in 10 min and empties in 50 min. Thus there is a 1-hr cycle time. After the switch is closed, the timer is set to open the contact in 10 min. If there is a failure in the mechanism, the alarm horn sounds. The operator then opens the switch to prevent the tank from overfilling and therefore rupturing.

Solution   A fault tree for the tank rupture is shown in Fig. 12.19. Notice how the analyst has used primary (i.e., basic), secondary, and command faults at several points in developing the tree. The operator's actions, a primary failure, are interpreted as the operator's failing to push the button when the alarm sounds. A secondary fault would occur, for example, if the operator is absent or unconscious when the alarm sounded, and the command fault for the operator would take place if the alarm does not sound.

The foregoing examples give some idea of the problems inherent in drawing fault trees. The reader should consult more advanced literature for

---

* Adapted from E. J. Henley and H. Kumamoto, *Reliability Engineering and Risk Assessment,* Prentice-Hall, Englewood Cliffs, NJ, 1981.

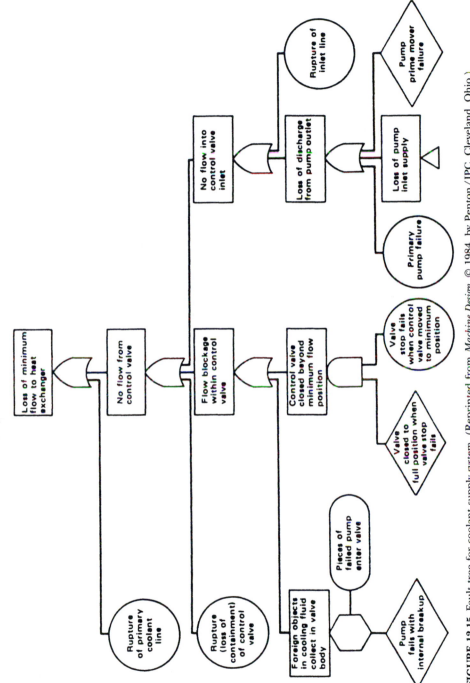

FIGURE 12.15 Fault tree for coolant supply system. (Reprinted from *Machine Design*, © 1984, by Penton/IPC, Cleveland, Ohio.)

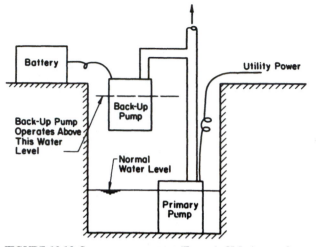

FIGURE 12.16 Sump pump system. (From A. H-S. Ang and
W. H. Tang, *Probability Concepts in Engineering Planning and De-
sign,* Vol. 2, p. 496. Copyright © 1984, by John Wiley and
Sons, New York. Reprinted by permission.)

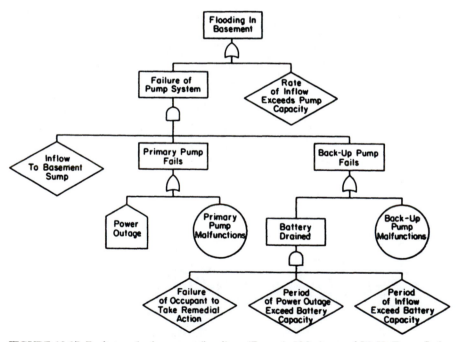

FIGURE 12.17 Fault tree for basement flooding. (From A. H-S. Ang and W. H. Tang, *Probability Concepts in Engineering Planning and Design,* Vol. 2, p. 496. Copyright © 1984, by John
Wiley and Sons, New York. Reprinted by permission.)

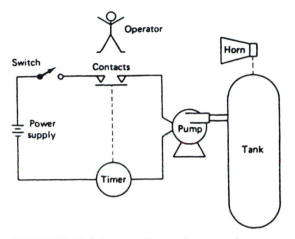

FIGURE 12.18 Schematic diagram for a pumping system. (From Ernest J. Henley and Hiromitsu Kumamoto, *Reliability Engineering and Risk Assessment*, p. 73, © 1981, with permission from Prentice-Hall, Englewood Cliffs, NJ.)

fault-tree constructions for more complex configurations, keeping in mind that the construction of a valid fault tree for any real system (as opposed to textbook examples) is necessarily a learning experience for the analyst. As the tree is drawn, more and more knowledge must be gained about the details of the system's components, its failure modes, the operating and maintenance procedures and the environment in which the system is to be located.

12.6 DIRECT EVALUATION OF FAULT TREES

The evaluation of a fault tree proceeds in two steps. First, a logical expression is constructed for the top event in terms of combinations (i.e., unions and intersections) of the basic events. This is referred to as qualitative analysis. Second, this expression is used to give the probability of the top event in terms of the probabilities of the primary events. This is referred to as quantitative analysis. Thus, knowing the probabilities of the primary events, we can calculate the probability of the top event. In these steps the rules of Boolean algebra contained in Table 12.2 are very useful. They allow us to simplify the logical expression for the fault tree and thus also to streamline the formula giving the probability of the top event in terms of the primary-failure probabilities.

In this section we first illustrate the two most straightforward methods for obtaining a logical expression for the top event, top-down and bottom-up evaluation. We then demonstrate how the resulting expression can be reduced in a way that greatly simplifies the relation between the probabilities of top and basic events. Finally, we discuss briefly the most common forms

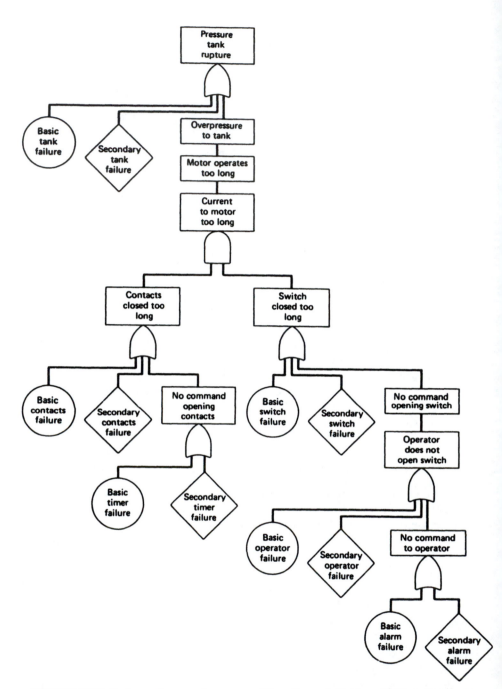

FIGURE 12.19 Fault tree for pumping system. (From Ernest J. Henley and Hiromitsu Kuma-moto, *Reliability Engineering and Risk Assessment*, p. 73, © 1981, with permission from Prentice-Hall, Englewood Cliffs, NJ.)

TABLE 12.2 Boolean Logic

A	B	$A \cap B$	$A \cup B$
0	0	0	0
1	0	0	1
0	1	0	1
1	1	1	1

that the primary-failure probabilities take and demonstrate the quantitative evaluation of a fault tree.

The so-named direct methods discussed in this section become unwieldy for very large fault trees with many components. For large trees the evaluation procedure must usually be cast in the form of a computer algorithm. These algorithms make extensive use of an alternative evaluation procedure in which the problem is recast in the form of so-called minimum cut sets, both because the technique is well suited to computer use and because additional insights are gained concerning the failure modes of the sytem. We define cut sets and discuss their use in the following section.

Qualitative Evaluation

Suppose that we are to evaluate the fault tree shown in Fig. 12.20. In this tree we have signified the primary failures by uppercase letters A through C. Note

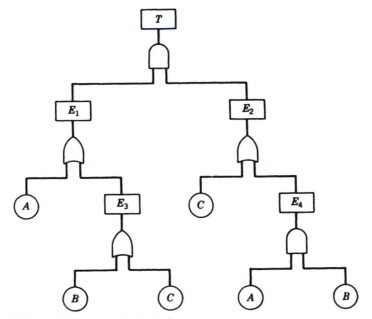

FIGURE 12.20 Example of a fault tree.

that the same primary failure may occur in more than one branch of the tree. This is typical of systems with m/N redundancy of the type discussed in Chapter 9. The intermediate events are indicated by E_i, and the top event by T.

Top Down To evaluate the tree from the top down, we begin at the top event and work our way downward through the levels of the tree, replacing the gates with the corresponding OR or AND symbol. Thus we have

$$T = E_1 \cap E_2 \tag{12.1}$$

at the highest level of the tree, and

$$E_1 = A \cup E_3; \qquad E_2 = C \cup E_4 \tag{12.2}$$

at the intermediate level. Substituting Eq. 12.2 into Eq. 12.1, we then obtain

$$T = (A \cup E_3) \cap (C \cup E_4). \tag{12.3}$$

Proceeding downward to the lowest level, we have

$$E_3 = B \cup C; \qquad E_4 = A \cap B. \tag{12.4}$$

Substituting these expressions into Eq. 12.3, we obtain as our final result

$$T = [A \cup (B \cup C)] \cap [C \cup (A \cap B)]. \tag{12.5}$$

Bottom Up Conversely, to evaluate this same tree from the bottom up, we first write the expressions for the gates at the bottom of the fault tree as

$$E_3 = B \cup C; \qquad E_4 = A \cap B. \tag{12.6}$$

Then, proceeding upward to the intermediate level, we have

$$E_1 = A \cup E_3; \qquad E_2 = C \cup E_4. \tag{12.7}$$

Hence we may substitute Eq. 12.6 into Eq. 12.7 to obtain

$$E_1 = A \cup (B \cup C) \tag{12.8}$$

and

$$E_2 = C \cup (A \cap B). \tag{12.9}$$

We now move to the highest level of the fault tree and express the AND gate appearing there as

$$T = E_1 \cap E_2. \tag{12.10}$$

Then, substituting Eqs. 12.8 and 12.9 into Eq. 12.10, we obtain the final form:

$$T = [A \cup (B \cup C)] \cap [C \cup (A \cap B)]. \tag{12.11}$$

The two results, Eqs. 12.5 and 12.11, which we have obtained with the two evaluation procedures, are not surprisingly the same.

Logical Reduction   For most fault trees, particularly those with one or more primary failures occurring in more than one branch of the tree, the rules of Boolean algebra contained in Table 2.1 may be used to simplify the logical expression for T, the top event. In our example, Eq. 12.11 can be simplified by first applying the associative and then the commutative law to write $A \cup (B \cup C) = (A \cup B) \cup C = C \cup (A \cup B)$. Then we have

$$T = [C \cup (A \cup B)] \cap [C \cup (A \cap B)]. \tag{12.12}$$

We then apply the distributive law with $X \equiv C$, $Y \equiv A \cup B$, and $Z \equiv A \cap B$ to obtain

$$T = C \cup [(A \cup B) \cap (A \cap B)]. \tag{12.13}$$

From the associative law we can eliminate the parenthesis on the right. Then, since $A \cap B = B \cap A$, we have

$$T = C \cup [(A \cup B) \cap B \cap A]. \tag{12.14}$$

Now, from the absorption law $(A \cup B) \cap B = B$. Hence

$$T = C \cup (B \cap A). \tag{12.15}$$

This expression tells us that for the fault tree under consideration the failure of the top system is caused by the failure of C or by the failure of both A and B. We then refer to $M_1 = C$ and $M_2 = A \cap B$ as the two failure modes leading to the top event. The reduced fault tree can be drawn to represent the system as shown in Fig. 12.21.

Quantitative Evaluation

Having obtained, in its simplest form, the logical expression for the top event in terms of the primary failures, we are prepared to evaluate the probability

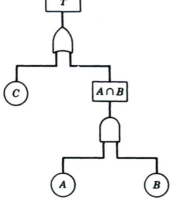

FIGURE 12.21 Fault-tree equivalent to Fig. 12.20.

that the top event will occur. The evaluation may be divided into two tasks. First, we must use the logical expression and the rules developed in Chapter 2 for combining probabilities to express the probability of the top event in terms of the probabilities of the primary failures. Second, we must evaluate the primary-failure probabilties in terms of the data available for component unreliabilities, component unavailabilities, and demand-failure probabilities.

Probability Relationships To illustrate the quantitative evaluation, we again use the fault tree that reduces to Eq. 12.15. Since the top event is the union of C with $B \cap A$, we use Eq. 2.10 to obtain

$$P\{T\} = P\{C\} + P\{B \cap A\} - P\{A \cap B \cap C\}, \tag{12.16}$$

thus expressing the top events in terms of the intersections of the basic events. If the basic events are known to be independent, the intersections may be replaced by the products of basic-event probabilities. Thus, in our example,

$$P\{T\} = P\{C\} + P\{A\}P\{B\} - P\{A\}P\{B\}P\{C\}. \tag{12.17}$$

If there are known dependencies between events, however, we must determine expression for $P\{A \cap B\}$, $P\{A \cap B \cap C\}$, or both through more sophisticated treatments such as the Markov models discussed in Chapter 11. Alternatively, we may be able to apply the β-factor treatment of Chapter 9 for common-mode failures.

Even where independent failures can be assumed, a problem arises when larger trees with many different component failures are considered. Instead of three terms as in Eq. 12.17, there may be hundreds of terms of vastly different magnitudes. A systematic way is needed for making reasonable approximations without evaluating all the terms. Since the failure probabilities are rarely known to more than two or three places of accuracy, often only a few of the terms are of significance. For example, suppose that in Eq. 12.17 the probabilities of A, B, and C are $\sim 10^{-2}$, 10^{-4}, and $\sim 10^{-6}$, respectively. Then the first two terms in Eq. 12.17 are each of the order 10^{-6}; in comparison the last term is of the order of 10^{-12} and may therefore be neglected.

One approach that is used in rough calculations for larger trees is to approximate the basic equation for $P\{X \cup Y\}$ by assuming that both events are improbable. Then, instead of using Eq. 2.10, we may approximate

$$P\{X \cup Y\} \approx P\{X\} + P\{Y\}, \tag{12.18}$$

which leads to a conservative (i.e., pessimistic) approximation for the system failure. For our simple example, we have, instead of Eq. 12.17, the approximation

$$P\{T\} \approx P\{C\} + P\{A\}P\{B\}. \tag{12.19}$$

The combination of this form of the rare-event approximation and the assumption of independence,

$$P\{X \cap Y\} = P\{X\}P\{Y\}, \tag{12.20}$$

often allows a very rough estimate of the top-event probability. We simply perform a bottom-up evaluation, multiplying probabilities at AND gates and adding them at OR gates. Care must be exercised in using this technique, for it is applicable only to trees in which basic events are not repeated—since repeated events are not independent—or to trees that have been logically reduced to a form in that primary failures appear only once. Thus we may not evaluate the tree as it appears in Fig. 12.20 in this way, but we may evaluate the reduced form in Fig. 12.21. More systematic techniques for truncating the prohibitively long probability expressions that arise from large fault trees are an integral part of the minimum cut-set formulation considered in the next section.

Primary-Failure Data   In our discussions we have described fault trees in terms of failure probabilities without specifying the particular types of failure represented either by the top event or by the primary-failure data. In fact, there are three types of top events and, correspondingly, three types of basic events frequently used in conjunction with fault trees. They are (1) the failure on demand, (2) the unreliability for some fixed period of time t, and (3) the unavailability at some time.

When failures on demand are the basic events, a value of p is needed. For the unreliability or unavailability it is often possible to use the following approximations to simplify the form of the data, since the probabilities of failure are expected to be quite small. If we assume a constant failure rate, the unreliability is

$$\tilde{R} \simeq \lambda t. \tag{12.21}$$

Similarly, the most common unavailability is the asymptotic value, for a system with constant failure and repair rates λ and ν. From Eq. 10.56 we have

$$\tilde{A}(\infty) = 1 - \frac{\nu}{\nu + \lambda}. \tag{12.22}$$

But, since in the usual case $\nu \gg \lambda$, we may approximate this by

$$\tilde{A}(\infty) \approx \lambda/\nu. \tag{12.23}$$

Often, demand failures, unreliabilities, and unavailabilities will be mixed in a single fault tree. Consider, for example, a very simple fault tree for the failure of a light to go on when the switch is flipped. We assume that the top event, T, is the failure on demand for the light to go on, which is due to

X = bulb burned out,

Y = switch fails to make contact,

Z = power failure to house.

Therefore $T = X \cup Y \cup Z$. In this case, X might be considered an unreliability of the bulb, with the time being that since it was originally installed; Y would be a demand failure, assuming that the cause was a random failure of the switch to make contact; and Z would be the unavailability of power to the circuit. Of course, the tree can be drawn in more depth. Is the random demand failure the only significant reason (a demand failure) for the switch not to make contact, or is there a significant probability that the switch is corroded open (an unreliability)?

12.7 FAULT-TREE EVALUATION BY CUT SETS

The direct evaluation procedures just discussed allow us to assess fault trees with relatively few branches and basic events. When larger trees are considered, both evaluation and interpretation of the results become more difficult and digital computer codes are invariably employed. Such codes are usually formulated in terms of the minimum cut-set methodology discussed in this section. There are at least two reasons for this. First, the techniques lend themselves well to the computer algorithms, and second, from them a good deal of intermediate information can be obtained concerning the combination of component failures that are pertinent to improvements in system design and operations.

The discussion that follows is conveniently divided into qualitative and quantitative analysis. In qualitative analysis information about the logical structure of the tree is used to locate weak points and evaluate and improve system design. In quantitative analysis the same objectives are taken further by studying the probabilities of component failures in relation to system design.

Qualitative Analysis

In these subsections we first introduce the idea of minimum cut sets and relate it to the qualitative evaluation of fault trees. We then discuss briefly how the minimum cut sets are determined for large fault trees. Finally, we discuss their use in locating system weak points, particularly possibilities for common-mode failures.

Minimum Cut-Set Formulation A minimum cut set is defined as the smallest combination of primary failures which, if they all occur, will cause the top event to occur. It, therefore, is a combination (i.e., intersection) of primary failures sufficient to cause the top event. It is the smallest combination in that all the failures must take place for the top event to occur. If even one of the failures in the minimum cut set does not happen, the top event will not take place.

The terms minimum cut set and failure mode are sometimes used interchangeably. However, there is a subtle difference that we shall observe hereafter. In reliability calculations a failure mode is a combination of component or other failures that cause a system to fail, regardless of the consequences of the failure. A minimum cut set is usually more restrictive, for it is the minimum combination of failures that causes the top event as defined for a particular fault tree. If the top event is defined broadly as system failure, the

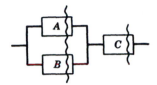

FIGURE 12.22 Minimum cut sets on a reliability block diagram.

two are indeed interchangeable. Usually, however, the top event encompasses only the particular subset of system failures that bring about a particular safety hazard.

The origin for using the term cut set may be illustrated graphically using the reduced fault tree in Fig. 12.21. The reliability block diagram corresponding to the tree is shown in Fig. 12.22. The idea of a cut set comes originally from the use of such diagrams for electric apparatus, where the signal enters at the left and leaves at the right. Thus a minimum cut set is the minimum number of components that must be cut to prevent the signal flow. There are two minimum cut sets, M_1, consisting of components A and B, and M_2, consisting of component C.

For a slightly more complicated example, consider the redundant system of Fig. 9.9, for which the equivalent fault tree appears in Fig. 12.8. In this system there are five cut sets, as indicated in the reliability block diagram of Fig. 12.23.

For larger systems, particularly those in which the primary failures appear more than once in the fault tree, the simple geometrical interpretation becomes problematical. However, the primary characteristics of the concept remain valid. It permits the logical structure of the fault tree to be represented in a systematic way that is amenable to interpretation in terms of the behavior of the minimum cut sets.

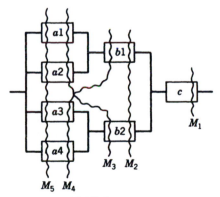

FIGURE 12.23 Minimum cut sets on a reliability block diagram of a seven-component system.

Suppose that the minimum cut sets of a system can be found. The top event, system failure, may then be expressed as the union of these sets. Thus, if there are N minimum cut sets,

$$T = M_1 \cup M_2 \cup \cdots \cup M_N. \tag{12.24}$$

Each minimum cut set then consists of the intersection of the minimum number of primary failures required to cause the top event. For example, the minimum cut sets for the system shown in Figs. 12.8 and 12.23 are

$$M_1 = c \qquad\qquad M_3 = a1 \cap a2 \cap b2$$

$$M_2 = b1 \cap b2 \qquad M_4 = a3 \cap a4 \cap b1 \tag{12.25}$$

$$M_5 = a1 \cap a2 \cap a3 \cap a4.$$

Before proceeding, it should be pointed out that there are other cut sets that will cause the top event, but they are not minimum cut sets. These need not be considered, however, because they do not enter the logic of the fault tree. By the rules of Boolean algebra contained in Table 2.1, they are absorbed into the minimum cut sets. This can be illustrated using the configuration of Fig. 12.23 again. Suppose that we examine the cut set $M_0 = b1 \cap c$, which will certainly cause system failure, but it is not a minimum cut set. If we include it in the expression for the top event, we have

$$T = M_0 \cup M_1 \cup M_2 \cup \cdots \cup M_N. \tag{12.26}$$

Now suppose that we consider $M_0 \cup M_1$. From the absorption law of Table 2.1, however, we see that

$$M_0 \cup M_1 = (b1 \cap c) \cup c = c. \tag{12.27}$$

Thus the nonminimum cut set is eliminated from the expression for the top event. Because of this property, minimum cut sets are often referred to simply as cut sets, with the minimum implied.

Since we are able to write the top event in terms of minimum cut sets as in Eq. 12.24, we may express the fault tree in the standardized form shown in Fig. 12.24. In this X_{mn} is the nth element of the mth minimum cut set. Note from our example that the same primary failures may often be expected to occur in more than one of the minimum cut sets. Thus the minimum cut sets are not generally independent of one another.

Cut-Set Determination   In order to utilize the cut-set formulations, we must express the top event as the union of minimum cut sets, as in Eq. 12.24. For small fault trees this can be done by hand, using the rules of Table 2.1, just as we reduced the top-event expression for T given by Eq. 12.11 to the two-cut-set expression given by Eq. 12.15. For larger trees, containing perhaps 20 or more primary failures, this procedure becomes intractable, and we must resort to digital computer evaluation. Even then the task may be prodigious, for a larger tree with a great deal of redundancy may have a million or more minimum cut sets.

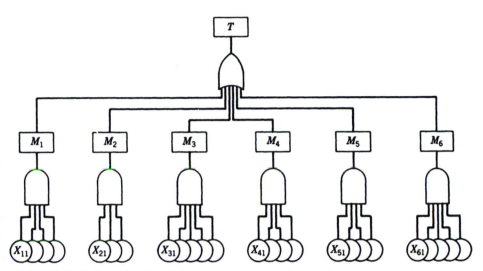

FIGURE 12.24 Generalized minimum cut-set representation of a fault tree.

The computer codes for determining the cut sets* do not typically apply the rules of Boolean algebra to reduce the expression for the top set to the form of Eq. 12.24. Rather, a search is performed for the minimum cut sets; in this, a failure is represented by 1 and a success by 0. Then each expression for the top event is evaluated using the outcome shown in Table 12.2 for the union and intersection of the events. A number of different procedures may be used to find the cut sets. In exhaustive searches, all single failures are first examined, and then all combinations of two primary failures, and so on. In general, there are 2^N, where N is the number of primary failures that must be examined. Other methods involve the use of random number generators in Monte Carlo simulation to locate the minimum cut sets.

When millions of minimum cut sets are possible, the search procedures are often truncated, for cut sets requiring many primary failures to take place are so improbable that they will not significantly affect the overall probability of the top event. Moreover, simulation methods must be terminated after a finite number of trials.

Cut-Set Interpretations   Knowing the minimum cut sets for a particular fault tree can provide valuable insight concerning potential weak points of complex systems, even when it is not possible to calculate the probability that either a particular cut set or the top event will occur. Three qualitative considerations, in particular, may be very useful: the ranking of the minimal cut sets by the number of primary failures required, the importance of particular component failures to the occurrence of the minimum cut sets, and the susceptibility of particular cut sets to common-mode failures.

---

* See, for example, N. J. McCormick, *Reliability and Risk Analysis*, Academic Press, New York, 1981.

Minimum cut sets are normally categorized as singlets, doublets, triplets, and so on, according to the number of primary failures in the cut set. Emphasis is then put on eliminating cut sets corresponding to small numbers of failures, for ordinarily these may be expected to make the largest contributions to system failure. In fact, the common design criterion, that no single component failure should cause system failure is equivalent to saying that all singlets must be removed from the fault tree for which the top event is system failure. Indeed, if component failure probabilities are small and independent, then provided that they are of the same order of magnitude, doublets will occur much less frequently than singlets, triplets much less frequently than doublets, and so on.

A second application of cut-set information is in assessing qualitatively the importance of a particular component. Suppose that we wish to evaluate the effect on the system of improving the reliability of a particular component, or conversely, to ask whether, if a particular component fails, the system-wide effect will be considerable. If the component appears in one or more of the low-order cut sets, say singlets or doublets, its reliability is likely to have a pronounced effect. On the other hand, if it appears only in minimum cut sets requiring several independent failures, its importance to system failure is likely to be small.

These arguments can rank minimum cut-set and component importance, assuming that the primary failures are independent. If they are not, that is, if they are susceptible to common-mode failure, the ranking of cut-set importance may be changed. If five of the failures in a minimum cut set with six failures, for example, can occur as the result of a common cause, the probability of the cut set's occurring is more comparable to that of a doublet.

Extensive analysis is often carried out to determine the susceptibility of minimum cut sets to common-cause failures. In an industrial plant one cause might be fire. If the plant is divided into several fire-resistant compartments, the analysis might proceed as follows. All the primary failures of equipment located in one of the compartments that could be caused by fire are listed. Then these components would be eliminated from the minimum cut sets (i.e., they would be assumed to fail). The resulting cut sets would then indicate how many failures—if any—in addition to those caused by the fire, would be required for the top event to happen. Such analysis is critical for determining the layout of the plant that will best protect it from a variety of sources of damage: fire, flooding, collision, earthquake, and so on.

Quantitative Analysis

With the minimum cut sets determined, we may use probability data for the primary failures and proceed with quantitative analysis. This normally includes both an estimate of the probability of the top event's occurring and quantitative measures of the importance of components and cut sets to the top event. Finally, studies of uncertainty about the top event's happening, because the probability data for the primary failures are uncertain, are often needed to assess the precision of the results.

Top-Event Probability To determine the probability of the top event, we must calculate

$$P\{T\} = P\{M_1 \cup M_2 \cup \cdots \cup M_N\}. \tag{12.28}$$

As indicated in Section 2.2, the union can always be eliminated from a probability expression by writing it as a sum of terms, each one of which is the probability of an intersection of events. Here the intersections are the minimum cut sets. Probability theory provides the expansion of Eq. 12.28 in the following form

$$P\{T\} = \sum_{i=1}^{N} P\{M_i\} - \sum_{i=2}^{N} \sum_{j=1}^{i-1} P\{M_i \cap M_j\}$$

$$+ \sum_{i=3}^{N} \sum_{j=2}^{i-1} \sum_{k=1}^{j-1} P\{M_i \cap M_j \cap M_k\} - \cdots \tag{12.29}$$

$$+ (-1)^{N-1} P\{M_1 \cap M_2 \cap \cdots \cap M_N\}.$$

This is sometimes referred to as the inclusion—exclusion principle.

The first task in evaluating this expression is to evaluate the probabilities of the individual minimum cut sets. Suppose that we let X_{im} represent the mth basic event in minimum cut set i. Then

$$P\{M_i\} = P\{X_{i1} \cap X_{i2} \cap X_{i3} \cap \cdots \cap X_{iM}\}. \tag{12.30}$$

If it may be proved that the primary failures in a given cut set are independent, we may write

$$P\{M_i\} = P\{X_{i1}\}P\{X_{i2}\} \cdots P\{X_{iM}\}. \tag{12.31}$$

If they are not, a Markov model or some other procedure must be used to relate $P\{M_i\}$ to the properties of the primary failures.

The second task is to evaluate the intersections of the cut-set probabilities. If the cut sets are independent of one another, we have simply

$$P\{M_i \cap M_j\} = P\{M_i\}P\{M_j\}, \tag{12.32}$$

$$P\{M_i \cap M_j \cap M_k\} = P\{M_i\}P\{M_j\}P\{M_k\}, \tag{12.33}$$

and so on. More often than not, however, these conditions are not valid, for in a system with redundant components, a given component is likely to appear in more than one minimum cut set: If the same primary failure appears in two minimum cut sets, they cannot be independent of one another. Thus an important point is to be made. Even if the primary events are independent of one another, the minimum cut sets are unlikely to be. For example, in the fault trees of Figs. 12.8 and 12.23 the minimum cut sets $M_1 = c$ and $M_2 = b1 \cap b2$ will be independent of one another if the primary failures of components $b1$

and $b2$ are independent of c. In this system, however, M_2 and M_3 will be dependent even if all the primary failures are independent because they contain the failure of component $b2$.

Although minimum cut sets may be dependent, calculation of their intersections is greatly simplified if the primary failures are all independent of one another, for then the dependencies are due only to the primary failures that appear in more than one minimum cut set. To evaluate the intersection of minimum cut sets, simply take the product of probabilities that appear in one or more of the minimal cut sets:

$$P\{M_i \cap M_j\} = P\{X_{1ij}\}P\{X_{2ij}\} \cdots P\{X_{Nij}\}, \tag{12.34}$$

where $X_{1ij}, X_{2ij}, \ldots, X_{Nij}$ is the list of the failures that appear in M_i, M_j, or both.

That the foregoing procedure is correct is illustrated by a simple example. Suppose that we have two minimal cut sets $M_1 = A \cap B$, $M_2 = B \cap C$, where the primary failures are independent. We then have

$$M_1 \cap M_2 = (A \cap B) \cap (B \cap C) = A \cap B \cap B \cap C, \tag{12.35}$$

but $B \cap B = B$. Thus

$$P\{M_1 \cap M_2\} = P\{A \cap B \cap C\} = P\{A\}P\{B\}P\{C\}. \tag{12.36}$$

In the general notation of Eq. 12.34 we would have

$$X_{112} = A, \qquad X_{212} = B, \qquad X_{312} = C. \tag{12.37}$$

With the assumption of independent primary failures, the series in Eq. 12.29 may in principle be evaluated exactly. When there are thousands or even millions of minimum cut sets to be considered, however, the task may be both prohibitive and unwarranted, for many of the terms in the series are likely to be completely negligible compared to the leading one or two terms.

The true answer may be bracketed by taking successive terms, and it is rarely necessary to evaluate more than the first two or three terms. If $P\{T\}$ is the exact value, it may be shown that[*]

$$P_1\{T\} \equiv \sum_{i=1}^{N} P\{M_i\} > P\{T\}, \tag{12.38}$$

$$P_2\{T\} \equiv P_1\{T\} - \sum_{i=2}^{N} \sum_{j=1}^{i-1} P\{M_i \cap M_j\} < P\{T\}, \tag{12.39}$$

$$P_3\{T\} \equiv P_2\{T\} + \sum_{i=3}^{N} \sum_{j=2}^{i-1} \sum_{k=1}^{j-1} P\{M_i \cap M_j \cap M_k\} > P\{T\}. \tag{12.40}$$

and so on, with $P_4\{T\} < P\{T\}$.

[*] W. E. Vesely, "Time Dependent Methodology for Fault Tree Evaluation," *Nucl. Eng. Design*, **13**, 337-357 (1970).

Often the first-order approximation $P_1\{T\}$ gives a result that is both reasonable and pessimistic. The second-order approximation might be evaluated to check the accuracy of the first. And rarely would more than the third-order approximation be used.

Even taking only a few terms in Eq. 12.38 may be difficult, and wasteful, if a million or more minimum cut sets are present. Thus, as mentioned in the preceding subsection, we often truncate the number of minimum cut sets to include only those that contain fewer than some specified number of primary failures. If all the failure probabilities are small, say <0.1, the cut-set probabilities should go down by more than an order of magnitude as we go from singlets to doublets, doublets to triplets, and so on.

Importance   As in qualitative analysis, it is not only the probability of the top event that normally concerns the analyst. The relative importance of single components and of particular minimum cut sets must be known if designs are to be optimized and operating procedures revised.

Two measures of importance* are particularly simple but useful in system analysis. In order to know which cut sets are the most likely to cause the top event, the cut-set importance is defined as

$$I_{M_i} = \frac{P\{M_i\}}{P\{T\}}$$

(12.41)

for the minimum cut set i. Generally, we would also like to determine the relative importance of different primary failures in contributing to the top event. To accomplish this, the simplest measure is to add the probabilities of all the minimum cut sets to which the primary failure contributes. Thus the importance of component X_i is

$$I_{X_i} = \frac{1}{P\{T\}} \sum_{X_i \in M_l} P\{M_l\}.$$

(12.42)

Other more sophisticated measures of importance have also found applications.

Uncertainty   What we have obtained thus far are point or best estimates of the top event's probability. However, there are likely to be substantial uncertainties in the basic parameters—the component failure rates, demand failures, and other data—that are input to the probability estimates. Given these considerable uncertainties, it would be very questionable to accept point estimates without an accompanying interval estimate by which to judge the precision of the results. To this end the component failure rates and other data may themselves be represented as random variables with a mean or best-estimate value and a variance to represent the uncertainty. The lognormal distribution has been very popular for representing failure data in this manner.

---

* See, for example, E. J. Henley and H. Kumamoto, op. cit., Chapter 10.

For small fault trees a number of analytical techniques may be applied to determine the sensitivity of the results to the data uncertainty. For larger trees the Monte Carlo method has found extensive use.*

Bibliography

Ang, A. H-S., and W. H. Tang, *Probability Concepts in Engineering Planning and Design*, Vol. 2, Wiley, NY, 1984.

Brockley, D., (ed.) *Engineering Safety*, McGraw-Hill, London, 1992.

Burgess, J. A., "Spotting Trouble Before It Happens," *Machine Design*, **42**, No. 23, 150 (1970).

Green, A. E., *Safety Systems Analysis*, Wiley, NY, 1983.

Guttman, H. R., unpublished lecture notes, Northwestern University, 1982.

Henley, E. J., and J. W. Lynn (eds.), *Generic Techniques in System Reliability Assessment*, Nordhoff, Leyden, Holland, 1976.

Henley, E. J., and H. Kumamoto, *Probabilistic Risk Assessment*, IEEE Press, New York, 1992.

————. *Reliability Engineering and Risk Assessment*, Prentice-Hall, Englewood Cliffs, NJ, 1981.

McCormick, E. J., *Human Factors in Engineering Design*, McGraw-Hill, NY, 1976.

McCormick, N. J., *Reliability and Risk Analysis*, Academic Press, NY, 1981.

PRA Procedures Guide, Vol 1. U.S. Nuclear Regulatory Commission, NUREG/CR-2300, 1983.

Rasmussen, J., "Human Factors in High Risk Technology," in *High Risk Technology*, A. E. Green (ed.), Wiley, NY, 1982.

Roberts, H. R., W. E. Vesley, D. F. Haast and F. F. Goldberg, *Fault Tree Handbook*, U.S. Nuclear Regulatory Commission, NUREG-0492, 1981.

D. H. Stamatis, Failure Modes and *Effect Analysis*, ASQC Quality Press, Milwaukee, WI, 1995

Swain, A. D., and H. R. Guttmann, *Handbook of Human Reliability Analysis with Emphasis on Nuclear Power Plant Applications*, U.S. Nuclear Regulatory Commission, NUREG/CR-1287, 1980.

Vesely, W. E., "Time Dependent Methodology for Fault Tree Evaluation," *Nucl. Eng. Design*, **13**, 1970.

EXERCISES

12.1 Classify each of the failures in Fig. 12.15 as (*a*) passive, (*b*) active, or (*c*) either.

12.2 Make a list of six population stereotypical responses.

12.3 Suppose that a system consists of two subsystems in parallel. Each has a mission reliability of 0.9.

---

* See, for example, E. J. Henley and H. Kumamoto, op. cit., Chapter 12.

(a) Draw a fault tree for mission failure and calculate the probability of the top event.

(b) Assume that there are common-mode failures described by the β-factor method (Chapter 9) with $\beta = 0.1$. Redraw the fault tree to take this into account and recalculate the top event.

12.4 Find the fault tree for system failure for the following configurations.

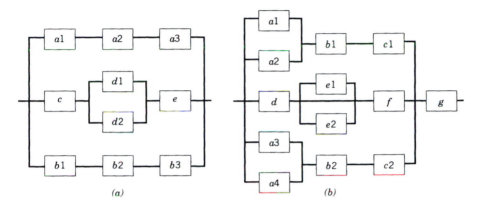

(a)   (b)

12.5 Find the minimum cut sets of the following fault tree.

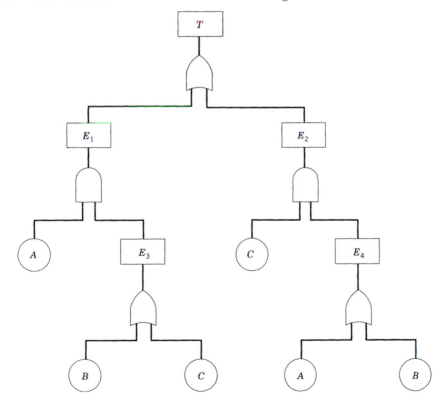

12.6 Draw a fault tree corresponding to the reliability block diagram in Exercise 9.37.

12.7 The following system is designed to deliver emergency cooling to a nuclear reactor.

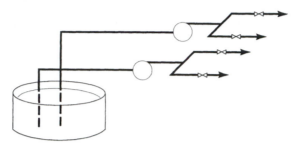

In the event of an accident the protection system delivers an actuation signal to the two identical pumps and the four identical valves. The pumps then start up, the valves open, and liquid coolant is delivered to the reactor. The following failure probabilities are found to be significant:

$p_{ps} = 10^{-5}$ the probability that the protection system will not deliver a signal to the pump and valve actuators.

$p_p = 2 \times 10^{-2}$ the probability that a pump will fail to start when the actuation signal is received.

$p_v = 10^{-1}$ the probability that a valve will fail to open when the actuation signal is received.

$p_r = 0.5 \times 10^{-5}$ the probability that the reservoir will be empty at the time of the accident.

(a) Draw a fault tree for the failure of the system to deliver any coolant to the primary system in the event of an accident.

(b) Evaluate the probability that such a failure will take place in the event of an accident.

12.8 Construct a fault tree for which the top event is your failure to arrive on time for the final exam of a reliability engineering course. Include only the primary failures that you think have probabilities large enough to significantly affect the result.

12.9 Suppose that a fault tree has three minimum cut sets. The basic failures are independent and do not appear in more than one cut set. Assume that $P\{M_1\} = 0.03$, $P\{M_2\} = 0.12$ and $P\{M_3\} = 0.005$. Estimate $P\{T\}$ by the three successive estimates given in Eqs. 12.38, 12.39, and 12.40.

12.10 Develop a logical expression for the fault trees in Fig. 12.13 in terms of the nine root causes. Find the minimum cut sets.

12.11 Suppose that for the fault tree given in Fig. 12.21 $P\{A\} = 0.15$, $P\{B\} = 0.20$, and $P\{C\} = 0.05$.

(a) Calculate the cut-set importances.
(b) Calculate the component importances.
   (Assume independent failures.)

12.12 The logical expression for a fault tree is given by

$$T = A \cap (B \cup C) \cap [D \cup (E \cap F \cap G)].$$

(a) Construct the corresponding fault tree.
(b) Find the minimum cut sets.
(c) Construct an equivalent reliability block diagram.

12.13 From the reliability block diagram shown in Figure 12.23, draw a fault tree for system failure in minimum cut-set form. Assume that the failure probabilities for component types *a, b,* and *c* are, respectively, 0.1, 0.02, and 0.005. Assuming independent failures, calculate

(a) $P\{T\}$, the probability of the top event;
(b) the importance of components $a1$, $b1$ and c;
(c) the importance of each of the five minimum cut sets.

12.14 Construct the fault trees for system failure for the low- and high-level redundant systems shown in Fig. 9.7. Then find the minimum cut sets.

APPENDIX A

Useful Mathematical Relationships

A.1 INTEGRALS

Definite Integrals

$$\int_0^\infty e^{-ax}\,dx = \frac{1}{a}, \qquad a > 0.$$

$$\int_0^\infty x^n e^{-ax}\,dx = \frac{n!}{a^{n+1}}, \qquad n = \text{integer} \geq 0,\, a > 0.$$

$$\int_0^\infty e^{-a^2x^2}\,dx = \frac{\sqrt{\pi}}{2a}, \qquad a > 0.$$

$$\int_0^\infty x e^{-x^2}\,dx = \tfrac{1}{2}.$$

$$\int_0^\infty x^2 e^{-x^2}\,dx = \frac{\sqrt{\pi}}{4}.$$

$$\int_0^\infty x^{2n} e^{-ax^2}\,dx = \frac{1 \cdot 3 \cdot 5 \cdots (2n-1)}{2^{n+1} a^n}\sqrt{\pi/a}, \qquad n = \text{integer} > 0,\, a > 0.$$

Integration by Parts

$$\int_a^b f(x)\,\frac{d}{dx}g(x)\,dx = f(b)g(b) - f(a)g(a) - \int_a^b g(x)\,\frac{d}{dx}f(x)\,dx.$$

Derivative of an Integral

$$\frac{d}{dc}\int_p^q f(x, c)\,dx = \int_p^q \frac{\partial}{\partial c} f(x, c)\,dx + f(q, c)\,\frac{dq}{dc} - f(p, c)\,\frac{dp}{dc}.$$

A.2 EXPANSIONS

Integer Series

$$1 + 2 + 3 + \cdots + n = \frac{n}{2}(n + 1).$$

$$1^2 + 2^2 + 3^2 + \cdots + n^2 = \frac{n}{6}(2n^2 + 3n + 1).$$

$$1^3 + 2^3 + 3^3 + \cdots + n^3 = \frac{n^2}{4}(n + 1)^2.$$

$$1 + 3 + 5 + \cdots + (2n - 1) = n^2.$$

Binomial Expansion

$$(p + q)^N = \sum_{n=0}^{N} C_n^N p^n q^{N-n}.$$

$$C_n^N \equiv \frac{N!}{(N - n)! n!}.$$

Geometric Progression

$$\frac{1 - p^n}{1 - p} = 1 + p + p^2 + p^3 + \cdots + p^{n-1}.$$

Infinite Series

$$e^x = 1 + \frac{x}{1!} + \frac{x^2}{2!} + \frac{x^3}{3!} + \cdots, \qquad x^2 < \infty.$$

$$\log(1 + x) = x - \frac{x^2}{2} + \frac{x^3}{3} - \frac{x^4}{4} + \cdots, \qquad x^2 < 1.$$

$$\frac{1}{1 - x} = 1 + x + x^2 + x^3 + x^4 + \cdots, \, x^2 < 1$$

$$\frac{1}{(1 - x^2)} = 1 + 2x + 3x^2 + 4x^3 + \cdots, \, x^2 < 1$$

$$\frac{1 + x}{(1 - x^2)^3} = 1 + 2^2 x + 3^2 x^2 + 4^2 x^3 + \cdots, \, x^2 < 1$$

A.3 SOLUTION OF A FIRST-ORDER LINEAR DIFFERENTIAL EQUATION

$$\frac{d}{dx} y(x) + \alpha(x) y(x) = S(x).$$

Note that

$$\frac{d}{dx} y(x) \exp\left[\int_{x_0}^{x} \alpha(x') \, dx'\right] = \left[\frac{d}{dx} y(x) + \alpha(x) y(x)\right] \exp\left[\int_{x_0}^{x} \alpha(x') \, dx'\right].$$

Thus, multiplying by the integrating factor $\exp[\int_{x_0}^{x} \alpha(x') \, dx']$, we have

$$\frac{d}{dx} y(x) \exp\left[\int_{x_0}^{x} \alpha(x') \, dx'\right] = S(x) \exp\left[\int_{x_0}^{x} \alpha(x') \, dx'\right].$$

Integrating between x_0 and x, we have

$$y(x) = y(x_0) \exp\left[-\int_{x_0}^{x} \alpha(x')\,dx'\right] + \int_{x_0}^{x} dx'\,S(x') \exp\left[-\int_{x'}^{x} \alpha(x'')\,dx''\right].$$

If α is a constant, then

$$y(x) = y(x_0) \exp[-\alpha(x - x_0)] + \int_{x_0}^{x} dx'\,S(x') \exp[-\alpha(x - x')].$$

Binomial Sampling Charts

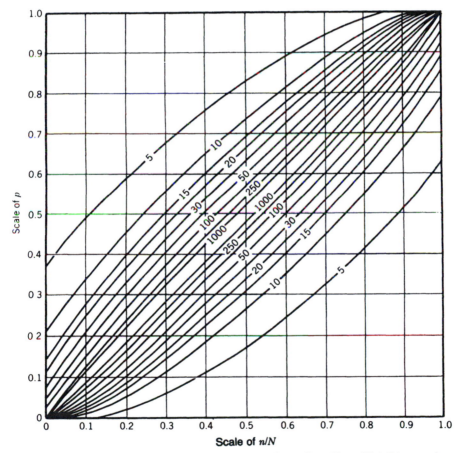

FIGURE B.1 An 80% confidence interval for binomial sampling. (From W. J. Dixon and F. J. Massey, Jr., *Introduction to Statistical Analysis*, 2nd ed., © 1957, with permission from McGraw-Hill Book Company, New York.)

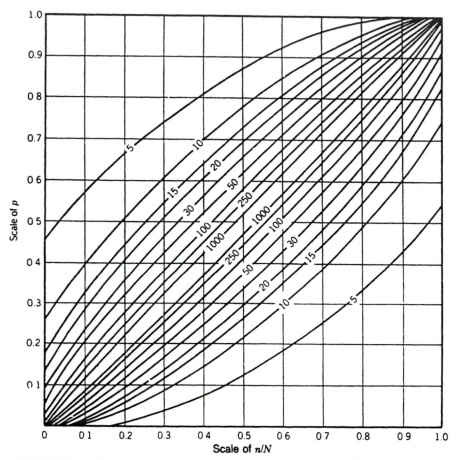

FIGURE B.2 A 90% confidence interval for binomial sampling. (From W. J. Dixon and F. J. Massey, Jr., *Introduction to Statistical Analysis,* 2nd ed., © 1957, with permission from McGraw-Hill Book Company, New York.)

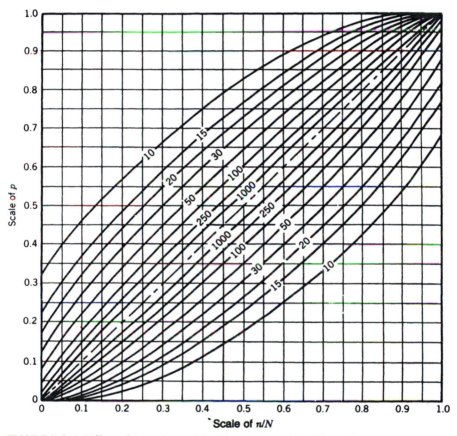

FIGURE B.3 A 95% confidence interval for binomial sampling. [From. E. S. Pearson and C. J. Clopper, "The Use of Confidence or Fiducial Limits Illustrated in the Case of the Binomial," *Biometrika*, **26**, 404 (1934). With permission of Biometrika.]

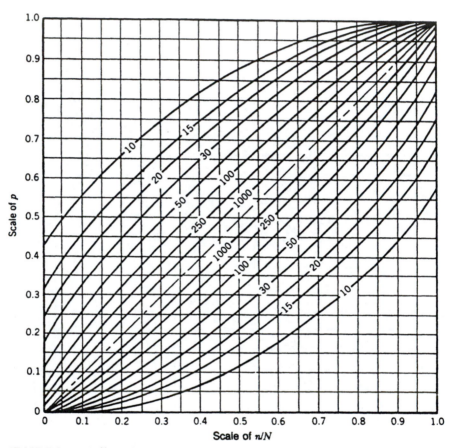

FIGURE B.4 A 99% confidence interval for binomial sampling. [From E. S. Pearson and C. J. Clopper, "The Use of Confidence or Fiducial Limits Illustrated in the Case of the Binomial," *Biometrika,* **26,** 404 (1934). With permission of Biometrika.]

APPENDIX C

Φ(z): Standard Normal CDF

z	.00	.01	.02	.03	.04	.05	.06	.07	.08	.09
− .0	.5000	.4960	.4920	.4880	.4840	.4801	.4761	.4721	.4681	.4641
− .1	.4602	.4562	.4522	.4483	.4443	.4404	.4364	.4325	.4286	.4247
− .2	.4207	.4168	.4129	.4090	.4052	.4013	.3974	.3936	.3897	.3859
− .3	.3821	.3783	.3745	.3707	.3669	.3632	.3594	.3557	.3520	.3483
− .4	.3446	.3409	.3372	.3336	.3300	.3264	.3228	.3192	.3156	.3121
− .5	.3085	.3050	.3015	.2981	.2946	.2912	.2877	.2343	.2810	.2776
− .6	.2743	.2709	.2676	.2643	.2611	.2578	.2546	.2514	.2483	.2451
− .7	.2420	.2389	.2358	.2327	.2297	.2266	.2236	.2206	.2177	.2148
− .8	.2119	.2090	.2061	.2033	.2005	.1977	.1949	.1922	.1894	.1867
− .9	.1841	.1814	.1788	.1762	.1736	.1711	.1685	.1660	.1635	.1611
−1.0	.1587	.1562	.1539	.1515	.1492	.1469	.1446	.1423	.1401	.1379
−1.1	.1357	.1335	.1314	.1292	.1271	.1251	.1230	.1210	.1190	.1170
−1.2	.1131	.1131	.1113	.1093	.1075	.1056	.1038	.1020	.1003	.09853
−1.3	.09680	.09510	.09342	.09176	.09012	.08851	.08691	.08534	.08379	.08226
−1.4	.08076	.07927	.07780	.07636	.07493	.07353	.07215	.07078	.06944	.06811
−1.5	.06681	.06552	.06426	.06301	.06178	.06057	.05938	.05821	.05705	.05592
−1.6	.05480	.05370	.05262	.05155	.05050	.04947	.04846	.04746	.04648	.04551
−1.7	.04457	.04363	.04272	.04182	.04093	.04006	.03920	.03836	.03754	.03673
−1.8	.03593	.03515	.03438	.03362	.03288	.03216	.03144	.03074	.03005	.02938
−1.9	.02872	.02807	.02743	.02680	.02619	.02559	.02500	.02442	.02385	.02330
−2.0	.02275	.02222	.02169	.02118	.02068	.02018	.01970	.01923	.01876	.01831
−2.1	.01786	.01743	.01700	.01659	.01618	.01578	.01539	.01500	.01463	.01426
−2.2	.01390	.01355	.01321	.01287	.01255	.01222	.01191	.01160	.01130	.01101
−2.3	.01072	.01044	.01017	$.0^2 9903$	$.0^2 9642$	$.0^2 9387$	$.0^2 9137$	$.0^2 8894$	$.0^2 8656$	$.0^2 8424$
−2.4	$.0^2 8198$	$.0^2 7976$	$.0^2 7760$	$.0^2 7549$	$.0^2 7344$	$.0^2 7143$	$.0^2 6947$	$.0^2 6756$	$.0^2 6569$	$.0^2 6387$
−2.5	$.0^2 6210$	$.0^2 6037$	$.0^2 5868$	$.0^2 5703$	$.0^2 5543$	$.0^2 5386$	$.0^2 5234$	$.0^2 5085$	$.0^2 4940$	$.0^2 4799$
−2.6	$.0^2 4661$	$.0^2 4527$	$.0^2 4396$	$.0^2 4269$	$.0^2 4145$	$.0^2 4025$	$.0^2 3907$	$.0^2 3793$	$.0^2 3681$	$.0^2 3573$
−2.7	$.0^2 3467$	$.0^2 3364$	$.0^2 3264$	$.0^2 3167$	$.0^2 3072$	$.0^2 2980$	$.0^2 2890$	$.0^2 2803$	$.0^2 2718$	$.0^2 2635$
−2.8	$.0^2 2555$	$.0^2 2477$	$.0^2 2401$	$.0^2 2327$	$.0^2 2256$	$.0^2 2186$	$.0^2 2118$	$.0^2 2052$	$.0^2 1988$	$.0^2 1926$
−2.9	$.0^2 1866$	$.0^2 1807$	$.0^2 1750$	$.0^2 1695$	$.0^2 1641$	$.0^2 1589$	$.0^2 1538$	$.0^2 1489$	$.0^2 1441$	$.0^2 1395$
−3.0	$.0^2 1350$	$.0^2 1306$	$.0^2 1264$	$.0^2 1223$	$.0^2 1183$	$.0^2 1144$	$.0^2 1107$	$.0^2 1070$	$.0^2 1035$	$.0^2 1001$
−3.1	$.0^3 9676$	$.0^3 9354$	$.0^3 9043$	$.0^3 8740$	$.0^3 8447$	$.0^3 8164$	$.0^3 7888$	$.0^3 7622$	$.0^3 7364$	$.0^3 7114$
−3.2	$.0^3 6871$	$.0^3 6637$	$.0^3 6410$	$.0^3 6190$	$.0^3 5976$	$.0^3 5770$	$.0^3 5571$	$.0^3 5377$	$.0^3 5190$	$.0^3 5009$
−3.3	$.0^3 4834$	$.0^3 4663$	$.0^3 4501$	$.0^3 4342$	$.0^3 4189$	$.0^3 4041$	$.0^3 3897$	$.0^3 3758$	$.0^3 3624$	$.0^3 3495$
−3.4	$.0^3 3369$	$.0^3 3248$	$.0^3 3131$	$.0^3 3018$	$.0^3 2909$	$.0^3 2803$	$.0^3 2701$	$.0^3 2602$	$.0^3 2507$	$.0^3 2415$
−3.5	$.0^3 2326$	$.0^3 2241$	$.0^3 2158$	$.0^3 2078$	$.0^3 2001$	$.0^3 1926$	$.0^3 1854$	$.0^3 1785$	$.0^3 1718$	$.0^3 1653$
−3.6	$.0^3 1591$	$.0^3 1531$	$.0^3 1473$	$.0^3 1417$	$.0^3 1363$	$.0^3 1311$	$.0^3 1261$	$.0^3 1213$	$.0^3 1166$	$.0^3 1121$
−3.7	$.0^3 1078$	$.0^3 1036$	$.0^4 9961$	$.0^4 9574$	$.0^4 9201$	$.0^4 8842$	$.0^4 8496$	$.0^4 8162$	$.0^4 7841$	$.0^4 7532$
−3.8	$.0^4 7235$	$.0^4 6948$	$.0^4 6673$	$.0^4 6407$	$.0^4 6152$	$.0^4 5906$	$.0^4 5669$	$.0^4 5442$	$.0^4 5223$	$.0^4 5012$
−3.9	$.0^4 4810$	$.0^4 4615$	$.0^4 4427$	$.0^4 4247$	$.0^4 4074$	$.0^4 3908$	$.0^4 3747$	$.0^4 3594$	$.0^4 3446$	$.0^4 3304$
−4.0	$.0^4 3167$	$.0^4 3036$	$.0^4 2910$	$.0^4 2789$	$.0^4 2673$	$.0^4 2561$	$.0^4 2454$	$.0^4 2351$	$.0^4 2242$	$.0^4 2157$
−4.1	$.0^4 2066$	$.0^4 1978$	$.0^4 1894$	$.0^4 1814$	$.0^4 1737$	$.0^4 1662$	$.0^4 1591$	$.0^4 1523$	$.0^4 1458$	$.0^4 1395$
−4.2	$.0^4 1335$	$.0^4 1277$	$.0^4 1222$	$.0^4 1168$	$.0^4 1118$	$.0^4 1069$	$.0^4 1022$	$.0^5 9774$	$.0^5 9345$	$.0^5 8934$
−4.3	$.0^5 8540$	$.0^5 8163$	$.0^5 7801$	$.0^5 7455$	$.0^5 7124$	$.0^5 6807$	$.0^5 6503$	$.0^5 6212$	$.0^5 5934$	$.0^5 5668$
−4.4	$.0^5 5413$	$.0^5 5169$	$.0^5 4935$	$.0^5 4712$	$.0^5 4498$	$.0^5 4294$	$.0^5 4098$	$.0^5 3911$	$.0^5 3732$	$.0^5 3561$
−4.5	$.0^5 3398$	$.0^5 3241$	$.0^5 3092$	$.0^5 2949$	$.0^5 2813$	$.0^5 2682$	$.0^5 2558$	$.0^5 2439$	$.0^5 2325$	$.0^5 2216$
−4.6	$.0^5 2112$	$.0^5 2013$	$.0^5 1919$	$.0^5 1828$	$.0^5 1742$	$.0^5 1660$	$.0^5 1581$	$.0^5 1506$	$.0^5 1434$	$.0^5 1366$
−4.7	$.0^5 1301$	$.0^5 1239$	$.0^5 1179$	$.0^5 1123$	$.0^5 1069$	$.0^5 1017$	$.0^6 9680$	$.0^6 9211$	$.0^6 8765$	$.0^6 8339$
−4.8	$.0^6 7933$	$.0^6 7547$	$.0^6 7178$	$.0^6 6827$	$.0^6 6492$	$.0^6 6173$	$.0^6 5869$	$.0^6 5580$	$.0^6 5304$	$.0^6 5042$
−4.9	$.0^6 4792$	$.0^6 4554$	$.0^6 4327$	$.0^6 4111$	$.0^6 3906$	$.0^6 3711$	$.0^6 3525$	$.0^6 3348$	$.0^6 3179$	$.0^6 3019$

z	.00	.01	.02	.03	.04	.05	.06	.07	.08	.09
.0	.5000	.5040	.5080	.5120	.5160	.5199	.5239	.5279	.5319	.5359
.1	.5398	.5438	.5478	.5517	.5557	.5596	.5636	.5675	.5714	.5753
.2	.5793	.5832	.5871	.5910	.5948	.5987	.6026	.6064	.6103	.6141
.3	.6179	.6217	.6255	.6293	.6331	.6368	.6406	.6443	.6480	.6517
.4	.6554	.6591	.6628	.6664	.6700	.6736	.6772	.6808	.6844	.6879
.5	.6915	.6950	.6985	.7019	.7054	.7088	.7123	.7157	.7190	.7224
.6	.7257	.7291	.7324	.7359	.7389	.7422	.7454	.7486	.7517	.7549
.7	.7580	.7611	.7642	.7673	.7703	.7734	.7764	.7794	.7823	.7852
.8	.7881	.7910	.7939	.7967	.7995	.8023	.8051	.8078	.8106	.8133
.9	.8159	.8186	.8212	.8238	.8264	.8289	.8315	.8340	.8365	.8389
1.0	.8413	.8438	.8461	.8485	.8508	.8531	.8554	.8577	.8599	.8621
1.1	.8643	.8665	.8686	.8708	.8729	.8749	.8770	.8790	.8810	.8830
1.2	.8849	.8869	.8888	.8907	.8925	.8944	.8962	.8980	.8997	.90147
1.3	.90320	.90490	.90658	.90824	.90988	.91149	.91309	.91466	.91621	.91774
1.4	.91924	.92073	.92220	.92364	.92507	.92647	.92785	.92922	.93056	.93189
1.5	.93319	.93448	.93574	.93699	.93822	.93943	.94062	.94179	.94295	.94408
1.6	.94520	.94630	.94738	.94845	.94950	.95053	.95154	.95254	.95352	.95449
1.7	.95543	.95637	.95728	.95818	.95907	.95994	.96080	.96164	.96246	.96327
1.8	.96407	.96485	.96562	.96638	.96712	.96784	.96856	.96926	.96995	.97062
1.9	.97128	.97193	.97257	.97320	.97381	.97441	.97500	.97558	.97615	.97670
2.0	.97725	.97778	.97831	.97882	.97932	.97982	.98030	.98077	.98124	.98169
2.1	.98214	.98257	.98300	.98341	.98382	.98422	.98461	.98500	.98537	.98574
2.2	.98610	.98645	.98679	.98713	.98745	.98778	.98809	.98840	.98870	.98899
2.3	.98928	.98956	.98983	$.9^20097$	$.9^20358$	$.9^20613$	$.9^20863$	$.9^21106$	$.9^21344$	$.9^21576$
2.4	$.9^21802$	$.9^22024$	$.9^22240$	$.9^22451$	$.9^22656$	$.9^22857$	$.9^23053$	$.9^23244$	$.9^23431$	$.9^23613$
2.5	$.9^23790$	$.9^23963$	$.9^24132$	$.9^24297$	$.9^24457$	$.9^24614$	$.9^24766$	$.9^24915$	$.0^25060$	$.0^25201$
2.6	$.9^25339$	$.9^25473$	$.9^25604$	$.9^25731$	$.9^25855$	$.9^25975$	$.9^26093$	$.9^26207$	$.0^26319$	$.0^26427$
2.7	$.9^26533$	$.9^26636$	$.9^26736$	$.9^26833$	$.9^26928$	$.9^27020$	$.9^27110$	$.9^27197$	$.0^27282$	$.0^27365$
2.8	$.9^27445$	$.9^27523$	$.9^27599$	$.9^27673$	$.9^27744$	$.9^27814$	$.9^27882$	$.9^27948$	$.0^28012$	$.0^28074$
2.9	$.9^28134$	$.9^28193$	$.9^28250$	$.9^28305$	$.9^28359$	$.9^28411$	$.9^28462$	$.0^28511$	$.0^28559$	$.0^28605$
3 0	$.9^28650$	$.9^28694$	$.9^28736$	$.9^28777$	$.9^28817$	$.9^28856$	$.9^28893$	$.9^28930$	$.9^28965$	$.9^28999$
3.1	$.9^30324$	$.9^30646$	$.9^30957$	$.9^31260$	$.9^31553$	$.9^31836$	$.9^32112$	$.9^32378$	$.9^32636$	$.9^32886$
3.2	$.9^33129$	$.9^33363$	$.9^33590$	$.9^33810$	$.9^34024$	$.9^34230$	$.9^34429$	$.9^34623$	$.9^34810$	$.9^34991$
3.3	$.9^35166$	$.9^35335$	$.9^35499$	$.9^35658$	$.9^35811$	$.9^35959$	$.9^36103$	$.9^36242$	$.9^36376$	$.9^36505$
3.4	$.9^36631$	$.9^26752$	$.9^36869$	$.9^36982$	$.9^37091$	$.9^37197$	$.9^37299$	$.9^37398$	$.9^37493$	$.9^37585$
3.5	$.9^37674$	$.9^37759$	$.9^37842$	$.9^37922$	$.9^37999$	$.9^38074$	$.9^38146$	$.9^38215$	$.9^38282$	$.9^38347$
3.6	$.9^38409$	$.9^38469$	$.9^38527$	$.9^38583$	$.9^38637$	$.9^38689$	$.9^38739$	$.9^38787$	$.9^38834$	$.9^38879$
3.7	$.9^38922$	$.9^38964$	$.9^40039$	$.9^40426$	$.9^40799$	$.9^41158$	$.9^41504$	$.9^41838$	$.9^42159$	$.9^42468$
3.8	$.9^42765$	$.9^43052$	$.9^43327$	$.9^43593$	$.9^43848$	$.9^44094$	$.9^44331$	$.9^44558$	$.9^44777$	$.9^44988$
3.9	$.9^45190$	$.9^45385$	$.9^45573$	$.9^45753$	$.9^45926$	$.9^46092$	$.9^46253$	$.9^46406$	$.9^46554$	$.9^46696$
4.0	$.9^46833$	$.9^46964$	$.9^47090$	$.9^47211$	$.9^47327$	$.9^47439$	$.9^47546$	$.9^47649$	$.9^47748$	$.9^47843$
4.1	$.9^47934$	$.9^48022$	$.9^48106$	$.9^48186$	$.9^48263$	$.9^48338$	$.9^48409$	$.9^48477$	$.9^48542$	$.9^48605$
4.2	$.9^48665$	$.9^48723$	$.9^48778$	$.9^48832$	$.9^48882$	$.9^48931$	$.9^48978$	$.9^50226$	$.9^50655$	$.9^51066$
4.3	$.9^51460$	$.9^51837$	$.9^52199$	$.9^52545$	$.9^52876$	$.9^53193$	$.9^53497$	$.9^53788$	$.9^54066$	$.9^54332$
4.4	$.9^54587$	$.9^54831$	$.9^55065$	$.9^55288$	$.9^55502$	$.9^55706$	$.9^55902$	$.9^56089$	$.9^56268$	$.9^56439$
4.5	$.9^56602$	$.9^56759$	$.9^56908$	$.9^57051$	$.9^57187$	$.9^57318$	$.9^57442$	$.9^57561$	$.9^57675$	$.9^57784$
4.6	$.9^57888$	$.9^57987$	$.9^58081$	$.9^58172$	$.9^58258$	$.9^58340$	$.9^58419$	$.9^58494$	$.9^58566$	$.9^58634$
4.7	$.9^58699$	$.9^58761$	$.9^58821$	$.9^58877$	$.9^58931$	$.9^58983$	$.9^60320$	$.9^60789$	$.9^61235$	$.9^61661$
4.8	$.9^62067$	$.9^62453$	$.9^62822$	$.9^63173$	$.9^63508$	$.9^63827$	$.9^64131$	$.9^64420$	$.9^64696$	$.9^64958$
4.9	$.9^65208$	$.9^65446$	$.9^65673$	$.9^65889$	$.9^66094$	$.9^66289$	$.9^66475$	$.9^66652$	$.9^66821$	$.9^66981$

From A. Hald, *Statistical Tables and Formulas,* Wiley, New York, 1952. Table II. Reproduced by permission. See also W. Nelson, *Applied Life Data Analysis,* Wiley, New York, 1982.

APPENDIX D

Probability Graph Papers

The general procedures used with all probability graph papers may be illustrated using the Weibull paper shown in Fig. D.1. The times to failure or other random variable are ranked (i.e., placed in ascending order): $t_1 \leq t_2 \leq t_3 \leq \ldots \leq t_N$. The CDF is then estimated at each time using Eq. 5.12,

$$\hat{F}(t_i) = \frac{i}{N+1}, \qquad i = 1, 2, 3, \cdots N, \tag{D.1}$$

and the appropriate probability paper is used to plot $\hat{F}(t_i)$ versus t_i. The points should fall roughly along a straight line if the random variable is described by the distribution. A straight line is drawn through the data, and the distribution parameters are estimated from the line.

Graph papers for the exponential, normal, lognormal, maximum extreme value, Weibull, and minimum extreme value distributions are given in Figs. D.2 through D.7. For plotting convenience the vertical and horizontal axes such papers are labeled with values of F and t. Observe, however, that the ordinate scales are nonlinear while the abscissa is either linear or logarithmic. These scales result from the rectification of the equation describing each distribution to the form

$$y(F) = \frac{1}{q}[x(t) - x(p)]. \tag{D.2}$$

The function $y(F)$ and $x(t)$ are derived for each distribution in Chapter 5 and summarized in Table D.1. The distribution parameters are expressed in terms of p and q also as indicated in the table.

The values of p and q, and hence the parameters, may be determined from the straight line drawn on the probability paper. Equation D.2 indicates that the condition $t_o = p$ satisfies

$$y[F(t_o)] = 0. \tag{D.3}$$

The value of F for which this holds is given in Table D.1 for each distribution. Thus for the Weibull plot in Fig. D.1, we note that at t_o, $F = 0.632$, and thus from the horizontal and vertical dashed lines drawn on Fig. D.1 $t_o = p = \theta = 46$ hr. To determine q, we find the values of $F(t_+)$ and $F(t_-)$ such that

$$y[F(t_\pm)] = \pm 1. \tag{D.4}$$

The corresponding values of $F(t_\pm)$ are tabulated for each distribution in Table D.1. Combining Eqs. D.2 and D.4, we obtain

$$x(t_\pm) - x(p) = \pm q, \tag{D.5}$$

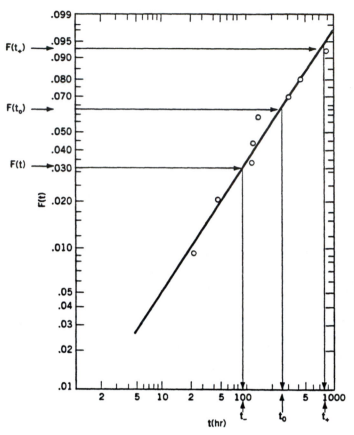

FIGURE D.1 Example Weibull probability plot.

TABLE D.1 Probability Graphing Information

distribution	$F(t)$	$y(F)$	$x(t)$	p	q	$F(t_o)$	$F(t_+)$	$F(t_-)$
exponential	$1 - e^{-t/\theta}$	$\ln[1/(1-F)]$	t	0	θ	0.632	—	—
normal	$\Phi\left(\dfrac{t-\mu}{\sigma}\right)$	$\Phi^{-1}(F)$	t	μ	σ	0.500	0.841	0.159
lognormal	$\Phi\left[\dfrac{1}{\omega}\ln(t/t_o)\right]$	$\Phi^{-1}(F)$	$\ln(t)$	t_o	ω	0.500	0.841	0.159
max. extreme val.	$\exp[-e^{-(t-u)/\theta}]$	$-\ln[\ln(1/F)]$	t	Θ	u	0.368	0.692	0.066
Weibull	$1 - e^{-(t/\theta)^m}$	$\ln[\ln[1/(1-F)]]$	$\ln(t)$	θ	$1/m$	0.632	0.934	0.308
min. extreme val.	$1 - \exp[-e^{(t-u)/\theta}]$	$\ln[\ln[1/(1-F)]]$	t	Θ	u	0.632	0.934	0.308

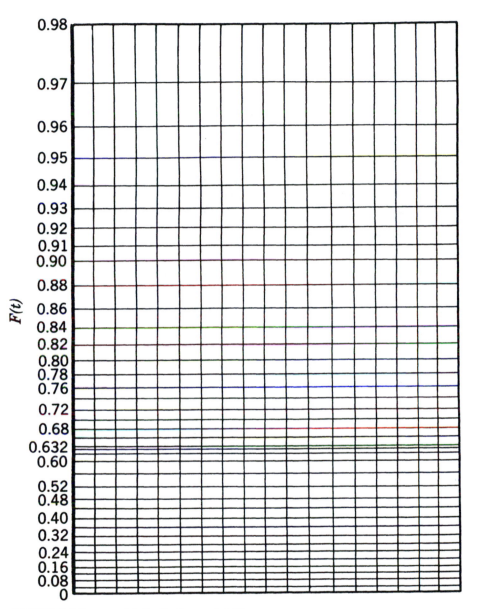

FIGURE D.2 Exponential distribution probability paper.

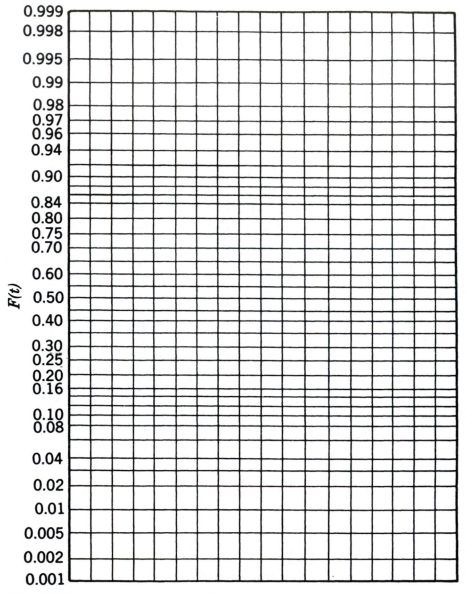

FIGURE D.3 Normal distribution probability paper.

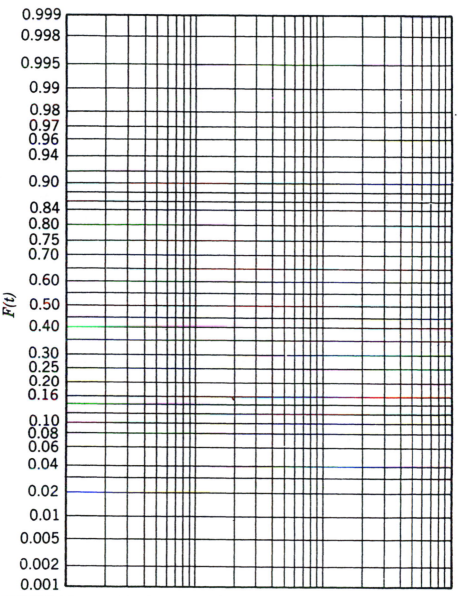

FIGURE D.4 Lognormal distribution probability paper.

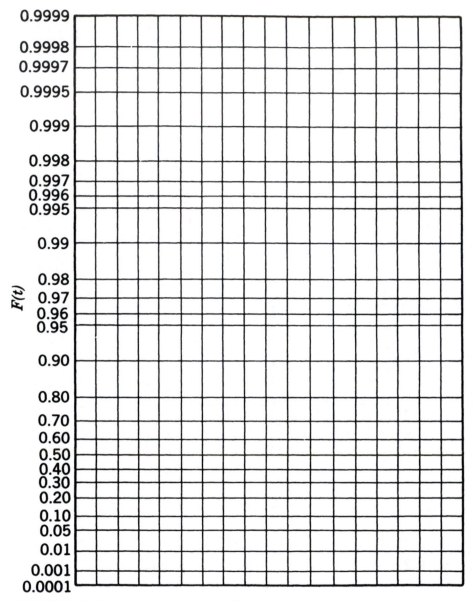

FIGURE D.5 Maximum extreme-value probability paper.

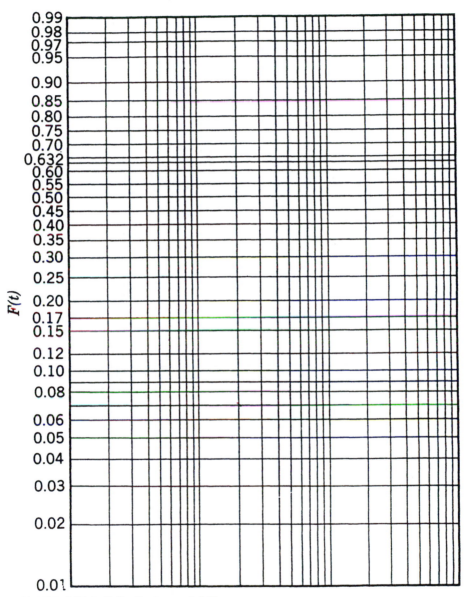

FIGURE D.6 Weibull distribution probability paper.

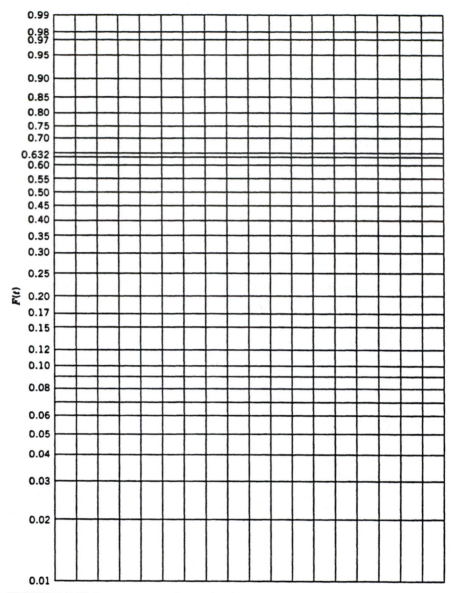

FIGURE D.7 Minimum extreme-value probability paper.

or with p eliminated between equations,

$$q = \tfrac{1}{2}[x(t_+) - x(t_-)].\qquad\text{(D.6)}$$

Finally, for the exponential normal and extreme value distributions, where $x(t) = t$, we have $q = (t_+ - t_-)/2$, while for the lognormal and Weibull distributions where $x(t) = \ln(t)$ we obtain $q = \ln(t_+/t_-)/2$. In our Weibull example, Table D.1 yields $F(t_+) = 0.934$ and $F(t_-) = 0.308$. Therefore from the horizontal and vertical dashed lines drawn on Fig. D.1 we obtain $t_+ = 800$ hrs and $t_- = 90$ hrs. Hence $m = 1/q = 2/\ln(8000/9000) = 0.92$.

Answers to Odd-Numbered Exercises

CHAPTER 2

2.1 (a) 0.72, (b) 0.115, (c) 0.59, (d) 0.165, (e) 0.115, (f) 0.425 (independent).

2.3 (a) 0.5, (b) 0.25, (c) 0.625, (d) 0.5.

2.5 (a) 0.7225, (b) 0.0225.

2.7 $R_{DL} = 0.9048$.

2.9 (a) $P\{X\} = 0.04$, (b) $P\{X_1|X_2\} = 0.25$.

2.11 (a) $C = 1/14$, (b) $F(1) = 1/14$, $F(2) = 5/14$, $F(3) = 1$, (c) $\mu \approx 2.57$ $\sigma = 2.10$.

2.13 $\mu \approx 1.53$, $\sigma^2 \approx 1.97$.

2.15 (a) 10, (b) 36, (c) 792, (d) 20.

2.17 0.0734.

2.19 $P_{NEW} = 0.0036$.

2.21 (a) 0.058, (b), 6.6×10^{-5}.

2.23 (a) 0.594, (b) 0.0166.

2.25 (a) 0.353, (b) 3.0.

2.27 0.0803.

2.29 (a) $1 - 1.2 \times 10^{-6}$, (b) 0.851.

2.31 230 consecutive starts.

2.33 (a) 2×10^{-4}, (b) 0.061, (c) 0.678.

2.35 0.140 ± 0.053, 0.140 ± 0.068.

2.37 415 units to test; no more than 18 failures to pass.

2.39 $\beta = 12\%$.

CHAPTER 3

3.1 $b = 6$, $\mu \approx 0.5$, $\sigma \approx 0.22$.

3.3 (a) $a = 18 \times 10^6$ hr^3, (b) 3000 hr.

3.5 (a) $f(x) = 0.04xe^{-0.2x}$, (b) $\mu \approx 10$, $\sigma^2 \approx 50$, (c) 0.0278

3.7 (a) 1 μm, (b) 80.8%, (c) 0.720 μm.

3.9 (a) $(e^{-x/\gamma} - 1)/(e^{-\tau/\gamma} - 1)$, (b) 0.168.

3.11 (a) $-$, (b) 8.32 cm, (c) 9.76 cm, (d) $-$.

3.13 $sk = \dfrac{\langle x^3 \rangle - 3\langle x^2 \rangle\langle x \rangle + 2\langle x \rangle^3}{(\langle x^2 \rangle - \langle x \rangle^2)^{3/2}}$.

3.15 (a) $f_y(y) =$
$$\frac{1}{b-a}\frac{1}{B}\left(\frac{y-a}{b-a}\right)^{r-1}\left(1 - \frac{y-a}{b-a}\right)^{t-r-1}$$
(b) $\mu_\gamma = (b-a)\dfrac{r}{t} + a.$

3.17 (a) 0.1056, (b) 1043 lbs, (c) 21.6 lbs.

3.19 7.44 hrs.

3.21 $\mu \approx 19.8$ kips, $\sigma \approx 1.676$ kips.

3.23 (a) $n = 5.58$, (b) $n = 1.57$.

3.25 (a) 0.026, (b) 0.308 yrs.

3.27 (a) 1.24×10^{-6}, (b) 0.037, (c) 0.311.

CHAPTER 4

4.1 (a) $\$125 \, x^2$, (b) $\$25$, (c) 0.056.

4.3 $L_o/3$.

4.5 (a) 0.463, (b) \$10, (c) 3.01.

4.7 0.0508

4.9 (a) 26.6 ppm, (b) 778 ppm.

4.11 (a) 0.86638, (b) 0.86638A, (c) 0.788, (d) $0.5515c^2$.

4.13 780 ppm.

4.15 0.0174.

4.17 (a) 2.00, (b) 0.0049 cm, (c) 0.680.

CHAPTER 5

5.1 (a) $\mu \approx 15061$, $\sigma^2 \approx 0.016935$, (b) graph.

5.3 (a) $\hat{\mu} = 20.3$, $\hat{\sigma}^2 = 142.8$, $\widehat{sk} = 0.794$, $\widehat{ku} = 0.716$ (b) $\mu = 20.3$, $\sigma^2 = 412$, sk = 2, ku = 7

5.5 $\hat{m} = 1.26$, $\hat{\theta} = 37$, $r^2 = 0.972$

5.7 $\hat{\mu} = 10.78$, $\hat{\sigma} = 6.28$.

5.9 1: $\mu = 49.8$, $\sigma = 0.80$, 2: $\mu = 50.5$, $\sigma = 1.53$.

5.11 $\hat{t}_0 = 17.0$, $\hat{\omega} = 0.824$, $r^2 = 0.957$.

5.13 (a) graph, (b) 103,419, (c) 2,507, (d) 0.987.

5.15 (a) graph, (b) 514 hr.

5.17 90%: 547, 95%: 651

5.19 103,421 ± 3150.

CHAPTER 6

6.1 (a) $16/(t + 4)^2$, (b) $2/(t + 4)$, (c) 4.

6.3 (a) 130 hr, (b) 256 hr, (c) 155 hr, (d) 513 hr.

6.5 (a) 0.966, (b) 0.980, (c) 0.975, (d) 0.990.

6.7 (a) 0.905, (b) 0.9275.

6.9 (a) 1.63, (b) 0.224.

6.11 47 days.

6.13 $\lambda = 0.105/\mathrm{hr}$.

6.15 MTTF $= \sqrt{\pi}\ \theta/2$.

6.17 0.04921.

6.19 28%.

6.21 (a) 1.667 hr, (b) 0.127 hr, (c) increases.

6.23 (a) 3.98 yr, (b) 3.14 yr.

6.25 2×10^6 cycles.

6.27 (a) 123 hr, (b) 6.3%, (c) 86%.

6.29 MTTF $= \sqrt{\pi}\ \theta/\sqrt{4N}$.

6.31 2.5%.

6.33 (a) 70.2 failures/yr, (b) nine flashlights.

6.35 (a) 0.939, (b) 1.87×10^{-3}, (c) 3.88×10^{-5}.

6.37 (a) 0.2856, (b) 0.1315, (c) 1.25.

6.39 (a) 1/15, (b) 0.00213.

CHAPTER 7

7.1 (a) 1.39×10^{-3}, (b) 721 V, (c) 2161 V.

7.3 $r = 1 + \dfrac{1}{a\gamma}(e^{-2a\gamma} - e^{-a\gamma})$.

7.5 $\tilde{R} = 0.2090$.

7.7 >10 strands.

7.9 15.7 Nm.

7.11 $c_0/l_0 = 4.64$.

7.13 9%.

7.15 (a) 0.269, (b) 0.00669.

7.17 (a) 9 cables, (b) 9 cables.

7.19 85.6 lbs.

7.21 0.0436.

7.23 10^{-15}.

7.25 0.670.

7.27 (a) 0.18, (b) 0.06, (c) 2.40 yr.

7.29 (a) 87 cycles,
(b) 1.25×10^6 cycles.

CHAPTER 8

8.1 (a) 0.647, (b) 0.999.

8.3 130 min.

8.5 (a) 74.4 min, (b) 129 min.

8.7 (a) graph, (b) $\alpha = 0.5011$.

8.9 $\hat{t}_0 = 96.4$ hr, $\hat{\omega} = 0.712$,
MTTF $= 124$ hr.

8.11 $\hat{t}_0 = 92.4$ hr, $\hat{\omega} = 0.657$,
MTTF $= 115$ hr.

8.13 $\hat{m} = 2.16$, $\hat{\sigma} = 110$ hr,
MTTF $= 97.5$ hr.

8.15 1.95 months.

8.17 $\mu = 48.1$, $\sigma^2 = 351.2$.

8.19 $m \approx 2.5$, $\theta \approx 130$.

8.21 (a) graph, (b) $\mu \approx 7000$ hr,
$\sigma \approx 3000$ hr, (c) 48%.

8.23 increasing with time.

8.25 $m \approx 2.4$, $\theta \approx 12$.

8.27 $\hat{R}(t_i) = \dfrac{N + 0.7 - i}{N + 0.4}$

8.29 143%.

8.31 MTTF $= 9.76$ months, 90% confidence limits: 6.54 & 16.61 months.

8.33 (a) 177 hr,
(b) $104 < \mu < 324$ hr.

8.35 33.8 days.

CHAPTER 9

9.1 $R' = 0.9289$.

9.3 6 units.

9.5 (a) 0.827, (b) 0.683, (c) 0.696.

9.7 (a) $1/4\lambda^2$, (b) $5/4\lambda^2$,
(c) parallel larger.

9.9 (a) $2e^{-(t/\theta)^m} - e^{-2(t/\theta)^m}$,
(b) $1 - (t/\theta)^{2m}$.

9.11 (a) 0.990, (b) 0.973.

9.13 0.629.

9.15 (a) $R = e^{-3\lambda t}$,
(b) $R = 1 - (1 - e^{-\lambda t})^3$
(c) $R = 2e^{-2\lambda t} - e^{-3\lambda t}$,
(d) graph.

9.17 (a) 30 days, (b) 27.3 days,
(c) 27.3 days.

9.19 $0.647 \sqrt{\pi}\,\theta$.

9.21 (a) 2.242×10^{-2}, (b) 0.1376.

9.23 (a) 0.9938, (b) 0.9960,
(c) 0.9798, b is best.

9.25 (a) $2R^2 - R^4$, (b) $(2R - R^2)^2$.

9.27 3.2×10^{-8}.

9.29 (a) 2/3 MTTF,
(b) 11/6 MTTF.

9.31 (a) 5 detectors, 7 amplifiers, 5 annunciators, (b) $30,800.

9.33 (a) 0.9867, (b) 0.9952.

9.35 (a) 0.9769, (b) 0.99978.

CHAPTER 10

10.1 (a) 0.885, (b) every 6300 hr,
(c) every 4275 hr.

10.3 No, maximum value is 0.934.

10.5 (a) 0.7225, (b) 0.8825,
(c) 0.7188.

10.7 (a) 4.04θ, (b) 455%.

10.9 1.044θ.

10.11 (a) 18.4 hr,
(b) 12.9 hr, 29.5 hr.

10.13 (a) 0.9315, (b) 20.4 hr.

10.15 0.980.

10.17 65.5 days.

10.19 2.2×10^{-4}/day.

10.21 (a) 0.897, (b) $\lambda = 0.013$/hr,
$\mu = 0.111$/hr,
(c) 2% difference.

10.23 (a) 0.968, (b) 0.946,
(c) every 18.6 days.

10.25 (a) 0.9594, (b) every 87.5 days.

10.27 every 1980 hr.

CHAPTER 11

11.1 (a) 0.058 MTTF, (b) 0.129
MTTF, (c) 0.182 MTTF.

11.3 (a) $1 - \lambda(2\lambda^* - \lambda)t^2$, (b) 1.56.

11.5 (a) $2/\lambda$, (b) $\lambda^2 t/(1 + \lambda t)$.

11.7 standby: $2/\lambda^2$,
active parallel: $5/4\lambda^2$.

11.9 (a) shared-load system,
(b) 1.063.

11.11 (a) proof, (b) $\approx 1 - \frac{3}{8}(\lambda t)^4$,
(c) active: 0.99990,
standby: 0.99996.

11.13 (a) $2(1 + \lambda t)e^{-\lambda t} - (1 + \lambda t)^2 e^{-2\lambda t}$,
(b) $1 - \frac{1}{4}\lambda^4 t^4$,
active parallel: $1 - \lambda^4 t^4$.

11.15 1.2×10^{-3}.

11.17 (a) 0.9998, (b) 0.9996.

11.19 0.09902.

11.21 with $\varepsilon \equiv \lambda/v$, (a)
$$\frac{1 + \varepsilon + \varepsilon^2 + \varepsilon^3}{1 + \varepsilon + \varepsilon^2 + \varepsilon^3 + \varepsilon^4},$$
(b) $\approx 1 - \varepsilon^4$,
(c) identical, $\approx 1 - 1.6 \times 10^{-7}$.

11.23 (a) 0.9961, (b) yes.

CHAPTER 12

12.1 passive-inlet line rupture,
either-valve closed when stop
fails, active-all other failures.

12.3 (a) 0.01, (b) 0.0185.

12.5 $A \cap B, A \cap C, B \cap C$.

12.7 (a) graph, (b) 9.15×10^{-4}.

12.9 0.12800, 0.12385, 0.12387.

12.11 (a) M_1: 0.382, M_2: 0.637,
(b) A: 0.382, B: 0.382, C: 0.637.

12.13 (a) 5.9×10^{-3}, (b) 0.0508,
0.1016, 0.847 (c) 0.847, 0.0678,
0.0339, 0.0339, 0.0169.

INDEX

Printed in the United States
61386LVS00001B/105